Molecular Biology of *Saccharomyces*

edited by
L.A. Grivell

Molecular Biology Section, University of Amsterdam, The Netherlands

Reprinted from *Antonie van Leeuwenhoek* 62: 1/2

SPRINGER SCIENCE+BUSINESS MEDIA, B.V.

Library of Congress Cataloging-in-Publication Data

Molecular biology of Saccharomyces / edited by L.A. Grivell.
p. cm.
ISBN 978-94-010-5104-0 ISBN 978-94-011-2504-8 (eBook)
DOI 10.1007/978-94-011-2504-8
1. Saccharomyces. 2. Yeast fungi. 3. Fungal molecular biology.
I. Grivell, L. A..
QK623.S23M644 1992
589.2'330488--dc20 92-13592

ISBN 978-94-010-5104-0

Printed on acid-free paper

Originally published by Kluwer Academic Publishers in 1992

Contents

Editorial 1

Nehrbass U. & Hurt E.C.: Nuclear transport and nuclear pores in yeast 3

Gasser S.M., Walter R., Dang Q. & Cardenas M.E.: Topoisomerase II: its functions and phosphorylation 15

Diffley J.F.X.: Global regulators of chromosome function in yeast 25

Brown J.D., Plumpton M. & Beggs J.D.: The genetics of nuclear pre-mRNA splicing: a complex story 35

Linder P.: Molecular biology of translation in yeast 47

Kunau W.-H. & Hartig A.: Peroxisome biogenesis in *Saccharomyces cerevisiae* 63

Gellissen G., Melber K., Janowicz Z.A., Dahlems U.M., Weydemann U., Piontek M., Strasser A.W.M. & Hollenberg C.P.: Heterologous protein production in yeast 79

Konopka J.B. & Fields S.: The pheromone signal pathway in *Saccharomyces cerevisiae* 95

Thevelein J.M.: The RAS-adenylate cyclase pathway and cell cycle control in *Saccharomyces cerevisiae* 109

Bolotin-Fukuhara M. & Grivell L.A.: Genetic approaches to the study of mitochondrial biogenesis in yeast 131

Antonie van Leeuwenhoek **62**: 1, 1992.

Editorial

The yeast *Saccharomyces* is one of the simplest eukaryotes known and also one of the world's most important commercial micro-organisms. As an experimental organism, there is much to recommend it: it grows rapidly; it can be manipulated by classical genetic and by recombinant DNA techniques; it is likely to be the first eukaryote for which a complete genomic sequence will be available and it offers unique opportunities for putting predictions of gene function to the test.

In this Special Issue, I have brought together a number of reviews dealing with topics in fundamental and applied aspects of yeast molecular biology and genetics. Although apparently diverse in scope, these articles strengthen the view held by an increasing number of research scientists that *Saccharomyces* can be regarded as a model eukaryote in a variety of respects, including the way it regulates gene expression, carries out housekeeping tasks necessary for cell growth and division, responds to signals from the environment and targets proteins to various destinations. The Special Issue thus not only tells us more about how this versatile micro-organism itself 'ticks', but how it can help us better to understand other eukaryotic cells.

My thanks are due to the contributors, who responded with such enthusiasm to my request to produce at short notice up-to-date readable surveys of recent developments in their respective fields. I believe that they have succeeded.

University of Amsterdam L.A. Grivell

Antonie van Leeuwenhoek **62**: 3–14, 1992.

Nuclear transport and nuclear pores in yeast

U. Nehrbass & E.C. Hurt
European Molecular Biology Laboratory, Postfach 10.2209, Meyerhofstraße 1, D-6900 Heidelberg, Germany

Key words: Saccharomyces cerevisiae, nuclear pore complex, nuclear transport, nuclear localization sequence, nucleoporins

Abstract

The central features of nuclear import have been conserved during evolution. In yeast the nuclear accumulation of proteins follows the same selective and active transport mechanisms known from higher eukaryotes. Yeast nuclear proteins contain nuclear localization sequences (NLS) which are presumably recognized by receptors in the cytoplasm and the nuclear envelope. Subsequent to this recognition step, nuclear proteins are translocated into the nucleus via the nuclear pore complexes. The structure of the yeast nuclear pore complex resembles that of higher eukaryotes. Recently, the first putative components of the yeast nuclear import machinery have been cloned and sequenced. The genetically amenable yeast system allows for an efficient structural and functional analysis of these components. Due to the evolutionary conservation potential insights into the nuclear import mechanisms in yeast can be transferred to higher eukaryotes. Thus, yeast can be considered as a eukaryotic model system to study nuclear transport.

1. Introduction

Nuclear proteins are synthesized in the cytoplasm and post-translationally transported across the nuclear envelope. Nuclear transport is controlled by the nuclear envelope which consists of the outer and inner nuclear membrane, nuclear pore complexes (NPCs), and an underlying nuclear lamina. The double membrane is constituted of two lipid bilayers that enclose the perinuclear space. The bilayers meet only at sites where nuclear pore complexes, large proteinaceous assemblies, traverse the envelope. The outer nuclear membrane is continuous with the endoplasmic reticulum and carries ribosomes, whilst at the inner membrane the nuclear lamina forms attachment points for chromatin and nucleoskeletal elements.

It has become clear that in order to be specifically transported into the nucleus, nuclear proteins require a common transport determinant. This implies the existence of a decoding system that selectively recognizes the determinant, and mediates the translocation across the nuclear membrane. During the past few years the coding signal of nuclear proteins, the nuclear localization sequence (NLS), has been identified, and the site of entry into the nucleus has been shown to be the nuclear pore complex. Recent work has focused on the 'decoding system' i.e. on NLS receptors in the cytoplasm or at the nuclear envelope. However, the molecular mechanisms underlying nuclear transport have remained largely unknown.

Progress in nuclear transport research on the molecular level requires a system which provides fast and systematic approaches to find components of the import machinery. In this respect, yeast has advanced into the limelight of scientific interest, being a versatile experimental system that allows for the investigation of nuclear transport using both *in vivo* and *in vitro* approaches. A widely used *in vivo* approach has been the analysis of the subcellular localization of mutated nuclear proteins or fusion proteins in yeast cells with the goal of identifying NLS. The genes encoding the nuclear proteins can be delivered into the cells on yeast shuttling plasmids and their expression can be regulated exploiting regulable yeast promoters. The fate of the *in vivo* expressed proteins is then followed by indirect immunofluorescence, electron microscopy, subcellular fractionation, or with the help of functional assays. For the identification of components of the nuclear import machinery, a variety of different approaches (genetic, biochemical or immunological) is currently applied. Once a component of the import machinery has been found, yeast offers a number of powerful tools for its functional analysis. In cases where it is possible to obtain conditionally lethal alleles of such components, yeast allows for the identification of extragenic suppressors, which themselves could turn out to encode for proteins of the import machinery.

To be considered as a model system for nuclear import, yeast should allow for a generalization of experimental findings. This, however, is only possible if yeast has an evolutionarily conserved import machinery sharing components and principal mechanisms with the nuclear import apparatus of higher eukaryotes. In this respect, the following chapters will try to provide an up to date picture of yeast nuclear transport in the context of corresponding data obtained from higher eukaryotes.

2. Targeting nuclear proteins

In the original view on the targeting mechanism of nuclear proteins for nuclear entry, specific signal sequences within the nuclear proteins were suggested to be responsible for nuclear targeting. These signals were postulated to meet two criteria: deleting or mutating them should abolish nuclear transport and fusing these signal sequences to non-nuclear proteins should induce accumulation of the reporter protein in the nucleus. First experimental evidence for the existence of short and confined nuclear targeting sequences came from work by Dingwall et al. (1982) who studied the targeting of nucleoplasmin after microinjection into *Xenopus* oocytes. Truncated nucleoplasmin lacking the C-terminal domain was not able to enter the nucleus, whereas the short C-terminal domain could mediate nuclear entry of attached colloidal gold particles. However, when the truncated nucleoplasmin lacking the C-terminal domain was injected into the nucleus directly it was retained, thus ruling out a retention model for nuclear transport. Subsequently, a nuclear localization signal consisting of the short amino acid sequence Pro-Lys-Lys-Lys-Arg-Lys-Val was found to be responsible for the nuclear targeting of SV40 large T antigen, and single amino acid exchanges within this signal (Lys128 to Thr or Asn) rendered a non-nuclear protein (Kalderon et al. 1984). Although nuclear transport is induced by the short heptapeptide, the context of this NLS within the protein is crucial for its optimal function. Rhis and Peters (1989) showed that for efficient nuclear transport of a SV40-lacZ fusion protein, the 15 amino acids preceeding the NLS were also required. Furthermore, the NLS has to be exposed on the surface of the protein in order to have access to components of the import machinery; when burried within the hydrophobic interior of pyruvate kinase, the SV40 large T antigen NLS was unable to mediate nuclear transport (Roberts et al. 1987; Nelson & Silver 1989). Until recently, the SV40 large T antigen NLS with its stretch of lysinal amino acid residues was considered to be the prototype of a NLS. Recent work by Robbins et al. (1991), however, showed that NLSs may have a more complex organization. In a detailed analysis applying site-directed mutagenesis of the nucleoplasmin NLS, two adjacent basic domains, separated by 10 amino acids, were found to be required for efficient nuclear targeting. The transport defect caused by a single mutation in one domain was amplified by a successive mutation in the other domain, indicating their interdependence. Although amino acid exchanges in the spacer sequence did not interfere with the functionality of the bipartite NLS, alterations in the length of the spacer element impaired nuclear transport. An extensive search in protein data libraries then revealed that many other nuclear proteins share the principle of a bipartite nuclear targeting motif, as initially found in nucleoplasmin.

Nuclear localization sequences from higher eukaryotes are functional in evolutionarily distant species such as yeast, showing that the mechanism of nuclear transport has been conserved during evolution. Nelson & Silver (1989) demonstrated that the SV40 large T antigen NLS sequence mediates nuclear accumulation of non-nuclear proteins in *Saccharomyces cerevisiae.* Two yeast ribosomal proteins, L3 (Moreland et al. 1985) and L29 (Underwood & Fried 1990), were shown in a fusion protein approach to contain NLS, which in case of L3 is contained within the first 21 amino-terminal amino acids, whilst nuclear transport of L29-lacZ fusion proteins can be mediated by two independent, internal, basic heptapeptides. Some of the nuclear localization signals identified in yeast proteins, however, reveal unique features. The yeast transcription factor GAL4 contains its nuclear localization signal within an extended stretch of the amino-terminal domain, not resembling the bipartite consensus NLS motif (Silver et al. 1984). The transcriptional repressor matα2 contains two independent NLSs being functionally distinct (Hall et al. 1984; Hall et al. 1990). In addition to a N-terminal sequence that resembles other NLS, a distinct and distant amino acid stretch within the homeodomain was able to mediate nuclear transport of a matα2-lacZ fusion protein. Deletion of the internal NLS reduces nuclear transport so that the nuclear protein accumulates around the nuclear periphery, most likely the nuclear pore complexes. Based on these observations, the N-terminal NLS signal was suggested to mediate association with the nuclear envelope, whilst the internal signal triggers translocation through the pore. Whether such a two step import mechanism is also true for nuclear proteins containing only a single NLS remains to be shown in the yeast system.

Surprisingly, proteins whose molecular weight is below the exclusion limit of nuclear pore complexes and thus could enter the nucleus by diffusion, can contain NLSs as shown for the yeast histone H2B (Moreland et al. 1987). The amino acids 28 to 33 (Gly-Lys-Lys-Arg-Ser-Lys-Ala) in histone H2B, which resemble the SV40 large T antigen NLS, will target a non-nuclear protein into the nucleus. A point mutation exchanging Lys31 for Met abolishes nuclear transport of the corresponding fusion protein. However, a H2B-lacZ fusion protein containing a mutated NLS, but in addition the H2A dimerization domain, was localized in the nucleus. This result suggested that H2A and H2B may be co-transported to the nucleus as a heterodimer. NLS mediated transport of small nuclear proteins has also been observed in mammalian cells e.g. in case of histone H1 (Breeuwer & Goldfarb 1990).

3. The import machinery

a) The decoding system: putative receptors

The existence of short and confined nuclear targeting sequences within a nuclear protein implies the existence of a decoding system, which specifically recognizes the NLS and catalyzes translocation through the nuclear pores. It was shown that nuclear transport may be divided into two steps, with a 'quick' binding step of the nuclear protein to the nuclear envelope being followed by the 'slow', ATP-dependent translocation through the nuclear pore channel (Newmeyer & Forbes 1988; Richardson et al. 1988). Moreover, nuclear transport is saturable, arguing that this is a receptor-mediated process with limiting amounts of receptor. There is no experimental data pointing to the exact mechanism of receptor recognition. The diversity of NLS sequences however, requires a complex recognition mechanism, unless what is recognized is not the primary sequence but rather a common determinant like the secondary structure. Each group of NLS could have a specific receptor or alternatively there could be adaptor molecules in the cytoplasm which recognize individual NLS and deliver them to a common receptor at the pore complex. *In vivo* studies indicate that NLS may first bind to such soluble receptors in the cytoplasm before the complex is delivered to the nuclear pores (Richardson et al. 1988; Breeuwer & Goldfarb 1990).

Approaches to identify NLS binding proteins were mainly biochemical, exploiting an assumed stable association of the receptor with its ligand in cell-free binding assays. Accordingly, ligand overlays were developed to measure binding of labeled

NLS-conjugates to cellular proteins immobilized on nitrocelluose. In higher eukaryotes two soluble cytoplasmic NLS binding proteins of 54 and 56 kDa have been identified by Adam and Gerace (1991). Evidence for an involvement of these proteins in nuclear uptake mechanisms comes from *in vitro* studies, where the authors could show that the purified 54 kDa and 56 kDa proteins can stimulate nuclear import.

In yeast, the 'NLS overlay' approach led to the identification of two putative NLS binding proteins (Lee & Melese 1989; Silver et al. 1989). Both proteins of 70 and 67 kDa specifically bound to the SV40 large T antigen NLS, but not to a mutated, import-incompetent form of this NLS. Although both proteins have been identified by a very similar approach, and have a similar molecular weight, they are not identical (Silver, personal communication).

The gene for the 67 kDa protein NSR1 has been cloned and sequenced (Lee et al. 1991). In gene disruption experiments it was demonstrated that *NSR1* is not essential, but resulting *nsr*$^-$ mutants grow poorly. NSR1 is tightly associated with nuclear structures since it can only be extracted with urea or high salt. In indirect immunofluorescence NSR1 gives an intranuclear staining typical of a nucleolar localization. Moreover, the amino acid sequence of NSR1 reveals a glycine/arginine-rich repetitive sequence and two putative RNA binding motifs.

Similarly, the 70 kDa NLS-binding protein is a nuclear protein, but less tightly associated with nuclei (Stochaj et al. 1991). In indirect immunofluorescence on whole yeast cells, anti-70 kDa antibodies give a strong nuclear and only a weak cytoplasmic staining. Binding of the nuclear reporter protein SV40-HSA (SV40 large T antigen NLS coupled to human serum albumin) to purified yeast nuclei is reduced both by antibodies against the 70 kD protein and by salt stripped nuclei, which removes this NLS binding protein (Stochaj et al. 1991). Binding can, however, be reconstituted by addition of a crude fraction enriched in NLS binding proteins.

The *in vivo* role of these two NLS binding proteins is not clear so far and accordingly their potential to participate in nuclear uptake mechanisms will have to await further clarification. This, however, should be possible in yeast if mutants of these genes can be generated. The *nsr*$^-$ mutant e.g. can be tested for a possible correlation of its sick phenotype to an impairment in nuclear transport, as would be expected if NSR1 was functioning as an NLS receptor *in vivo*.

Surprisingly, NSR1 is not the only nucleolar protein having NLS binding properties. For example, NO38 can be purified from nucleolar extracts by affinity chromatography using a synthetic peptide corresponding to the SV40 large T antigen nuclear localization signal (Goldfarb 1988). Another protein, p140 from rat liver, initially identified in cell extracts by chemical crosslinking to NLS peptides, was later shown to be a nucleolar shuttling protein (Meier & Blobel 1990). At the moment, however, it seems difficult to reconcile these data with a single cellular process in nuclear transport. It is speculated that these NLS binding proteins might function to chaperone karyophiles through the pore complexes, before they return to the cytoplasm. A final evaluation, however, whether NO38 and p140 are involved in NLS dependent nuclear transport, will also have to await experimental evidence that these proteins function as NLS receptors *in vivo*.

b) Nuclear pores and nucleoporins

Proteins destined for the nucleus have to cross the nuclear membrane. Proteins smaller than the functional pore diameter of 9 nm can passively diffuse through sites in the double membrane. Moreover, these sites of entry have to provide a mechanism that allows for larger proteins to be translocated in a specific, signal-dependent manner. The structural entities of the nuclear envelope designed to meet these requirements are the nuclear pore complexes.

The structure of the nuclear pore complex

Since their discovery in the early fifties, the nuclear pores always attracted structural biologists and therefore were subjected to extensive structural studies. Their involvement in nuclear transport, however, remained debated until Feldherr and coworkers devised assays which allowed microinjection of colloidal gold and ferritin coated with nuclear proteins into the cytoplasm of cells. This approach led to the identification of pore complexes as the entry and exit sites of nuclear transport (reviewed by Peters 1986). The structural analysis of the pore complexes went far beyond the functional characterization. From electron microscopy it became clear that that NPCs have a conserved principal building device: on the cytoplasmic and the nucleoplasmic faces of the pore complex are two coaxial rings consisting of eight subunits (Akey 1989). These rings sandwich a large spoke assembly made up of eight subunits that have attachments to both rings at the edge of the complex and include a central structure referred to as the central plug, central granule or transporter. As implied by its name, this transporter is thought to be part of the transport channel. Moreover, there are fibrils extending from the pore rings into the nucleo- and cytoplasm (Richardson et al. 1988). The cytoplasmic fibrils have been reported to be involved in a possible receptor function for nuclear proteins. Some of these cytoplasmic pore fibrils may also interact with intermediate filaments (Carmo-Fonseca et al. 1987).

The nuclear pore complex has an overall diameter of 120 nm, a relative molecular mass of 124 MD and is estimated to contain more than 100 different proteins (Featherstone et al. 1988). The number of nuclear pores varies with cell type, but in most cases is quite constant (Franke 1974; Maul 1977). Yeast nuclei have been found to contain 10–15 pores per μm^2 of nuclear surface, with the total number of nuclear pores adding up to 200 per nucleus (Moor & Mühlethaler 1963; Jordan et al. 1977). The distribution of nuclear pores is reported to be uneven, particularly in cells derived from starved cultures (Moor & Mühlethaler 1963). The absolute number of pores varies during the yeast cell cycle with increases in early G_0 phase and before nuclear division (Jordan et al. 1977). The pore density however remains relatively constant due to a rapid growth of the nuclear envelope during the peaks of nuclear pore assembly.

Most of our knowledge on the nuclear pore architecture stems from electron microscopic studies of NPC from higher eukaryotes, especially from *Xenopus* nuclear membranes. Less is known about the structural organization of nuclear pore complexes from yeast. This is partly due to the fact that yeast is not easily amenable to electron microscopic techniques.

However, the most recent and detailed analysis of the yeast nuclear pore complex confirms that the blueprint of nuclear pore complex architecture had already been elaborated in the early eukaryotic evolution (Allen & Douglas 1989). Upon subnuclear fractionation of isolated yeast nuclei by sequential extraction with detergent, nucleases and salt, the yeast NPC reveals an organization similar to the NPC of higher eukaryotes. Isolated NPCs are still associated with a remaining filamentous structure, that resembles the nuclear lamina of higher eukaryotes. Although putative analogues of the lamin A, B and C from higher eukaryotes have been identified in yeast using immunological approaches (Georgatos et al. 1989), there is, however, no biochemical evidence for the existence of nuclear lamina as yet. The organization of pore complexes in this preparation resembles a wagon wheel, with the pore complex sitting at the hub of the wheel. One type of filament (5–9 nm) radiates from the pore complex to the rim of the wheel, which might be constituted by another filament. The radially emanating filaments seem to be organized by a 30 nm ring that becomes visible in partially disassembled pore complexes lacking the central plug or transporter. These stripped pore complexes still reveal a set of spokes that are organized by the 30 nm ring in an eight-fold symmetry. In situ, in isolated yeast nuclei, the NPC extend short filaments into the cytosol, again in analogy to higher eukaryotes. These filaments could be involved in mediating an interaction of the NPC with cytoskeletal elements.

Components of the nuclear pore complex

Despite the large size and structual complexity of the pore complex, only a small number of pore complex proteins were so far identified. The most detailed work on nuclear pore complex components and their structural and functional analysis has been done with rat liver nuclei, using immunological, biochemical and recently also recombinant DNA techniques.

One prominent family of 8–10 novel NPC constituents of higher eukaryotes are the nucleoporins, which play a crucial part in nuclear transport (see below). They have in common an O-glycosylation with N-acetylglucosamine (GlcNAc) on serine and threonine residues and thus can bind the lectin wheat germ agglutinin (Hart et al. 1989). Nucleoporins were identified with the help of monoclonal antibodies raised against rat liver nuclear preparations and by their ability to bind to WGA (Davis & Blobel 1986; Park et al. 1987; Snow et al. 1987; Davis & Blobel 1987). In immuno-electron microscopic studies, the nucleoporins were either found to be labelled along filamentous structures at the marginal rings of the pore complex (Snow et al. 1987) or in other studies were found to be part of the central transporter (Akey & Goldfarb 1989). This contradiction in the localization of nucleoporins could be reconciled if the central transporter in the nuclear pore complex preparations of Akey and Goldfarb (1989) were the result of collapsed filamentous structures plugging the pores.

Since the basic building device of nuclear pores has been conserved during evolution it was reasonable to assume that yeast may contain the same class of nuclear pore proteins. In fact, when Western blots of whole cell yeast extracts were probed with monoclonal antibodies against the rat liver nucleoporin p62, three cross-reactive bands of 95 kDa, 110 kDa and 130 kDa were recognized (Aris & Blobel 1989; Davis & Fink 1990). Immunofluorescence localization using yeast cells and isolated nuclei showed a punctate and patchy staining pattern of the nuclear periphery, reminiscient of nuclear pore staining (Aris & Blobel 1989).

The nuclear pore complex protein NUP1

Using mammalian anti-nucleoporin monoclonal antibodies for expression screening of a yeast genomic library, the gene *NUP1* coding for a 130 kDa yeast nucleoporin was cloned (Davis & Fink 1990). In gene disruption experiments, the *NUP1* gene was found to be essential. Overexpression of NUP1 under the strong GAL10 promoter led to cell death in an often multi-budded state, indicating that a correct stoichiometry of NUP1 is of functional relevance. The localization of the NUP1 protein was shown by epitope tagging, since all antibodies available against NUP1 recognized other yeast nuclear pore proteins including NSP1. Therefore, a short influenza haemagglutinin epitope was inserted into the NUP1 protein. In indirect immunofluorescence, NUP1 could be localized to the nuclear periphery, giving a heterogeneous staining pattern highly similar to that obtained with anti-nucleoporin antibodies. From the amino acid sequence of NUP1, a three domain structure was deduced: a charged and very acidic amino-terminal domain, a middle portion consisting of 28 degenerate 9 amino acid long repeats separated by short and highly charged peptide stretches of varying length, and a short and basic C-terminal domain. In the middle, repetitive domain NUP1 shares common structural motifs with the yeast nucleoporin NSP1 (Hurt 1988; Nehrbass et al. 1990). It was shown that epitopes within this central, repetitive domain are responsible for the cross-reactivity of the nucleoporin antibodies (Davis & Fink 1990). So far it has not been investigated whether NUP1 is involved in nuclear transport mechanisms.

The nuclear pore complex protein NSP1

NSP1 was initially identified with the help of antibodies raised against the insoluble nucleoskeleton preparation of isolated yeast nuclei (Hurt 1988). One affinity-purified polyclonal antibody present in these immun-sera recognizes NSP1 as a 100 kDa protein on Western blots. Gene disruption experiments show that NSP1 is an essential gene and overexpression of NSP1 under the GAL10 promoter on a high copy number plasmid leads to cell death (Hurt 1989). In indirect immunofluorescence on whole yeast cells, NSP1 shows a heterogeneous punctate and occasionally blobby distribution at the nuclear periphery. Immunoelectron microscopic studies on frozen thin sections of whole yeast cells or isolated nuclei clearly demonstrated that NSP1 is localized at nuclear pores (Nehrbass et al. 1990; Hurt et al. 1992). This finding is in analogy with the localization of mammalian nucleoporins, and suggests a potential function in nuclear transport mechanisms or nuclear pore structure and/or assembly.

As deduced from the primary amino acid sequence, NSP1 can be divided into three domains, an amino terminal-, a central repetitive, and a carboxy-terminal domain. The C-terminal domain comprising one fourth of the protein contains all the essential function of NSP1 (Nehrbass et al. 1990). One of these functions is the correct localization of NSP1 at the nuclear periphery. This has been demonstrated by a gene fusion approach (Hurt 1990): cytosolic DHFR attached to the C-terminal domain of NSP1 was targeted to the nuclear envelope as shown by indirect immunofluorescence using antibodies against the DHFR moiety. The ability of the C-terminal domain to mediate nuclear pore localization might in molecular terms be accounted for by a heptad repeat motif stretching over the whole of the C-terminal domain. Heptad repeats of that organization, in general, are characterized by the occurence of a hydrophobic amino acid residue on position 1 and 4 of a repeating 7 amino acid stretch. They are found in intermediate filament proteins (e.g. cytokeratins and lamins) where they are responsible for the formation of α-helical coiled-coil dimers (Steinert et al. 1985; Steinert & Roop 1988; Cohen & Parry 1990). In case of NSP1, the heptad repeat motif could mediate interaction with a potential partner protein at the nuclear pore complex of similar secondary structure and thus target NSP1 to the pores. Experimental evidence for the functional significance of the heptad repeat organization comes from mutational analysis of the C-terminal domain (Nehrbass et al. 1990; U. Nehrbass, unpublished data).

The NSP1 C-terminal domain has been conserved during evolution. Sequence comparison with the C-terminus of human nucleoporin p62 (Carmo-Fonseca et al. 1991) reveals about 30% identity and 50% similarity, and is slightly less conserved in rat p62 (Starr & Hanover 1991) and *Xenopus* p68 (Cordes et al. 1991). Moreover, the C-terminal region of p62 also shows a heptad repeat organization similar to NSP1. In analogy to what has been discussed for NSP1, this heptad repeat motif could be responsible for the localization of p62 at the nuclear pore complex.

The central domain of NSP1 is characterized by 22 repetitive, highly conserved nonapeptides which are precisely spaced by stretches of ten hydrophilic amino acid residues (Hurt 1988). The repetitive nonapeptides generally correspond to peaks of high amphiphilicity when aligned to a hydrophobic moment plot for a predicted β-sheet structure. Within the nonapeptide repeat, a central 'FSF' motif is often found which gives rise to the amphiphilic moment. A functional analysis of the repetitive domain has not been reported so far and is difficult to do since this domain is not essential for NSP1 function (Nehrbass et al. 1990). However, the fact that the nonapeptide repeats are so highly conserved within NSP1 points to a specific, selectable function that allows for this conservation to occur. The functional importance of this domain is also indicated by the fact that this repeat structure is shared by other pore complex proteins like NUP1 and another yeast nuclear protein of about 95 kDa, tentatively called NSP2, which cross-reacts with polyclonal antibodies against the NSP1 middle repetitive domain (Aris & Blobel 1989; Nehrbass et al. 1990).

The cross-reactivity of antibodies directed against mammalian nucleoporin p62 with yeast nucleoporins suggests an evolutionary conservation of the repeat units and their secondary structure. The N-terminal half of p62 shows a repetitive building device with a conserved, repeating motif similar to that found in yeast nucleoporins (Carmo-Fonseca et al. 1991; Starr & Hanover 1991). Accordingly, truncated forms of NSP1 (Davis & Fink 1990) and mammalian p62 (Carmo-Fonseca et al. 1991) lacking the repetitive domain are no longer reactive with anti-nucleoporin monoclonal antibodies. Conversely, anti-peptide antibodies raised against a consensus repeat unit of NSP1 cross-react with several mammalian proteins including p62 and stain the nuclear pores in indirect immunofluorescence.

In summary, yeast contain a family of non-glycosylated nucleoporins exhibiting a common repetitive structural motif. The resulting abundance of structurally similar repeat units in various nuclear pore proteins may suggest a functional redundance. Indeed, the loss of the repetitive units of the NSP1 middle domain may not affect the essential function of this nucleoporin since it gets functionally compensated by repetitive units of other yeast nucleoporins. The genetic analysis of yeast mutated in more than one nucleoporin will allow to test such a hypothesis and might thereby lead us to the function of these conserved repetitive units.

c) Nucleoporins and nuclear transport

The structural conservation of nucleoporins from yeast to man suggests a functional conservation. In higher eukaryotes, the nucleoporins, in particular p62, have been demonstrated to play a crucial role in nuclear transport:

(i) antibodies specific for nucleoporins block nuclear import of proteins and RNA export (Featherstone et al. 1988);

(ii) the lectin WGA which binds to the sugar moiety of nucleoporins, blocks nuclear uptake of proteins both using reconstituted nuclei (Finlay et al. 1987; Newmeyer & Forbes 1988) and after microinjection into cells (Yoneda et al. 1987; Dabauvalle et al. 1988); furthermore, WGA is also demonstrated to inhibit export of ribosomal particles from the nucleus (Bataillé et al. 1990). The inhibition was shown not to be due to steric hindrance since WGA bound to the pore complexes still allowed dextrans to diffuse freely through the pores (Finlay et al. 1987).

(iii) pore complexes that have been immuno-depleted of a nucleoporin complex composed of p62, p58, and 54 using polyclonal antibodies are defective for nuclear import (Finlay et al. 1991). The amount of this complex present in partially depleted nuclei could be linearly correlated to the amount of nuclear transport of which these nuclei are capable.

The data on a possible involvement of nucleoporins in nuclear pore assembly are conflicting. The role of nucleoporins was investigated in *in vitro* reconstituted nuclei from *Xenopus* egg extracts (Finlay & Forbes 1990). After immuno-depletion of nucleoporins the extracts were still found competent of forming nuclear envelopes with nuclear pore complexes, but were unable to translocate nuclear proteins. In contrary Dabauvalle et al. (1990) found in similar reconstitution experiments that nucleoporins are required for nuclear pore assembly.

A role of yeast nucleoporins in nuclear transport mechanisms is currently under investigation. In an *in vitro* binding assay, anti-peptide antibodies against both the 70 kDa NLS-binding proteins and the repetitive units of NSP1 inhibited binding of SV40-HSA conjugates to isolated yeast nuclei (Stochaj et al. 1991). In our laboratory, we find that in a ts nsp1 mutant the nuclear reporter protein matα2-lacZ accumulates in the cytoplasm at restrictive conditions (Nehrbass et al. submitted).

What specific step in nuclear uptake could nucleoporins be involved in? Nucleoporins were speculated to be constituents of short filaments emanating from the pore complex (Snow et al. 1987; Park et al. 1987), filaments, that were later suggested to be involved in a binding function for karyophiles (Richardson et al. 1988). If this model were correct, nucleoporins could have a receptor function, mediating one of the early steps in translocation. One could envisage two principal possibilities: either the nucleoporins would directly bind nuclear proteins, or they could act as a docking site for NLS-binding proteins loaded with their nuclear import substrates. The latter hypothesis is backed by crosslinking experiments where the 70 kD yeast NLS-binding protein could be crosslinked to NSP1 on isolated yeast nuclei (Pam Silver, personal communications).

According to data from (Akey 1990), nucleoporins could be constituents of the central transporter. He proposes a model for the transporter in which two aligned rings of eight subunits form a flexible 'double iris diaphragm'. Inhibition of transport by WGA and anti-nucleoporin antibodies would then be due to crosslinking and immobilization of the subunits.

4. Regulated nuclear transport

Selective nuclear transport does not imply the rigid and permanent separation of nuclear and cytoplasmic proteins, but allows for the regulated nuclear import of some karyophiles. For a number of transcription factors controlled nuclear localization has been adopted by cells as a very immediate and powerful means of regulating their nuclear activity.

Selective nuclear transport can e.g. be achieved by phosphorylation of the nuclear protein. Rihs et al. (1991) found that the efficient import of a SV40-lacZ fusion protein is dependent both on the nuclear localization sequence and an adjacent phosphorylation site containing sequence. Whilst the specificity is determined by the NLS, the rate of uptake is controlled by the phosphorylation site containing sequence. The phosphorylation of a single residue in a casein kinase II site, 14 amino acids upstream of the NLS, was found sufficient to elicit the maximum transport enhancement. Efficient transport of the SV40 large T antigen thus requires phosphorylation. The opposite effect was found in the regulated, cell cycle dependent nuclear import of SWI5 in yeast (Moll et al. 1991). SWI5, a transcription factor required for the expression of HO, is transcribed during the S, G2 and M phase of the cell cycle (Nasmyth et al. 1987). The protein is only translocated to the nucleus at the end of anaphase, as cells enter G1 (Nasmyth et al. 1990). Within, or close to the NLS, three phosphorylated serine residues were mapped, the mutation of which led to constitutive nuclear import (Moll et al. 1991). These residues are phosphorylated in a cell-cycle dependent manner *in vivo*, being phosphorylated when SWI5 is in the cytoplasm and dephosphorylated when it is in the nucleus. Since all three residues are phosphorylated by purified CDC28 kinase, a model is proposed in which CDC28 kinase acts directly to control cell cycle dependent nuclear import of SWI5 (Moll et al. 1991).

Another way to selectively regulate nuclear transport is the reversible masking of the NLS by interaction of the karyophile with a partner protein. The binding of such a partner component, and thus the activity of the NLS, can again be regulated by phosphorylation, as observed for the transcription factor NF-κB (Bäuerle and Baltimore 1988). In the cytoplasm NF-κB occurs in a complex with IκB. The association of NK-κB and IκB, however, can be disrupted by phosphorylation, thereby exposing the NLS of NF-κB which is then translocated into the nucleus (Ghosh & Baltimore 1990). In case of the glucocorticoid receptor, selective nuclear uptake is achieved by dissociation of a heatshock protein (hsp90) from the NLS upon binding of hormone, thereby leading to the NLS dependent nuclear translocation of the hormone-receptor complex (Sanchez et al. 1985; Picard & Yamamoto 1987).

5. Experimental approaches to nuclear transport in yeast

As described above, one major approach to study nuclear transport in yeast has been investigating the subcellular localization of manipulated nuclear proteins in living cells. This led to the identification of NLSs within e.g. matα2 (Hall et al. 1984; Hall et al. 1990), GAL4 (Silver et al. 1984), the ribosomal proteins L3 (Moreland et al. 1985) and L29 (Underwood & Fried 1990). Applying biochemical and immunological methods, it has been possible to identify two putative NLS receptors (Lee & Melese 1989; Silver et al. 1989) and two essential components of the NPC (Davis & Fink 1990; Nehrbass et al. 1990).

Exploiting genetic approaches mutants defective in nuclear transport have been identified. In one elegant approach, a nuclear localization sequence was fused to a mitochondrial protein. Since the NLS was dominant the mitochondrial protein was mislocalized to the nucleus and thus could no longer perform its essential function in the mitochondrion. Cells therefore became respiration deficient and unable to use glycerol as a non-fermentable carbon source. This deficiency was exploited to screen for temperature-sensitive mutants, which regained the ability to grow on glycerol at the permissive temperature due to a mutation in the nuclear import machinery. The gene for one of such npl mutants, *NPL1,* was isolated and found to be homologous to the *E. coli* heat shock like protein DnaJ. Intriguingly, *npl1-1* is allelic to a previously identified secretion mutant *sec63,* which is required for protein translocation across the ER membrane (Rothblatt et al. 1989). So far it is not clear why npl1 is involved in both nuclear transport and translocation into the ER. One possible explanation would be that npl1 is required for insertion of membrane proteins associated with the NPC into the nuclear membrane. Such integral membrane proteins, like gp210, have been shown to be constituents of the nuclear pore complex in higher eukaryotes (Wozniak et al. 1989; Greber et al. 1990).

Another approach for the identification of nuclear import mutants in yeast is the screening for strains that can afford to express toxic nuclear proteins. One such possible screen has been designed by Barnes and Rine (1985), who expressed the *E. coli* restriction endonuclease Eco RI under control of the regulable GAL10 promoter in rad52 yeasts, deficient in double strand break repair. Only strains impaired in nuclear import should survive the toxic effects of the restriction enzyme on galactose containing medium. So far, this elegant screen has not been used successfully to find nuclear import mutants.

In the past three years several groups have independently set up *in vitro* systems to study nuclear transport, using purified yeast nuclei and various nuclear substrates (Kalinich & Douglas 1989; Garcia-Bustos et al. 1991; Stochaj et al. 1991). Yeast nuclei can be efficiently isolated by a variety of methods based on the enzymic digestion of the cell wall followed by lysis in a buffer supplemented with Ficoll as an osmotic stabilizer (Mann & Mecke 1980; Mann & Mecke 1982; Hurt et al. 1988; Aris & Blobel 1988; Amati & Gasser 1988; Allen & Douglas 1989; Kalinich & Douglas 1989; Cardenas et al. 1990). In the future such nuclear import assays may be potentially used to test the inhibitory effect of antibodies against possible components of the import machinery, or to test cytosol or nuclei of potential import mutant strains for their *in vitro* translocation competence.

6. Conclusions

Yeast as a well characterized genetically amenable system provides a number of powerful tools to study the mechanisms and components of nuclear transport. During the past years, researchers identified the first genes for such putative components of the import machinery in yeast. These could constitute a starting point for a stepwise, genetic and biochemical analysis of the nuclear pore complex leading to the identification of further components of the import aparatus. Moreover, the principle mechanisms of nuclear transport have been conserved during evolution from yeast to human. Thus, yeast seems to have advanced to a suitable model system for the investigation of the molecular mechanisms underlying nuclear transport.

References

Adam SA & Gerace L (1991) Cytosolic proteins that specifically bind nuclear localization signals are receptors for nuclear import. Cell 66: 837–847

Akey CW (1989) Interactions and structure of the nuclear pore complex revealed by cryo-electron microscopy. J. Cell Biol. 109: 955–970

— (1990) Visualization of transport-related configurations of the nuclear pore transporter. Biophys. J. 58: 341–355

Akey CW & Goldfarb DS (1989) Protein import through the nuclear pore complex is a multistep process. J. Cell Biol. 109: 971–982

Allen JL & Douglas MG (1989) Organization of the nuclear pore complex in *Saccharomyces cerevisiae.* J. Ultrastruc. Molec. Struct. Res. 102: 95–108

Amati BB & Gasser SM (1988) Chromosomal ARS and CEN elements bind specifically to the yeast nuclear scaffold. Cell 54: 967–978

Aris, JP & Blobel G (1988) Identification and characterization of a yeast nucleolar protein that is similar to a rat liver nucleolar protein. J. Cell Biol. 107: 17–31

— (1989) Yeast nuclear envelope proteins cross-react with an antibody against mammalian pore complex proteins. J. Cell Biol. 108: 2059–2067

Bäuerele PA & Baltimore D (1988) IκB: A specific inhibitor of NF-κB transription factor. Science 242: 540–546

Barnes G & Rine J (1985) Regulated expression of endonuclease EcoRI in *Saccharomyces cerevisiae:* nuclear entry and biological consequences. Proc. Natl. Acad. Sci. USA. 82: 1354–1358

Bataillé N, Helser T & Fried HM (1990) Cytoplasmic transport of ribosomal subunits microinjected into the *Xenopus* laevis oocyte nucleus: a generalized, facilitated process. J. Cell Biol. 111: 1571–1582

Breeuwer M & Goldfarb M (1990) Facilitated nuclear transport of histone H1 and other small nucleophilic proteins. Cell 60: 999–1008

Cardenas ME, Laroche T & Gasser SM (1990) The composition and morphology of yeast nuclear scaffolds. J. Cell Sci. 96: 439–450

Carmo-Fonseca M, Cidadao AJ & David-Ferreira JF (1987) Filamentous cross-bridges link intermediate filaments to the nuclear pore complexes. Eur. J. Cell Biol. 45: 282–290

Carmo-Fonseca M, Kern H & Hurt EC (1991) Human nucleoporin p62 and the essential yeast nuclear pore protein NSP1 show sequence homology and a similar domain organization. Eur. J. Cell Biol. 55: 17–30

Cohen C & Parry DAD (1990) α-helical coiled coils and bundles: how to design an α-helical protein. Proteins 7: 1–15

Cordes V, Waizenegger I & Krohne G (1991) Nuclear pore complex glycoprotein p62 of *Xenopus* laevis and mouse: cDNA cloning and identification of its glycosylated region. Eur. J. Cell Biol. 55: 31–47

Dabauvalle MC, Schulz B, Scheer U & Peters R (1988) Inhibition of nuclear accumulation of karyophilic proteins in living cells by microinjection of the lectin wheat germ agglutinin. Exp. Cell Res. 174: 291–296

Dabauvalle M-C, Loos K & Scheer U (1990) Identification of a soluble precursor complex essential for nuclear pore assembly. Chromosoma 100: 56–66

Davis LI & Blobel (1986) Identification and characterization of a nuclear pore complex protein. Cell 45: 699–709

— (1987) Nuclear pore complex contains a family of glycoproteins that includes p62: glycosylation through a previously unidentified cellular pathway. Proc. Natl. Acad. Sci. USA 84: 7552–7556

Davis LI & Fink GR (1990) The NUP1 gene encodes an essential component of the yeast nuclear pore complex. Cell 61: 965–978

Dingwall C, Sharnick SV & Laskey RA (1982) A polypeptide domain that specifies migration of nucleoplasmin into the nucleus. Cell 30: 449–458

Featherstone C, Darby MK & Gerace L (1988) A monoclonal antibody against the nuclear pore complex inhibits nucleocytoplasmic transport of protein and RNA *in-vivo*. J. Cell Biol. 107: 1289–1297

Feldherr CM, Kallenbach E & Schultz N (1984) Movement of a karyophilic protein through the nuclear pores of oocytes. J. Cell Biol. 99: 2216–2222

Finlay DR & Forbes DJ (1990) Reconstitution of biochemically altered nuclear pores: transport can be eliminated and restored. Cell 60: 17–29

Finlay DR, Newmeyer DD, Price TM & Forbes DJ (1987) Inhibition of *in-vitro* nuclear transport by a lectin that binds to nuclear pores. J. Cell Biol. 104: 189–200

Finlay DR, Meier E, Bradley P, Horecka J & Forbes DJ (1991) A complex of nuclear pore proteins required for pore functions. J. Cell Biol. 114: 169–183

Franke WW (1974) Structure, biochemistry and functions of the nuclear envelope. Int. Rev. Cytol. 4: 72–236

Garcia-Bustos JF, Wagner P & Hall MN (1991) Yeast cell-free nuclear protein import requires ATP hydrolysis. Exp. Cell Res. 192: 213–219

Georgatos SD, Maroulakou I & Blobel G (1989) Lamin A, Lamin B, and Lamin B receptor analogues in yeast. J. Cell Biol. 108: 2069–2082

Goldfarb D (1988) Karyophilic peptides: applications to the study of nuclear transport. Cell Biol. Inter. Reports. 12: 809–832

Ghosh S & Baltimore D (1990) Activation *in vitro* of NF-κB by phosphorylation of its inhibitor IκB. Nature 344: 678–682

Greber UF, Senior A & Gerace L (1990) A major glycoprotein of the nuclear pore complex is a membrane-spanning polypeptide with a large luminal domain and a small cytoplasmic tail. EMBO J. 9: 1495–1502

Hall MN, Hereford L & Herskowitz I (1984) Targeting of *E. coli* b-galactosidase to the nucleus in yeast. Cell 36: 1057–1065

Hall MN, Craik C & Hiraoka Y (1990) Homeodomain of yeast repressor alpha 2 contains a nuclear localization signal. Proc. Natl. Acad. Sci. USA 87: 6954–6958

Hart GW, Haltiwanger RS, Holt GD & Kelly WG (1989) Glycosylation in the nucleus and cytoplasm. Ann. Rev. Biochem. 58: 841–874

Hurt EC (1988) A novel nucleoskeletal-like protein located at the nuclear periphery is required for the life cycle of *Saccharomyces cerevisiae*. EMBO J. 7: 4323–4334

— (1989) NSP1, a yeast protein located at the nuclear periphery, is required for the cell cycle of *Saccharomyces cerevisiae*. J. Cell Sci. 12: 243–252

— (1990) Targeting of a cytosolic protein to the nuclear periphery. J. Cell Biol. 111: 2829–2837

Hurt EC, McDowall A & Schimmang T (1988) Nucleolar and nuclear envelope proteins of the yeast *Saccharomyces cerevisiae*. Eur. J. Cell Biol. 46: 554–563

Hurt EC, Mutvei A & Carmo-Fonseca M (1992) The nuclear envelope of the yeast *S. cerevisiae*. Int. Rev. Cytol. (in press)

Jordan EG, Severs NJ & Williamson DH (1977) Nuclear pore formation and the cell cycle in *Saccharomyces cerevisiae*. Exp. Cell Res. 104: 446–449

Kalderon D, Roberts BL, Richardson WP & Smith AE (1984) A short amino acid sequence able to specify nuclear location. Cell 39: 499–509

Kalinich JF & Douglas MG (1989) *In vitro* translocation through the yeast nuclear envelope. J. Biol. Chem. 264: 17979–17989

Lee WC & Melese T (1989) Identificaton and characterization of a nuclear localization sequence-binding protein in yeast. Proc. Natl. Acad. Sci. USA 86: 8808–8812

Lee WC, Xue Z & Melese T (1991) The *NSR1* gene encodes a protein that specifically binds nuclear localization sequences and has two RNA recognition motifs. J. Cell Biol. 113: 1–12

Mann K & Mecke D (1980) Isolation and characterization of nuclei and nuclear membranes from *Saccharomyces cerevisiae* protoplasts. FEBS Lett. 122: 95–99

Mann K-H & Mecke D (1982) The isolation of *Saccharomyces*

cerevisiae nuclear membranes with nuclease and high-salt treatment. Biochim. Biophys. Acta 687: 57–62

Maul GG (1977) The nuclear and cytoplasmic pore complex: structure, dynamics, distribution and evolution. Int. Rev. Cytol. 6: 75–186

Meier UT & Blobel G (1990) A nuclear localization signal binding protein in the nucleolus. J. Cell Biol. 111: 2235–2245

Moll T, Tebb G, Surana U, Robitsch H & Nasmyth K (1991) The role of phosphorylation and the CDC28 protein kinase in cell-cycle regulated nuclear import of the *S. cerevisiae* transcription factor SWI5. Cell 66: 743–758

Moor H & Mühlethaler K (1963) Fine structure in frozen-etched yeast cells. J. Cell Biol. 17: 609–627

Moreland RB, Nam HG, Hereford LM & Fried HM (1985) Identification of a nuclear localization signal of a yeast ribosomal protein. Proc. Natl. Acad. Sci. USA 82: 6561–6565

Moreland RB, Langevin GL, Singer RH, Garcea RL & Hereford LM (1987) Amino acid sequences that determine the nuclear localization of yeast histone 2B. Mol. Cell Biol. 7: 4048–4057

Nasmyth K, Seddon A & Ammerer G (1987) Cell cycle regulation of SW15 is required for mother-cell-specific HO transcription in yeast. Cell 49: 549–558

Nasmyth K, Adolf G, Lydall D & Seddon A (1990) The identification of a second cell cycle control on the HO promoter in yeast: cell cycle regulation of SWI5 nuclear entry. Cell 62: 631–647

Nehrbass U, Kern H, Mutvei A, Horstmann H, Marshallsay B & Hurt EC (1990) NSP1: A yeast nuclear envelope protein localized at the nuclear pores exerts its essential function by its carboxy-terminal domain. Cell 61: 979–989

Nelson M & Silver P (1989) Context affects nuclear protein localization in *Saccharomyces cerevisiae*. Mol. Cell Biol. 9: 384–389

Newmeyer DD & Forbes DJ (1988) Nuclear import can be separated into distinct steps *in vitro*: nuclear pore binding and translocation. Cell 52: 641–653

Park MK, D'Onofrio M, Willingham MC & Hanover JA (1987) A monoclonal antibody against a family of nuclear pore proteins (nucleoporins): O-linked N-acetylglucosamine is part of the immunodeterminant. Proc. Natl. Acad. Sci. USA 84: 6462–6466

Peters R (1986) Fluorescence microphotolysis to measure nucleocytoplasmic transport and intracellular mobility. Biochim. Biophys. Acta 864: 305–359

Picard D & Yamamoto KR (1987) Two signals mediate hormone-dependent nuclear localization of the glucocorticoid receptor. EMBO J. 6: 3333–3340

Richardson WD, Mills AD, Dilworth SM, Laskey RA & Dingwall C (1988) Nuclear protein migration involves two steps: rapid binding at the nuclear envelope followed by slower translocation through nuclear pores. Cell 52: 655–664

Rihs H-P & Peters R (1989) Nuclear transport kinetics depend on phosphorylation site-containing sequences flanking the karyophilic signal of the SV40 T-antigen. EMBO J. 8: 1479–1484

Rihs H-P, Jans DA, Fan H & Peters R (1991) The rate of nuclear cytoplasmic protein transport is determined by the casein kinase II site flanking the nuclear localization sequence of the SV40 T-antigen. EMBO J. 10: 633–639

Robbins J, Dilworth SM, Laskey RA & Dingwall C (1991) Two interdependent basic domains in nucleoplasmin nuclear targeting sequence: Identification of a class of bipartite nuclear targeting sequence. Cell 64: 615–623

Roberts BL, Richardson WD & Smith AE (1987) The effect of protein context on nuclear location signal function. Cell 50: 465–475

Rothblatt JA, Deshaies SL, Sanders SL, Daum G & Schekman R (1989) Multiple genes are required for proper insertion of secretory proteins into the endoplasmic reticulum in yeast. J. Cell Biol. 109: 2641–2652

Sadler I, Chiang A, Kurihara T, Rothblatt J, Way J & Silver P (1989) A yeast gene important for protein assembly into the endoplasmic reticulum and the nucleus has homology to DnaJ, an *Escherichia coli* heat shock protein. J. Cell Biol. 109: 2665–2675

Sanchez ER, Schlesinger MJ & Pratt WB (1985) The 90 kD non-steroid-binding phosphoprotein that binds to the untransformed glucocorticoid receptor in molybdate stabilized L-cell cytosol is the murine 90 kD heat shock protein. J. Biol. Chem. 260: 12398–12401

Silver P, Sadler I & Osborne MA (1989) Yeast proteins that recognize nuclear localization sequences. J. Cell Biol. 109: 983–989

Silver PA, Keegan LP & Ptashne M (1984) Amino terminus of the yeast GAL4 gene product is sufficient for nuclear localization. Proc. Natl. Acad. Sci. USA 81: 5951–5955

Snow CM, Senior A & Gerace L (1987) Monoclonal antibodies identify a group of nuclear pore complex glycoproteins. J. Cell Biol. 104: 1143–1156

Starr CM & Hanover JA (1991) A common structural motif in nuclear pore proteins (nucleoporins). BioEssays 13: 145–146

Steinert PM & Roop DR (1988) Molecular and cellular biology of intermediate filaments. Ann. Rev. Biochem. 57: 593–625

Steinert PM, Steven AC & Roop DR (1985) The molecular biology of intermediate filaments. Cell 42: 411–419

Stochaj U, Osborne M, Kurihara T & Silver P (1991) A yeast protein that binds nuclear localization signals: Purification, localization, and antibody inhibition of binding activity. J. Cell Biol. 113: 1243–1254

Underwood MR & Fried HM (1990) Characterization of nuclear localization sequences derived from the yeast ribosomal protein L29. EMBO J. 9: 91–99

Wozniak RK, Bartnik E & Blobel G (1989) Primary structure analysis of an integral membrane glycoprotein of the nuclear pore. J. Cell Biol. 108: 2083–2092

Yoneda Y, Imamoto-Sonobe N, Yamaizumi M & Uchida T (1987) Reversible inhibition of protein import into the nucleus by wheat germ agglutinin injected into cultured cells. Exp. Cell Res. 173: 586–595

Antonie van Leeuwenhoek **62**: 15–24, 1992.

Topoisomerase II: its functions and phosphorylation

Susan M. Gasser, Robin Walter, Qi Dang[1] & Maria E. Cardenas[2]
Swiss Institute for Experimental Cancer Research (ISREC), CH-1066 Epalinges s/Lausanne, Switzerland
Present addresses: [1] University of California Medical School, San Francisco, CA, USA;
[2] Rockefeller University, New York, NY 10021, USA

Key words: chromosomes, nuclear scaffold, phosphorylation, topoisomerase II, yeast

Abstract

The gene encoding topoisomerase II in yeast is unique and essential, required for both mitotic and meiotic proliferation. The use of temperature-sensitive mutants in topoisomerase II have demonstrated roles in the relaxation of tortional stress, reduction of recombination rates, and in the separation of sister chromatids after replication. In vertebrate cells, topoisomerase II was shown to be the most abundant component of the metaphase chromosomal scaffold, and has been shown to play a role in chromosome condensation *in vitro*. The cell cycle control of chromosome condensation may well require phosphorylation of topoisomerase II, since the enzyme is more highly phosphorylated in metaphase than in G1. Recent studies have identified casein kinase II as the major enzyme phosphorylating topoisomerase II in intact yeast cells. The target sites of CKII are exclusively in the C-terminal 400 amino acids of topoisomerase II, the region that is most divergent among the eukaryotic type II enzymes and which is absent in the bacterial gyrase homologues.

Abbreviations: topoII – topoisomerase II; CKII – Casein Kinase II; SV40 – Simian Virus 40

Introduction

Eukaryotic chromosomes undergo major structural changes that reflect their varied functions through the cell cycle. These changes are mediated in part by DNA topoisomerases. To relieve both positive and negative supercoiling, eukaryotic organisms have two types of topoisomerases: DNA topoisomerase I and II. The type I enzymes alter the DNA linking number in steps of one by means of a transient single strand cut, while topoisomerase II makes a double stranded break, passes the same or another DNA molecule through the break and reseals the strands, changing the linking number in steps of two (reviewed by Wang 1985).

Because each has the capacity to relax supercoiled DNA, topoisomerases I and II can substitute for each other in many biological functions. Studies of yeast mutants suggest that both topoisomerases I and II are able to relieve DNA torsional stress or supercoiling introduced by chain elongation during transcription (Brill & Sternglanz 1988) and DNA replication (Kim & Wang 1989a; Brill et al. 1987). Similarly, studies with topoisomerase I-topoisomerase II double mutants demonstrated that either enzyme can suppress high levels of mitotic recombination in the ribosomal DNA repeated units (Christman et al. 1988; Kim & Wang 1989b).

Although both topoisomerases appear to compensate for each other in many instances, topoisomerase II is an essential enzyme while topoisomerase I is not. This reflects the unique ability of topoisomerase II to decatenate covalently closed DNA circles or intertwined helices, rendering the enzyme essential for the proper disjunction of sister chromatids in mitosis (Dinardo et al. 1984;

Holm et al. 1985; Uemura & Yanagida 1984, 1986) and for the segregation of chromosomes in meiosis (Rose et al. 1990). The requirement of topoisomerase II at mitosis may also reflect a role in chromosome condensation and the higher order folding of chromatin (Uemura et al. 1987).

Studies of chromosome structure suggest that topoisomerase II is involved in the organization of looped chromatin domains. Antisera raised against the most abundant component of the metaphase scaffold isolated from either human or chick chromosomes, a 170 kDa protein called SC-1, showed unequivocally that SC-1 is identical to topoisomerase II (Earnshaw et al. 1985; Gasser et al. 1986). Topoisomerase II was immunolocalized to the axial core of mitotic chromosomes (Earnshaw & Heck 1985; Gasser et al. 1986) and to the synaptonemal complex in pachytene cells from species as varied as chicken, mouse and yeast (Moens & Earnshaw 1989; Klein et al. 1992). Topoisomerase II also cofractionates with the interphase nuclear matrix fraction of *Drosophila* and yeast cells (Berrios et al. 1985; Berrios & Fisher 1988; Cardenas et al. 1990). This does not depend upon the enzymatic activity of the enzyme, since temperature-sensitive mutants in topoisomerase II show the same fractionation pattern in yeast at either permissive or non-permissive temperatures (Cardenas et al. 1990). Electron microscopic studies have suggested that the enzyme preferentially recognizes sites where two DNA molecules cross over each other, in either positively or negatively supercoiled DNA (Zechiedrich & Osheroff 1990). The role of topoisomerase II in chromosome condensation, its binding sites in AT-rich DNA, and its post-translational modification are discussed below.

A role in chromatin condensation

A study of a cold sensitive topoisomerase II mutant in *Schizosaccharomyces pombe* provided the first functional evidence that topoisomerase II was required for proper chromosome condensation (Uemura et al. 1987). In a cold-sensistive (cs) β-tubulin mutant at nonpermissive temperature, the microtubules are destabilized and one observes by DAPI fluorescence the highly condensed *S. pombe* mitotic chromosomes (Umesono et al. 1983). When this mutation is coupled with a top2 cs mutant, chromosomes do not condense properly, remaining in an extended organization reminiscent of prophase (Uemura et al. 1987).

This observation was carried further by Laemmli and coworkers, who have shown that in a *Xenopus* egg extract the conversion of nuclei into condensed mitotic chromosomes requires the presence of topoisomerase II, either as a component of the substrate nuclei, or as a component of the egg extracts in which the mitotic events occur (Adachi et al. 1991). While HeLa nuclei, which have high levels of topoII, condense properly in topoII-depleted extracts, nuclei which have low levels of endogenous topoisomerase II are unable to form fully condensed chromosomes in the depleted extracts. Addition of exogenous yeast topoisomerase II, purified from an overexpressing strain, restores chromosome condensation. Although the requirement of topoII activity was not examined in this assay, earlier studies of nuclear disassembly in *Xenopus* extracts had noted the inhibition of chromosome condensation in the presence of VM-26, a specific inhibitor of topoisomerase II (Newport and Spann 1987).

Similar results have been obtained using an extract from somatic cells blocked in mitosis (Wood & Earnshaw 1990). Again it was observed that the degree of condensation of exogenously added nuclei in a topoII-depleted extract appeared to depend on the level of endogenous topoisomerase II. However, since chick erythrocyte nuclei, with <300 copies of topoII per nucleus, showed a significant condensation of chromatin without resolution into chromosomes, these authors stress the multiplicity of events involved in chromosome condensation. Among the obvious ones are histone H1 and H3 phophorylation and lamin disassembly.

A link has been proposed to exist between topoisomerase II activity and the ability of the $p34^{cdc2}$ kinase to phosphorylate histone H1 and disassemble the nuclear lamina (Peter et al. 1990). The topoII inhibitor VM-26 was shown to block the activation of $p34^{cdc2}$ kinase at G2/M, and the cells' progression into mitosis (Roberge et al. 1990). This might suggest that decatenation by topoisomerase II is required before entry into mitosis. An alternate explanation of these results, however, is that DNA strand breakage is induced by the inhibition of topoisomerase II, and that the damage stimulates a feedback pathway that blocks the cell in G2. There is ample evidence for such a pathway in *S. cerevisiae* (Hartwell & Weinert 1989).

While chromosome condensation in mammalian cells may be more complex than it is in yeast cells, the studies on topoisomerase II underscore the similarities in this important cell cycle event among eukaryotes. As described above, observations originally made in yeast mutants have been confirmed and extended using mammalian and vertebrate systems. In addition it is clear that *Drosophila* topoisomerase II can complement a yeast null mutant (Wyckoff & Hsieh 1988) and that in reconstituted systems from vertebrates, topoisomerase II from yeast can replace the endogenous enzyme (Adachi et al. 1991). Thus genetic and cellular approaches can be combined in the study of the mitotic function of topoisomerase II.

DNA binding properties

Many laboratories have tried to identify the preferred binding sites of topoisomerase II by mapping double strand breaks induced by the enzyme either *in vitro* or in intact cells. In some cases the sites mapped *in vitro* agree with those found *in vivo*, and in others they do not (reviewed in Liu 1989). Some of the discrepancy may be due to a certain promiscuity of the enzyme brought about by treatment with drugs (VM-26, m-AMSA, VP-16) that impede the religation step of the enzymatic reaction and thus enhance the detection of cleavages. However, it is likely that the primary cause of discrepancy is that topoisomerase II interacts differently with naked DNA in solution than with chromatin. It was noted that in reconstituted SV-40 chromatin and in the SV-40 chromosome *in vivo*, topoisomerase II cleaved primarily in linkers between nucleosomes or in DNase 1 hypersensitive sites, which generally did not coincide with preferred cleavage sites mapped on naked DNA *in vitro* (Yang et al. 1985; Capranico et al. 1990). In intact cells it was also observed that topoisomerase II cleavage sites fell either into DNase 1 hypersensitive regions, indicating an open, non-nucleosomal organization, or into nucleosomal spacers (Rowe et al. 1986; Reitman & Felsenfeld 1990; Udvardy & Schedl 1991). Although there appears no strict consensus for cleavage sites *in vivo*, sequencing of several cleavage sites found in nucleosomal linkers upstream of the *Drosophila* Histone H1 gene shows a preferential cut at GC rich stretches that are flanked by AT-rich regions (Kaes & Laemmli 1992).

In vitro the preference of topoisomerase II for AT-rich DNA is quite striking. Sander and Hsieh (1985) earlier defined a consensus for *Drosophila* topoisomerase II based on its cleavage of plasmid DNA *in vitro*. This AT-rich, 15bp sequence was but a loosely defined consensus; nonetheless, it was striking that this consensus for topoisomerase II occurred frequently in the fragments of DNA that have a high affinity for the nuclear scaffold (Gasser & Laemmli 1986a, b; Cockerill & Garrard 1986). These SARs or MARs (scaffold or matrix-attached regions) are candidates for sequences that organize the genome into looped domains, since they show a high affinity for a histone depleted nuclear and chromosomal substructure (Mirkovitch et al. 1984, 1986). In yeast these scaffold-associated regions are coincident with autonomously replicating sequences, commonly known as ARS elements (Amati & Gasser 1988, 1990). Consistent with studies in other species that map topoisomerase II cleavage sites to SARs (e.g. Udvardy et al. 1985), we show in Fig. 1 that yeast topoisomerase II also cuts preferentially in ARS sequences *in vitro*. As shown in Figure 1 for the rDNA repeat, transcribed regions have far fewer topoII cleavage sites.

An advance was made in understanding this interaction by showing that topoisomerase II binds preferentially to SAR sequences *in vitro*, whether it is induced to cleave the DNA or not (Adachi et al. 1989; Sperry et al. 1989). In addition, Adachi et al. (1989) showed that topoisomerase II binds in a cooperative manner and is capable of precipitating the DNA, probably as a result of aggregation of the protein-DNA complexes. Since the binding of to-

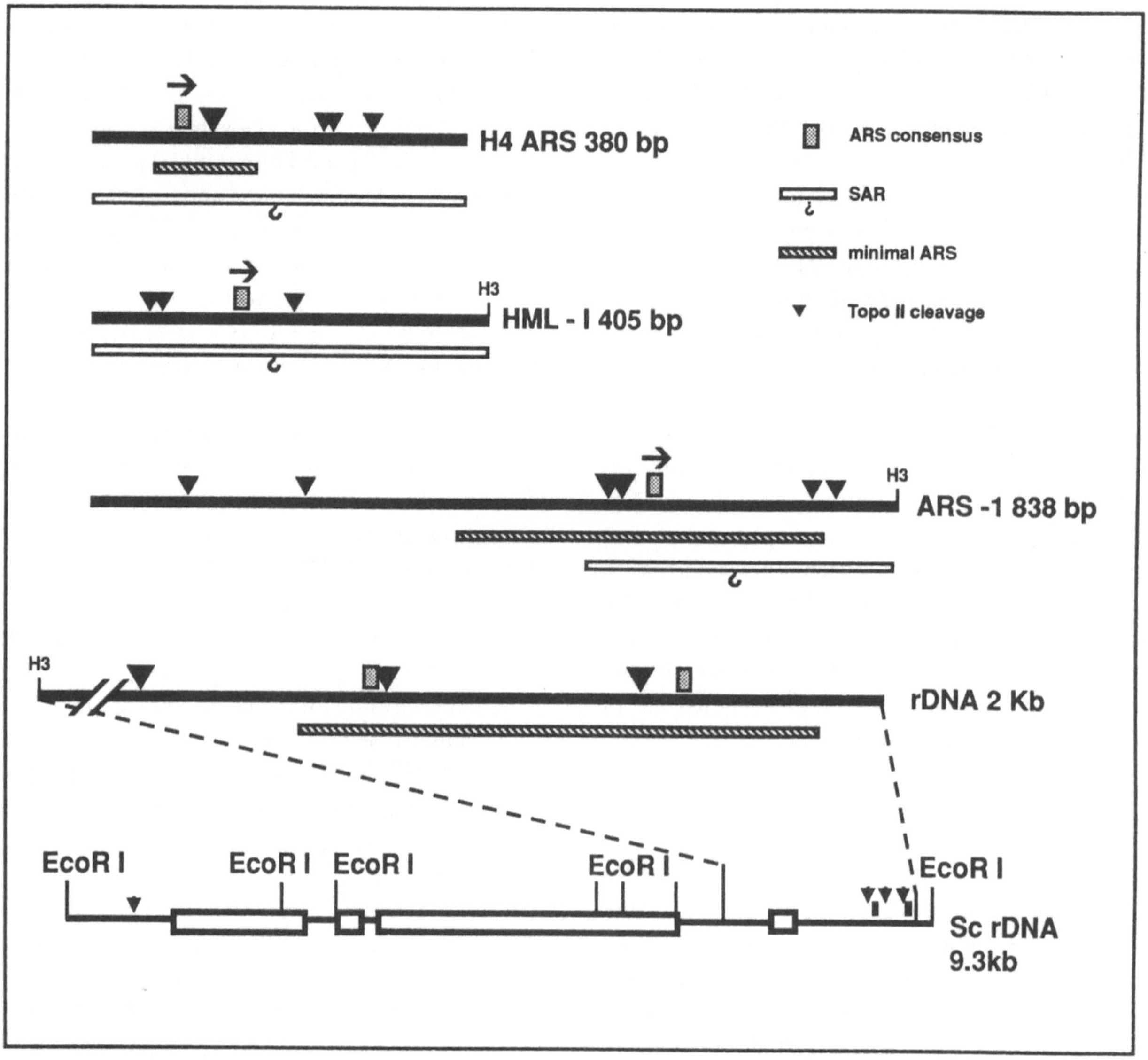

Fig. 1. Mapping of preferred sites of topoisomerase II cleavage in yeast ARS sequences.

Purified fragments of DNA containing the histone H4 ARS (H4 ARS), the HMLI silencer-ARS (HML-I), ARS-1, and the rDNA repeat (Sc rDNA), all from *S. cerevisiae*, were excised, end-labelled and incubated with purified yeast topoisomerase II. Trapping of the cleavage complex was achieved by the addition of 1% SDS in the absence of anti-topoII drugs. This leaves topoII covalently bound to the each strand of the cleaved double helix. Determination of the precise cleavage site was done on a sequencing gel, as described in Cardenas et al. (1992b). Cleavage sites are indicated by arrowheads. In the cases in which a minimal ARS domain has been identified, this is indicated by a hatched box. The location of the 11bp ARS consensus is indicated by a filled box with the arrows indicating the 5′ to 3′ direction of the T-rich strand of the consensus. The minimal fragments showing association with the nuclear scaffold of yeast is indicated below the maps of H4 ARS, ARS-1 and HML-I. The size and name of each fragment is indicated at the right.

poII is inhibited by distamycin, the interaction of topoisomerase II with SARs apparently reflects the enzyme's preference for the narrow minor groove of oligo dT-rich DNA (Nelson et al. 1987). Assuming that this interaction can occur *in vivo*, it suggests a means by which topoisomerase II could mediate chromosome condensation. If topoisomerase II were to able to complex other proteins and/or itself while binding SARs, or the nucleosomal linkers in SARs, the bases of loops might be brought together in clusters, forming a subunit of a radially organized chromatid. One would expect any such function of topoisomerase II to be regulated through the cell cycle.

Post-translational modification of topoisomerase II

In vitro phosphorylation experiments have shown that topoisomerase II is a substrate for casein kinase II (Ackerman et al. 1985), protein kinase C (Rottman et al. 1987; Sahyoun et al. 1986), Ca^{2+}/calmodulin-dependent protein kinase and $p34^{cdc2}$ kinase (Cardenas et al. 1992a). In all cases phosphorylation *in vitro* enhances topoisomerase II activity, as measured by the kinetoplast decatenation assay. In a variety of species ranging from yeast to mammalian cells, topoisomerase II has been shown to be phosphorylated *in vivo* (Ackerman et al. 1988; Rottman et al. 1987; Heck et al. 1989; Saijo et al. 1990; Cardenas et al. 1992a). Cell cycle phosphorylation studies in chicken lymphoblastoid cells have shown that topoisomerase II phosphorylation increases as cells enter mitosis (Heck et al. 1989). Similarly, studies with cell cycle-arrested yeast cells show that topoisomerase II is 6–10 times more highly phosphorylated in mitosis than in G1 (Cardenas et al. 1992a). Consistent with a role of phosphorylation in stimulating topoisomerase II activity at mitosis is the observation that soluble extracts from mitotic Hela cells contain more DNA decatenation activity than extracts obtained from S-phase cells (Estey et al. 1987). Taken together, these observations strongly suggest that *in vivo* the activity of topoisomerase II is modulated by phosphorylation. Our own studies have analyzed the pattern of modification obtained in both mitosis and G1 phase, and have identified the major kinase involved in the modification.

In studies using phosphate-labelled yeast cells, two dimensional phosphopeptide analysis of topoisomerase II showed that the phosphorylation patterns of the enzyme in G1 phase and at mitosis are related but not identical (Cardenas et al. 1992a). Among major phosphate accepting peptides, all but two comigrate with those phosphorylated by purified casein kinase II *in vitro*. Nonetheless, there are qualitative differences between the CKII-modified phosphopeptides observed in mitotic and in G1 phase cells. Some CKII sites appear to be preferentially targeted in mitosis, while others are weak in mitosis and relatively strong in G1 phase cells. Barring a non-physiological effect of the synchronization procedure, this suggests that the interaction of CKII with topoisomerase II varies through the cell cycle, the significance of which remains to be understood.

Genetic evidence using a temperature-sensitive casein kinase II mutant confirms the importance of casein kinase II in the modification of topoisomerase II. At non-permissive temperature in the casein kinase II mutant, the phosphate incorporated into topoisomerase II *in vivo* dropped to an almost undetectable level (Cardenas et al. 1992a). This role for casein kinase II in phosphorylation of topoII is likely to be conserved across species, since both topoisomerase II and CKII show high degrees of conservation in both sequence and function. The catalytic α subunit of CKII complexes the regulatory β subunit in an $\alpha_2\beta_2$ heterodimer cofiguration. The β subunit appears to be encoded by a single gene, while in yeast there are two α-subunits encoded by different genes (Padmanabha & Glover 1987; Chen-wu et al. 1988; Padmanabha et al. 1990). The disruption of both α subunits of CKII is lethal, and the *S. cerevisiae* cka1-cka2ts mutant used to study topoisomerase II phosphorylation arrests preferentially at the non-permissive temperature as large budded cells, a phenotype consistent with a G2-M arrest point (Glover CVC, unpublished results). This arrest point may be due in part to the requirement of CKII to activate topoisomerase II, which in its turn is required for the proper traversal of mitosis.

From the analysis of topoII phosphorylated *in vivo*, at least two significant phospho-acceptor sites could be identified that do not correspond to CKII sites *in vitro* (Cardenas et al. 1992a). These may indicate a difference in the specificity of CKII *in vitro* as compared to *in vivo*, or may represent acceptor sites for another kinase. In the latter case, there are two plausible explanations why non-CKII

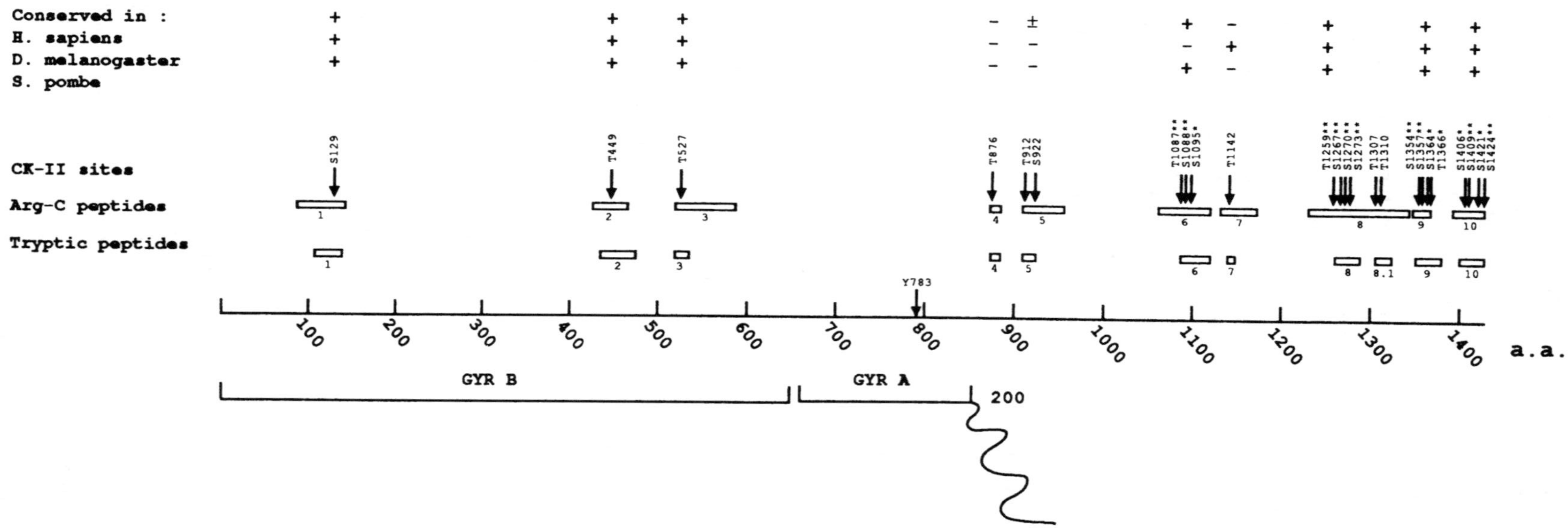

Fig. 2. Schematic representation of yeast topoisomerase II, its phoshopeptides and CKII target sites.

Topoisomerase II from *S. cerevisiae* is portrayed linearly as amino acids 1–1429 (a.a.). The tyrosine at a.a. 783 contacts DNA during catalysis. Regions of homology to the bacterial gyrase subunits A (GYR A) and B (GYR B) are shown. The amino acid residues that best fit the casein kinase II consensus for phosphoacceptor sites are indicated by arrows above the map, and the a.a. corresponding to the target serine or threonine is given. The corresponding tryptic peptides or the arginine C peptides are indicated as open bars. Single stars indicate sites which are likely to be phosphorylated by CKII *in vitro*, and two stars indicate sites are that are definitely phosphorylated by CKII. Above the corresponding peptides it is indicated whether or not the target sites are conserved (+) or not conserved (-) in topoII from *H. sapiens*, *D. melanogaster*, or *S. pombe*.

sites might not be significantly phosphorylated in the ck2 mutant. First, casein kinase II may be required to modify and activate the second kinase. Alternatively, the CKII modification may be required to render topoisomerase II a proper substrate for the second kinase. One intriguing possibility is that CKII phosphorylation of topoisomerase II is a necessary prerequisite for modification by a cell-cycle specific kinase, such as the $p34^{cdc2}$ kinase. Indeed, *in vitro* yeast topoisomerase II is a good substrate for $p34^{cdc2}$ kinase (Labbé et al. 1989; Cardenas et al. 1992b). Phosphorylation by $p34^{cdc2}$ kinase would be G2/M specific in higher eukaryotic cells, and might modify topoisomerase II in such a way to help catalyze the condensation of chromatin into metaphase chromosomes. While this question can probably not be addressed in yeast alone, a combined study of yeast and vertebrate cells may lead us to an understanding of the fine-tuning of topoisomerase II activities through the cell cycle by phosphorylation.

Mapping phosphoacceptor sites in topoisomerase II

The location of putative CKII phosphorylation sites in yeast topoisomerase II was determined by computer analysis using the minimal consensus of T/S-x-x-E/D, as described by Pinna (1990). While many potential sites could be identified, the phosphoamino acid analysis of the various of tryptic and arginine-C derived phosphopeptides showed that only sites in the C-terminus of topoisomerase II are labelled to a significant level in either G1 or M phase (Fig. 2). In fact, each of the four modified peptides contains multiple casein kinase II sites, which results in a characteristic series of diagonally aligned phosphopeptide derivatives after two dimensional peptide analysis (Cardenas et al. 1992a).

The C-terminal region of topoisomerase II diverges completely from the corresponding domain in the bacterial gyrase, suggesting that the regulation of activity by phosphorylation is an exclusive property of the eukaryotic topoisomerase II. The importance of this C-terminal domain is supported by the study of a truncated version of the yeast topoisomerase II. Truncation at a.a. 979, which removes all of the C-terminal CKII phospho-acceptor sites, renders the enzyme unable to complement a *S. cerevisiae* top2 ts mutant, even though the truncated form of the enzyme contains all regions of homology to gyrase A and B subunits including the active site tyrosine residue (MEC, unpublished results). Among the various eukaryotic type II topoisomerases that are cloned, despite significant sequence variation in this C-terminal region, sequence comparison shows that the presence and relative position of strong casein kinase II recognition sites are conserved among the *S. cerevisiae, S. pombe, Drosophila*, and human enzymes (Fig. 2). No tryptic peptide mapping data are available from species other than *S. cerevisiae*, but at some sites multiply phosphorylated peptides are predicted from the C-terminal domain, as we have observed in yeast. It will be necessary to modify the sites of phosphorylation by site-directed mutagenesis to confirm their localization and to provide further evidence concerning their proposed regulatory role.

The effects of phosphorylation on topoisomerase II

Many important nuclear proteins are substrates of casein kinase II. Among these are c-ErbA, E1A, the E7 protein of human papilloma virus (HPV), SV-40 T antigen, the avian c-Myb, c-Fos, p53, cAMP response element binding factor, Serum Response Factor, and two nucleolar substrates, B23 and nucleolin (reviewed in Pinna 1990; Meisner & Czech 1991). The effects of phosphorylation by CKII are not uniform among this diverse set of substrates; in the case of HPV E7, CKII phosphorylation is proposed to correlate with the transforming potential of the virus, while in the large T antigen, mutagenesis of the acceptor serine had little effect on the transforming potential of SV40 (Barbosa et al. 1990). On the other hand, mutation of the phospho-acceptor site decreases the rate of nuclear-cytoplasmic transport of β-galactosidase-T antigen fusion polypeptides (Rihs et al. 1991). In the case of yeast topoisomerase II, phosphorylation by CKII does not seem to be a prerequisite for nuclear targetting, since in the CKII ts mutants at non-permissive temperature, topoisomerase II is properly localized to the nucleus (MEC and SMG, unpublished results).

The most clear cut effect of topoII modification by CKII is a dramatic stimulation of its decatenation activity. Dephosphorylation of topoisomerase II renders the enzyme unable to decatenate kinetoplast DNA and its affinity for DNA is greatly diminished, as measured by a gel retardation assay. Phosphorylation by purified CKII suffices to reactivate alkaline phosphatase-treated topoisomerase II (Cardenas et al. 1992b). Thus, an active yeast topoisomerase II appears to require at least a basal level of phosphorylation. On the other hand, hyperphosphorylation of topoII by casein kinase II *in vitro* has little effect on the efficiency with which topoII could be trapped in its cleavable complex (Cardenas et al. 1992b). Thus the stimulation of decatention activity by casein kinase II most likely reflects an enhanced affinity for DNA and perhaps an enhanced efficiency at a step in the catalytic process other than DNA cleavage. Consistent with a role in chromosome condensation is the enhancement of protein-protein interactions and aggregation upon hyperphosphorylation *in vitro* (QD, in preparation). To determine which of the minor or the major phosphoacceptor sites on topoisomerase II plays a role in the observed stimulation of activity requires *in vitro* mutagenesis of the acceptor sites, work that is currently in progress.

Acknowledgements

Research was supported by grants to SMG from the Swiss National Science Foundation, the Swiss Cancer League, and the Human Frontiers Research Program. MEC was supported by the Roche Research Foundation (Switzerland).

References

Ackerman P, Glover CVC & Osheroff N (1985) Phosphorylation of DNA topoisomerase II by casein kinase II: modulation of eukaryotic topoisomerase II activity *in vitro*. Proc. Natl. Acad. Sci. USA 82: 3164–3168

Ackerman P, Glover CVC & Osheroff N (1988) Phosphorylation of DNA topoisomerase II *in vivo* and in total homogenates of *Drosophila* Kc cells. J. Biol. Chem. 263: 12653–12660

Adachi Y, Kaes E & Laemmli UK (1989) Preferential, cooperative binding of DNA topoisomerase II to scaffold-associated regions. EMBO J. 8: 3997–4006

Adachi Y, Luke M & Laemmli UK (1991) Chromosome assembly *in vitro*: topoisomerase II is required for condensation. Cell 64: 137–148

Amati B & Gasser SM (1988) Chromosomal ARS and CEN elements bind specifically to the yeast nuclear scaffold. Cell 54: 967–978

Amati B & Gasser SM (1990) Drosophila scaffold-attached regions bind nuclear scaffolds and can function as ARS elements in both budding and fission yeast. Mol. Cell. Biol. 10: 5442–5454

Barbosa MS, Edmonds C, Fisher C, Schiller JT, Lowy DR & Vousden KH (1990) The region of the HPV E7 oncoprotein homologous to adenovirus E1a and SV40 large T antigen contains separate domains for Rb binding and casein kinase II phosphorylation. EMBO J. 9: 153–160

Berrios M, Osheroff N & Fisher PA (1985) In situ localization of DNA topoisomerase II, a major polypeptide component of the Drosophila nuclear matrix fraction. Proc. Natl. Acad. Sci. USA 82: 4142–4146

Berrios S & Fisher PA (1988) Thermal stabilization of putative karyoskeletal protein-enriched fractions from S. cerevisiae. Mol. Cell. Biol. 8: 4573–4575

Brill SJ & Sternglanz R (1988) Transcription-dependent DNA supercoiling in yeast DNA topoisomerase mutants. Cell 54: 403–411

Brill SJ, Dinardo S, Voelkel Meiman K & Sternglanz R (1987) Need for DNA topoisomerase activity as a swivel for DNA replication for transcription of ribosomal RNA. Nature 326: 414–416

Capranico G, Jaxel C, Roberge M, Kohn KW & Pommier Y (1990) Nucleosome positioning as a critical determinant for the DNA cleavage sites of mammalian DNA topoisomerase II in reconstituted Simian virus 40 chromatin. Nucleic Acids Res. 18: 4553–4559

Cardenas ME, Laroche T & Gasser SM (1990) The composition and morphology of yeast nuclear scaffolds. J. Cell Sci. 96: 439–450

Cardenas ME, Dang Q, Glover CVC & Gasser SM (1992a) Casein kinase II phosphorylates the eukaryotic specific C-terminal domain of topoisomerase II *in vivo*. EMBO J. 11 (in press)

Cardenas ME, Hanna DE, Walter R & Gasser SM (1992b) Topoisomerase II requires phosphorylation for efficient DNA binding and DNA decatenation: involvement of casein kinase II (submitted to J. Cell Sci.)

Chen-wu JL, Padmanabha R & Glover CVC (1988) Isolation, sequencing and disruption of the CKA1 gene encoding the alpha subunit of yeast casein kinase II. Mol. Cell. Biol. 8: 4981–4990

Christman MF, Dietrich FS & Fink GR (1988) Mitotic recombination in the rDNA of *S. cerevisiae* is suppressed by the combined action of DNA topoisomerase I and II. Cell 55: 413–425

Cockerill PN & Garrard WT (1986) Chromosomal loop anchorage of the kappa imunoglobulin gene occurs next to the enhancer in a region containing topoisomerase II sites. Cell 44: 273–282

Dinardo S, Voelkel K & Sternglanz R (1984) DNA topoisomerase II mutant of Saccharomyces cerevisiae: topoisomerase II is required for segregation of daughter molecule at the termination of DNA replication. Proc. Natl. Acad. Sci. USA 82: 2616–2620

Earnshaw WC & Heck MMS (1985) Localization of topoisomerase II in mitotic chromosomes. J. Cell Biol. 100: 1716–1722

Earnshaw WC, Halligan B, Cooke CA, Heck MMS & Liu LF (1985) Topoisomerase II is a structural component of mitotic chromosome scaffolds. J. Cell Biol. 100: 1706–17150

Estey E, Adlakha RC, Hittelman WN & Zwelling LA (1987) Cell cycle stage dependent variations in drug-induced topoisomerase II mediated DNA cleavage and cytotoxicity. Biochemistry 26: 4338–4344

Gasser SM & Laemmli UK (1986a) The organization of chromatin loops: characterization of a scafffold attachment site. EMBO J. 5: 511–518

Gasser SM & Laemmli UK (1986b) Cohabitation of scaffold binding regions with upstream/enhancer elements of three developmentally regulated genes of D. melanogaster. Cell 46: 521–530

Gasser SM & Laemmli UK (1987) A glimpse at chromosomal order. Trends Genet. 3: 16–22

Gasser SM, Laroche T, Falquet J, Boy de la Tour E & Laemmli UK (1986) Metaphase chromosome structure involvement of topoisomrase II. J. Mol. Biol. 188: 613–629

Hartwell LH & Weinert TA (1989) Checkpoints: controls that ensure the order of cell cycle events. Science 246: 629–634

Heck MMS & Earnshaw WC (1986) Topoisomerase II, a specific marker for cell proliferation. J. Cell Biol. 103: 569–2581

Heck MMS, Hittelman WN & Earnshaw WC (1989) *in vivo* phosphorylation of the 170-kDa form of eukaryotic DNA topoisomerase II. J. Biol. Chem. 264: 15161–15164

Holm C, Goto T, Wang JC & Botstein D (1985) DNA topoisomerase II is required at the time of mitosis in yeast. Cell 41: 553–563

Kaes E & Laemmli UK (1992) *In vivo* topoisomerase II cleavage of the Drosophila histone and satellite III repeats: DNA sequence and structural requirements. EMBO J. 11: 705–716

Kim RA & Wang JC (1989a) A subthreshold level of DNA topoisomerases leads to the excision of yeast rDNA as extrachromosomal rings. Cell 57: 975–985

Kim RA & Wang JC (1989b) Function of DNA topoisomerases as replication swivels in Saccharomyces cerevisiae. J. Mol. Biol. 208: 257–267

Klein F, Laroche T, Cardenas ME, Hofmann JFX, Schweizer D & Gasser SM (1992) Localization of RAPI and Topoisomerase II in Nuclei and Meiotic Chromosomes of Yeast. J. Cell Biol. 117 (in press)

Labbé JC, Capony JP, Caput D, Cavadore JC, Derancourt J, Kaghad M, Lelias JM, Picard A & Dorée M (1989) MFP from starfish oocytes at first meiotic metaphase is a heterodimer containing one molecule of cdc2 and one molecule of cyclin B. EMBO J. 8: 3053–3058

Liu LF (1989) DNA topoisomerase poisons as anti-tumor drugs. Ann. Rev. Biochem. 58: 351–375

Meisner H & Czech PM (1991) Phosphorylation of transcriptional factors and cell-cycle-dependent proteins by casein kinase II. Curr. Op. in Cell Biol. 3: 474–483

Mirkovitch J, Mirault ME & Laemmli UK (1984) Organization of the higher-order chromatin loop: specific DNA attachment sites on nuclear scaffold. Cell 39: 223–232

Mirkovitch J, Gasser SM & Laemmli UK (1988) Scaffold attachment of DNA loops in metaphase chromosomes. J. Mol. Biol. 200: 101–110

Moens B & Earnshaw WC (1989) Anti-topoisomerase II recognizes meiotic chromosomal cores. Chromosoma 98: 317–322

Nelson HCM, Finch JT, Luisi BF & Klug A (1987) The structure of an oligo(dA).oligo(dT) tract and its biological implications. Nature 330: 221–226

Newport J & Spann T (1987) Disassembly of the nucleus in mitotic extracts: membrane vesicularization, lamin disassembly, and chromosome condensation are independent processes. Cell 48: 219–230

Padmanabha R & Glover CVC (1987) Casein kinase II of yeast contains two distinct polypeptides and an unusually large subunit. J. Biol. Chem. 262: 1829–1835

Padmanabha R, Chen-wu JL, Hanna DE & Glover CVC (1990) Isolation, sequencing and disruption of the yeast CKA2 gene: casein kinase II is essential for viability in *S. cerevisiae*. Mol. Cell. Biol. 10: 4089–4099

Peter M, Nakagawa J, Dorée M, Labbé J C & Nigg EA (1990) *in vitro* disassembly of the nuclear lamina and M phase-specific phosphorylation of lamins by cdc2 kinase. Cell 61: 591–602

Pinna LA (1990) Casein kinse II: an 'eminence grise' in cellular regulation? Biochim. Biophys. Acta 1054: 267–284

Reitman M & Felsenfeld G (1990) Developmental regulation of topoisomerase II sites and DNase I-hypersensitive sites in the chicken b-globin locus. Mol. Cell. Biol. 10: 2774–2786

Rihs HP, Jans DA, Fan H & Peters R (1991) The rate of nuclear cytoplasmic protein transport is determined by the casein kinase II site flanking the nuclear localization sequence of the SV40 T-antigen. EMBO J. 10: 633–639

Roberge M, Th'ng J, Hamaguchi J & Bradbury EM (1990) The topoisomerase II inhibitor VM-26 induces marked changes in histone H1 kinase activity, histones H1 and H3 phorphorylation and chromosome condensation in G2 phase and mitotic BHK cells. J. Cell Biol. 111: 1753–1762

Rose D, Thomas W & Holm C (1990) Segregation of recombined chromosomes in meiosis I requires topoisomerase II. Cell 60: 1009–1017

Rottman M, Schröder HC, Gramzow M, Renneisen K, Kurelec B, Dorn A, Friese U & Müller WEG (1987) Specific phosphorylation of proteins in pore-complex-laminae from the sponge *Geodia cydonium* by the homologous aggregation factor and phorbol ester. Role of protein kinase C in the phosphorylation of DNA topoisomerase II. EMBO J. 6: 3939–3944

Rowe TC, Wang JC & Liu LF (1986) *In vivo* localization of DNA topoisomerase II cleavage sites on Drosophila heat shock chromatin. Mol. Cell. Biol. 6: 985–992

Sahyoun N, Wolf M, Besterman J, Hsieh TS, Sander M, Levine III H, Chang KJ & Cuatrecasas P (1986) Protein kinase C phosphorylates topoisomerase II: topoisomerase activation and its possible role in phorbol ester-induced differentiation of HL-60 cells. Proc. Natl. Acad. Sci. USA 83: 1603–1607

Saijo M, Enamoto T, Hanaoka F & Ui M (1990) Purification and characterization of type II DNA topoisomerase from mouse FM3A cells: phosphorylation of topoisomerase II and modification of its activity. Biochemistry 29: 583–590

Sander M & Hsieh TS (1985) Drosophila topoisomerase II double-strand DNA cleavage: analysis of DNA sequence homology at the cleavage site. Nucleic Acids Res. 13: 1057–1072

Sperry AO, Blasquez VC & Garrard WT (1989) Dysfunction of chromosomal loop attachement sites: illegitimate recombination linked to matrix association regions and topoisomerase II. Proc. Natl. Acad. Sci. USA 86: 5497–5501

Tsai-Pflugfelder M, Liu L, Liu A, Tewey KM, Whang-Peng J, Knutsen T, Huebner K, Croce CM & Wang JC (1988) Cloning and sequencing of cDNA encoding human DNA topoisomerase II and localization of the gene to chromosome region 17q21–22. Proc. Natl. Acad. Sci. USA 85: 7177–7181

Udvardy A & Schedl P (1991) Chromatin structure, not DNA sequence specificity, is the primary determinant of topoisomerae II sites of action *in vivo*. Mol. Cell. Biol. 11: 4973–4984

Udvardy A, Schedl P, Sander M & Hsieh TS (1985) Novel partitioning of DNA cleavage sites for Drosophila topoisomerase II. Cell 40: 933–941

Uemura T & Yanagida M (1984) Isolation of type I and II DNA topoisomerase mutants from fission yeast: single and double mutants show different phenotypes in cell growth and chromatin organization. EMBO J. 3: 1737–1744

Uemura T & Yanagida M (1986) Mitotic spindle pulls but fails to separate chromosomes in type II DNA topoisomerase mutants: uncoordinated mitosis. EMBO J. 5: 1003–1010

Uemura T, Ohkura H, Adachi Y, Morino K, Shiozaki K & Yanagida M (1987) DNA topoisomerase II is required for condensation and separation of mitotic chromosomes in *S. pombe*. Cell 50: 917–925

Umesono K, Hiraoka Y, Toda T & Yanagida M (1983) Visualization of chromosomes in mitotically arrested cells of the fission yeast Schizosaccharomyces pombe. Curr. Genet. 7: 123–128

Wang JC (1985) DNA topoisomerases. Ann. Rev. Biochem. 54: 665–697

Wood ER & Earnshaw WC (1990) Mitotic chromain condensation *in vitro* using somatic cell extracts and nuclei with variable levels of endogenous topoisomerase II. J. Cell Biol. 111: 2839–2850

Wyckoff E & Hsieh TS (1988) Functional expression of a Drosophila gene in yeast: genetic complementation of DNA topoisomerase II. Proc. Natl. Acad. Sci. USA 85: 6272–6276

Yang L, Rowe TC, Nelson EM & Liu LF (1985) *in vivo* mapping of DNA topoisomerase II-specific cleavage sites on SV40 chromatin. Cell 41: 127–132

Zechiedrich EL & Osheroff N (1990) Eukaryotic topoisomerases recognize nucleic acid topology by preferentially interacting with DNA crossovers. EMBO J. 9: 4555–4562

Antonie van Leeuwenhoek **62**: 25–33, 1992.

Global regulators of chromosome function in yeast

John F.X. Diffley
Imperial Cancer Research Fund, Clare Hall Laboratories, Blanche Lane, South Mimms, Potters Bar, Herts. EN6 3LD, England

Key words: ARS, centromere, global transcriptional activators, telomeres

Abstract

In the yeast *Saccharomyces cerevisiae*, several abundant, sequence-specific DNA binding proteins are involved in multiple aspects of chromosome function. In addition to functioning as transcriptional activators of a large number of yeast genes, they are also involved in transcriptional silencing, the initiation of DNA replication, centromere function and regulation of telomere length. This review will consider each of these proteins, focusing on what is known about the mechanisms of their multiple functions.

RAP1

The best characterized of these global yeast factors is the product of the *RAP1* gene, known as RAP1, TUF and GRFI. RAP1 binds within elements involved in silencing the yeast mating type loci *HML* and *HMR* (Buchman et al. 1988a; Shore et al. 1987), within the promoters of many yeast genes (Brindle et al. 1990; Buchman et al. 1988b; Capieaux et al. 1989; Giesman et al. 1991; Hamil et al. 1988; Huet & Sentenac 1987; Kurtz & Shore 1991; Moehle & Hinnebusch 1991; Santangelo & Tornow 1990; Shore et al. 1987; Stanway et al. 1987; Vignais et al. 1987) and within yeast telomeres (Buchman et al. 1988a; Buchman et al. 1988b). The *RAP1* gene has been cloned and sequenced and is essential for viability in yeast (Shore & Nasmyth 1987). The RAP1 protein is 827 amino acids long(Shore & Nasmyth 1987) and contains a domain of 235 amino acids near the centre of the protein which is both necessary and sufficient for binding to the RAP1 consensus sequence with high affinity (Henry et al. 1990) (see Fig. 1). This rather large DNA binding domain does not contain any of the structural motifs usually associated with DNA binding domains like the zinc finger or the helix-turn-helix.

RAP1 was initially identified as a factor that bound between 200–400 bp upstream of most of the ribosomal protein genes at upstream activating sequences known as UASrpg (Huet et al. 1985). Since that time, RAP1 binding sites have been found in the promoters of a large number of yeast genes, in many cases coinciding with UAS elements. Additionally, oligonucleotides containing RAP1 binding sites, when cloned upstream of a UAS-less reporter gene, efficiently activate transcription (Brand et al. 1987). Thus, in the appropriate context, RAP1 can function as a transcriptional activator.

RAP1 can also act as a transcriptional repressor. In the yeast *Saccharomyces cerevisiae*, mating type is dictated by the expression of either α or a information from the *MAT* locus on chromosome III (for a review, see Nasmyth & Shore 1987). Two other loci on this chromosome, *HML* and *HMR*, also contain complete cassettes of α and a mating type information respectively but are maintained in a repressed or silent state by the concerted action of four *cis*-elements and a number of *trans*-acting factors (see Fig. 2).

RAP1 binds to two of the four sequences involved in silencing (*HMR E* and *HML E*) and the

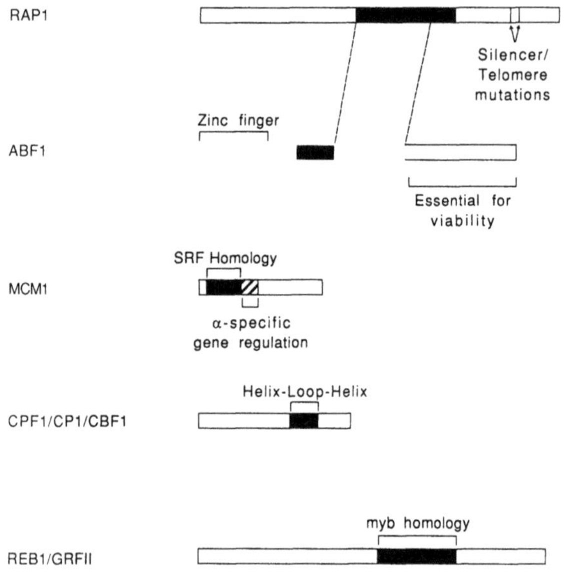

Fig. 1. A summary of the structures of the yeast global regulators. Some of the features described in the text are depicted here and references for these features can be found in the text. Proteins are drawn to scale and DNA binding domains are indicated as filled boxes. In the case of RAP1, ABF1 and MCM1, these boxes are based on experimental observation. In the case of CPF1/CP1/CBF1 the helix-loop-helix domain is shown while in the case of REB1/GRFII, the region of homology to the DNA binding domain of *myb* is shown. In both cases these motifs fall within the characterized DNA binding domains. The region of similarity between the DNA binding domains of ABF1 and RAP1 is indicated by broken lines. The data for the structure of the ABF1 protein is unpublished (Diffley & Stillman).

RAP1 binding sites within these sequences have been shown to have a role in silencing transcription (Brand et al. 1987; Mahoney et al. 1991). Each one of these transcriptional 'silencers' can also function as an autonomously replicating sequence (ARS) and the 11bp ARS core consensus element within the silencers appears to have a role in silencing (Brand et al. 1987; Mahoney et al. 1991), suggesting a causal link between the initiation of DNA replication and transcriptional silencing. Furthermore, the re-establishment of the repressed state, after derepression of a *SIR3*^{ts} mutant, requires passage through S phase (Miller & Nasmyth 1984). Plasmids containing the silencers as the sole ARS exhibit an extremely high plasmid stability when compared to other ARSs, and this enhanced plasmid stability is dependent upon the *SIR2–4* genes, suggesting that these genes may act directly through the silencers (Kimmerly et al. 1988; Kimmerly & Rine 1987).

RAP1 also binds within the yeast telomeric repeat *in vivo* (Conrad et al. 1990) and recent genetic analysis has demonstrated that RAP1 is involved in regulating telomere length. Overproduction of RAP1 results in a heterogeneous increase in telomere length (Conrad et al. 1990) while growth of a *rap1*^{ts} strain at the semi-permissive temperature results in a decrease in average telomere length (Lustig et al. 1990). Overproduction of a C-terminal portion of the protein (Conrad et al. 1990), which does not contain an intact DNA binding domain (Henry et al. 1990) interferes in telomere function, resulting in a dramatic increase in telomere length and suggesting that RAP1 may normally function by interacting with a cellular factor that limits telomere length.

Recently, several point mutations in the *RAP1* gene (*rap1*^{s}) that specifically affect transcriptional silencing without affecting viability have been isolated (Sussel & Shore 1991). In addition to their effect on transcriptional silencing, these mutants also show a significant increase in average telomere length indicating a mechanistic link between the two processes. Such a link is strengthened by the finding that several polII transcribed genes, when placed near a telomere, show an epigenetically inherited repression (Gottschling et al. 1990), very similar to that seen at the silent mating loci (Mahoney et al. 1991; Pillus & Rine 1989). These mutants show no defect in growth rate and, therefore, presumably activate transcription normally (Sussel & Shore 1991) suggesting that RAP1 may contain a domain specifically involved in transcriptional silencing/telomere length regulation. Since overexpression of *SIR4* rescues this defect in the *rap1*^{s} background but has the opposite result (i.e. derepression of the silent loci) in a *RAP1*$^{+}$ background (Marshall et al. 1987), interaction between RAP1 and SIR4 has been postulated (Sussel & Shore 1991). Based on its sequence similarity to the nuclear lamins, SIR4 had previously been postulated to associate with the nuclear lamina (Diffley & Stillman 1989b), where telomeres are often found (Agard & Sedat 1983). Taken together, these results suggest that RAP1 and SIR4 may mediate

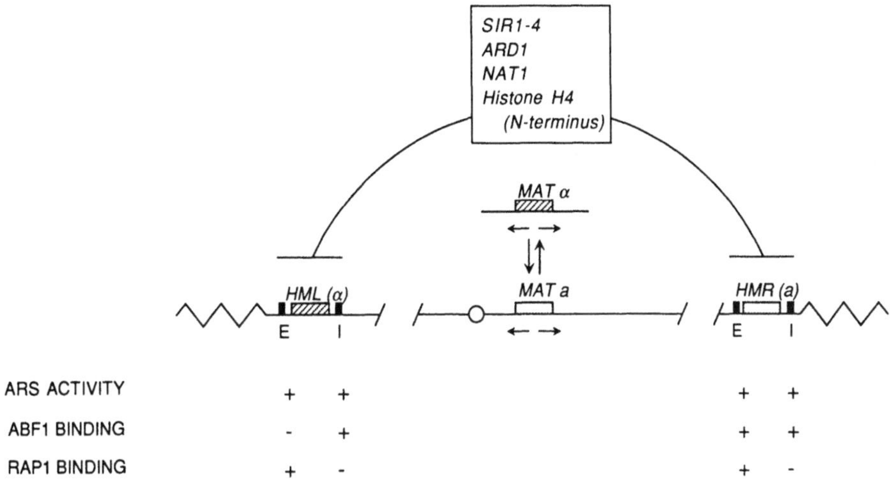

Fig. 2. A summary of yeast mating type control.

This figure summarizes the control of repression of themating type loci, HML and HMR. Details can be found in (Nasmyth and Shore 1987) and in the text.

interactions between telomeres or silencers and the nuclear lamina. This may help to explain the high mitotic stability of silencer containing plasmids: plasmids may be tethered to the lamina via a RAP1-SIR complex during the mitotic cycle in much the same way that REP3 has been postulated to enhance the mitotic stability of the 2-micron circle (Wu et al. 1986).

ABF1

A second well-characterized factor is the ARS binding factor 1 (ABF1). Initially identified as a factor that bound to three of the four silencers as well as several other origins of DNA replication (Buchman et al. 1988a; Diffley & Stillman 1988; Shore et al. 1987), ABF1 has subsequently been shown to bind within the promoters of a large number of yeast genes (see a list of computer predicted sites in Della Seta et al. 1990; also see Brindle et al. 1990; Buchman et al. 1988a; Buchman & Kornberg 1990; Chambers et al. 1990; Della Seta et al. 1990; Dorsman et al. 1991; Halfter et al. 1989; Hamil et al. 1988), in many cases coinciding with upstream activating sequences. The gene encoding ABF1 has been cloned and sequenced (Diffley & Stillman 1989a; Halfter et al. 1989; Rhode et al. 1989) and, like the *RAP1* gene, the *ABF1* gene is essential for viability.

The predicted sequence of the ABF1 gene contains three cysteine residues, all located in the amino-terminus of the protein which have been proposed to contribute to an unusual metal binding domain, somewhat related to the zinc finger (Diffley & Stillman 1989; Halfter et al. 1989; Rhode et al. 1989). Site directed mutagenesis revealed that at least one of these cysteine residues is essential for DNA binding (Halfter et al. 1989). Additionally, DNA binding by ABF1 is sensitive to p-hydroxy-mercuri-phenylsulfonic acid, a reversible alkylating agent specific for cysteine residues (Diffley & Stillman 1989a). After treatment of the modified protein with chelating agents to remove divalent cations, DNA binding could be restored only by the combined reversal of alkylation and addition of exogenous zinc indicating that ABF1 contains a metal binding domain essential for DNA binding (Diffley & Stillman 1989a).

The predicted amino acid sequence of ABF1 exhibits a low but significant degree of similarity to RAP1 over much of the protein (30% identity, 40% conservation over approximately 60% of the ABF1 protein, Diffley & Stillman 1989). As noted above, the DNA binding domain of RAP1 lies roughly in the middle of the protein, while ABF1 requires a metal binding domain located at the N-terminus for DNA binding, suggesting that ABF1 and RAP1 may contain unrelated DNA binding domains. This, in fact, is only partially correct. Recent analysis has shown that, in addition to the metal binding domain in the amino terminus, ABF1 also requires a second domain for high affinity, sequence-specific DNA binding. This domain is located near the centre of the protein and is related to the DNA binding domain of RAP1. Furthermore, this domain, when expressed and purified from *E.coli*, exhibits low affinity, sequence-specific DNA binding. Thus, the role of the N-terminal metal binding domain appears to be in increasing the overall affinity of ABF1 for DNA and extending the recognition sequence (Diffley & Stillman, unpublished results).

Although oligonucleotides containing an ABF1 binding site can activate transcription (Brand et al. 1987) when cloned upstream of a UAS-less promoter, this activation is far less than that seen by RAP1 oligonucleotides. Buchman & Kornberg (1990), however, have shown that, while ABF1 is, by itself, only a weak transcriptional activator, it can combine synergistically with other weak activators (Buchman & Kornberg 1990) to generate a strong transcriptional activator. Thus, ABF1, like RAP1, can activate transcription *in vivo*.

Several instances of overlap between ABF1 and RAP1 function exist. Most interesting perhaps are the ribosomal protein genes. While a large majority of the ribosomal genes are regulated by RAP1 via the UAS_{rpg}, several, such as TCM1 (Hamil et al. 1988) and S33 (Vignais et al. 1987) do not contain this sequence, but, instead, appear to regulated by ABF1 binding sites. The fact that all of these genes are carefully and coordinately regulated in response to growth conditions suggests some common regulation of ABF1 and RAP1.

In different contexts, ABF1 can repress as well as activate transcription. Like RAP1, ABF1 binds to silencer sequences, in this case, *HML I*, *HMR E* and *HMR I*. And, like RAP1, the ABF1 binding site within at least one of these sequences, *HMR E*, is important for silencer function (Brand et al. 1987). Interestingly, the ABF1 and RAP1 binding sites within *HMR E* are at least partially redundant, suggesting again that these two proteins may act by some common pathway.

In addition to its role in silencing the *HML* and *HMR* loci, ABF1 may also be involved in the repression of other genes. Perhaps the best characterized of these is the *ENO2* gene, encoding the glycolytic enzyme enolase. Like other glycolytic enzymes, *ENO2* is positively regulated by a gene known as *GCR1* (Baker 1986; Holland et al. 1990; Holland et al. 1987). In a *GCR1*$^+$ strain, an ABF1 site upstream of the *ENO2* gene lies within one of two classical UAS elements (Brindle et al. 1990; Holland et al. 1990). This same sequence acts to repress *ENO2* transcription in a *gcr1*$^-$ null mutant. Removal of the UAS containing the ABF1 site restores *ENO2* transcription to nearly wild type levels in a *gcr1*$^-$ background. An ABF1 binding site alone, however, when used to replace the entire *ENO2* promoter, functions as an activator in both a *GCR1*$^+$ and a *gcr1*$^-$ background. Thus, the UAS-/repressor sequence may be more complex than just an ABF1 site or, alternatively, ABF1 may exercise its repressor function through the second UAS normally present in the *ENO2* gene. Interestingly, this second UAS coincides with a RAP1 site (Brindle et al. 1990), suggesting that repression by ABF1 may occur either by competition with RAP1 for some limiting factor or by interaction with RAP1. Finally, this organization of binding sites is not specific to the *ENO2* gene, but is also seen in the upstream sequences of other genes encoding glycolytic enzymes like *PGK* (phosphoglycerate kinase) and *PYK* (pyruvate kinase) (Chambers et al. 1990).

ABF1 also appears to have a role in the initiation of DNA replication. In addition to the silencer/ARSs, ABF1 binds to several other ARS elements including ARS1 (Buchman et al. 1988a; Diffley & Stillman 1988; Shore et al. 1987; Sweder et al. 1988) and ARS121 (Eisenberg et al. 1988). Furthermore, the ABF1 binding sites at ARS1 (Diffley & Stillman 1988), ARS121 (Walker et al. 1989) and *HMR E* (Brand et al. 1987; Kimmerly et al. 1988) have been shown to be important for fully efficient ARS function.

Sequence analysis of yeast ARSs has revealed three distinct types of elements (reviewed in Diffley & Stillman 1990). The first of these is an 11bp core sequence which is absolutely required for ARS function. This sequence serves *in vivo* as a binding site for some cellular factor(s) (Diffley, unpublished), which may include the previously described ACBP/ssARS T protein (Hofmann & Gasser 1991; Kuno et al. 1990; Schmidt et al. 1991). The second type of element is sensitive to both distance and orientation relative to the ARS consensus sequence and has been correlated to nuclear scaffold attachment sites, DNA unwinding elements and additional binding sites for ACBP/ssARS T. The final type of element is insensitive to distance and orientation relative to the core sequence and has been correlated with binding sites for ABF1 (Walker et al. 1990).

The ability of ABF1 to act in a distance and orientation independent manner in both DNA replication and transcription suggests that the mechanism of its action may be similar in both processes. Whether this mechanistic similarity involves some common interface between transcriptional activators and either replication or transcription complexes (Mohr et al. 1990) or involves a less direct effect in which the transcriptional activator somehow alters local chromatin structure of the replication origin/promoter (Workman et al. 1991) is currently unclear.

MCM1

MCM1 was initially identified as one of a group of genes required for the efficient maintenance of ARS-containing minichromosomes (Maine et al. 1984). Subsequently, MCM1 was shown to be the same as PRTF (Jarvis et al. 1989; Passmore et al. 1989; Passmore et al. 1988), a factor involved in regulating haploid specific genes (for a review, see Dolan & Fields 1991). *MCM1* is an essential gene encoding a sequence-specific DNA binding protein (Passmore et al. 1989) closely related at the level of primary amino acid sequence to the human serum response factor (Norman et al. 1988) and the yeast *ARG80* gene product (Dubois et al. 1987). The primary similarities lie within an 80 amino acid domain near the amino terminus over which MCM1 and SRF are 70% identical. Expression of this domain of MCM1 alone is sufficient to prevent the loss of viability, the minichromosome maintenance defect, the defect in the repression of a-specific genes in *mcm1*$^-$ strains and the defect in transcriptional activation. A 23 amino acid acidic region located immediately adjacent to this domain is required for α-specific gene regulation while the C-terminal 166 amino acids, rich in glutamine, has only a minor role in transcriptional activation (Christ & Tye 1991).

MCM1 has a direct role in the function of at least some ARSs. ARS120, a sub-telomeric ARS, contains multiple MCM1 binding sites and removal of these sites compromises the activity of this ARS (V. Chang & B.-K. Tye, personal communication). It is tempting to speculate that the role of MCM1 in ARS120 function is similar to that of ABF1 in ARS1, ARS121 and *HMR E* function, although the distance dependence of the MCM1 sites at ARS120 has not yet been examined.

CPF1/CP1/CBF1

Yeast centromeres are composed of three structural elements designated CDEI, II and III. Mutational analysis has shown that CDEI is important but not essential for centromere function. A protein factor designated CPF1, CP1 and CBF1 binds to this sequence and binding sites related to the CDEI site have been found upstream of several yeast genes (Bram & Kornberg 1987). The gene encoding this protein, a member of the helix-loop-helix family, has been cloned by several groups (Baker & Masison 1990; Cai & Davis 1990; Mellor et al. 1990).

Although the *CPF1/CEP1/CBF1* gene is not essential, the null mutant has several clear phenotypes consistent with the locations of its binding sites. Null mutants exhibit a 4–10 fold increase in minichromosome loss (Baker & Masison 1990; Cai & Davis 1990; Mellor et al. 1990), similar to the effect of a CDEI deletion. Additionally, null mutants are auxotrophic for methionine, consistent with the fact that the promoters of several genes involved in methionine and S-adenosyl methionine biosynthesis (*MET2, MET3, MET25* and *SAM2*) contain CDEI-like elements. In one of these promoters, *MET25*, the CDEI element was shown to correlate with a UAS element (Thomas et al. 1989), and *in vivo* footprinting demonstrated the binding of CPF1/CP1/CBF1 (Mellor et al. 1990). Thus, there is a good correlation between CPF1/CP1/CBF1 binding in a promoter and transcriptional activation, although it should be pointed out that direct demonstration of a CDEI element oligonucleotide acting as a UAS element in a test promoter is lacking.

Furthermore, the story is not straightforward; mutations that eliminate the ability of CPF1/CP1/CBF1 to bind to its recognition sequence *in vitro* can still complement null mutants for the methionine auxotrophy but not the high loss rate of minichromosomes (Mellor et al. 1991) and null mutants still express wild type levels of *MET25*. The reasons for this are presently unclear. One possibility is that CPF1 is required for the activation of an as yet unidentified gene that is crucial for methionine biosynthesis and that this function can be performed by the non DNA-binding mutant. Another favoured model is that CPF1 interacts with a second cellular factor, perhaps through its helix-loop-helix, to establish methionine prototrophy by some means other than direct transcriptional activation.

REB1/GRFII

The final factor described here has been alternatively designated GRFII, Y protein, QBP and REB1 (Brandl & Struhl 1990; Chasman et al. 1990; Fedor et al. 1988; Morrow et al. 1990; Wang et al. 1990). Binding sites for this factor have been found within the promoters of a number of RNA polymerase II-transcribed genes, within both the enhancer and promoter of the RNA polymerase I-transcribed 35S rRNA gene, within the X and Y subtelomeric repeats, and near one centromere, CEN4. The gene encoding REB1 has been cloned and sequenced. This gene is essential for viability and contains a region similar to the DNA binding domain of the *myb* oncogene (Ju et al. 1990).

Like ABF1 (Buchman & Kornberg 1990), oligonucleotides containing REB1/GRFII binding sites only weakly activate a test promoter but can act synergistically with another weak transcriptional activator (Chasman et al. 1990). In other contexts, REB1/GRFII can act to repress transcription (Wang et al. 1990). The small, positive effect of the REB1/GRFII site alone can be mimicked in a cell-free system, but the synergistic effect cannot. Additionally, there is a strong distance dependence between the REB1/GRFII binding sites and its synergistic partner in transcriptional activation (Brandl & Struhl 1990; Chasman et al. 1990).

REB1/GRFII affects local chromatin structure, creating a short nucleosome-free region and positioning nucleosomes outside of this region (Fedor et al. 1988) which may explain its effects *in vivo*, but does not explain its effects *in vitro* since the cell-free transcription system does not involve chromatin or chromatin assembly. Thus, REB1/GRFII may act in transcription by two mechanisms; nucleosome exclusion/positioning and direct interaction with the transcriptional machinery.

Summary

Even among the small number of factors considered here, several different routes toward multifunctionality have been followed. In RAP1 and MCM1, specific subdomains within each protein are dedicated to different functions; a specific domain required for telomere/silencer function in RAP1 and one required for α-specific gene expression in MCM1. On the other hand, some domains have not been readily sub-dividible, for instance the transcriptional activation and replication domains of ABF1 and MCM1. Some of these factors, like ABF1 and RAP1 can act at distances while others, such as REB1/GRFII, apparently cannot. Thus, mechanistically, it is probably incorrect to group these factors together. Nonetheless, evolution has chosen to consolidate multiple functions in each of these proteins. Is this simply a matter of genetic economy or are there some subtle features of the regulation of these proteins that makes them particularly suitable for their tasks? In this regard, several of these factors exist *in vivo* as phosphoproteins (Francesconi & Eisenberg 1991; Ju et al. 1990;

Tsang et al. 1990) and it will be interesting to see how these phosphorylations are controlled and affect the functions of these factors. How does binding site context dictate function? That is, why does RAP1 interact with the telomere apparatus when bound at a telomere and the transcriptional apparatus when bound at a UAS? Finally, do related proteins exhibit a similar multi-functionality in higher eucaryotes? Identification and characterization of the yeast gene products that interact with these global factors in their various functions and further characterization of replication origins, telomeres and centromeres in higher eucaryotes over the next several years should begin to answer these questions.

Acknowledgements

Thanks to Bik Tye & Simon Dowell for their contributions to the sections on MCM1 and CPF/CP1/CBF1, respectively.

References

Agard DA & Sedat JW (1983). Three-dimensional architecture of a polytene nucleus. Nature 302: 676–681

Baker H (1986). Glycolytic gene expression in *Saccharomyces cerevisiae*: nucleotide sequence of *GCR1*, null mutants and evidence for expression. Mol. Cell. Biol. 6: 3774–3784

Baker RE & Masison DC (1990). Isolation of the gene encoding the *Saccharomyces cerevisiae* centromere-binding protein CP1. Mol. Cell. Biol. 10: 2458–2467

Bram RJ & Kornberg RD (1987). Isolation of a *Saccharomyces cerevisiae* centromere DNA-binding protein, its human homolog, and its possible role as a transcription factor. Mol. Cell. Biol. 7: 403–409

Brand AH, Micklem G & Nasmyth K (1987). A yeast silencer contains sequences that can promote autonomous plasmid replication and transcriptional activation. Cell 51: 701–719

Brandl CJ & Struhl K (1990). A nucleosome-positioning sequence is required for GCN4 to activate transcription in the absence of a TATA element. Mol. Cell. Biol. 10: 4256–4265

Brindle PK, Holland JP, Willet CE, Innis MA & Holland MJ (1990). Multiple factors bind the upstream activation sites of the yeast enolase genes *ENO1* and *ENO2*: ABFI protein, like repressor activator protein RAP1, binds *cis*-acting sequences which modulate repression or activation of transcription. Mol. Cell. Biol. 10: 4872–4885

Buchman AR & Kornberg RD (1990). A yeast *ARS*-binding protein activates transcription synergistically in combination with other weak activating factors. Mol. Cell. Biol. 10: 887–897

Buchman AR, Kimmerly WJ, Rine J & Kornberg RD (1988a). Two DNA-binding factors recognize specific sequences at silencers, upstream activating sequences, autonomously replicating sequences, and telomeres in *Saccharomyces cerevisiae*. Mol. Cell. Biol. 8: 210–225

Buchman AR, Lue NF & Kornberg RD (1988b). Connections between transcriptional activators, silencers, and telomeres as revealed by functional analysis of a yeast DNA-binding protein. Mol. Cell. Biol. 8: 5086–5099

Cai M & Davis RW (1990). Yeast centromere binding protein CBF1, of the helix-loop-helix protein family, is required for chromosome stability and methionine prototrophy. Cell 61: 437–446

Capieaux E, Vignais M-L, Sentenac A & Goffeau A (1989). The yeast H^+-ATPase gene is controlled by the promoter binding factor TUF. J. Biol. Chem. 264: 7437–7446

Chambers A, Stanway C, Tsang JSH, Henry Y, Kingsman AJ & Kingsman SM (1990). ARS binding factor 1 binds adjacent to RAP1 at the UASs of the yeast glycolytic genes *PGK* and *PYK1*. Nucleic Acids Res. 18: 5393–5399

Chasman DI, Lue NF, Buchman AR, LaPointe JW, Lorch Y & Kornberg RD(1990). A yeast protein that influences chromatin structure of UAS_G and functions as a powerful auxiliary gene activator. Genes & Dev. 4: 503–514

Christ C & Tye B-K (1991). Functional domains of the yeast transcription/replication factor MCM1. Genes & Dev. 5: 751–763

Conrad MN, Wright JH, Wolf AJ & Zakian VA (1990). RAP1 protein interacts with yeast telomeres *in vivo*: overproduction alters telomere structure and decreases chromosome stability. Cell 63: 739–750

Della Seta F, Ciafre S-A, Marck C, Santoro B, Presutti C, Sentenac A & Bozzoni I (1990). The ABF1 factor is the transcriptional activator of the L2 ribosomal protein genes in *Saccharomyces cerevisiae*. Mol. Cell. Biol. 10: 2437–2441

Della Seta F, Treich I, Buhler J-M & Sentenac A (1990). ABF1 binding sites in yeast RNA polymerase genes. J. Biol. Chem. 265: 15168–15175

Diffley JFX & Stillman B (1988). Purification of a yeast protein that binds to origins of DNA replication and a transcriptional silencer. Proc. Natl. Acad. Sci. USA 85: 2120–2124

Diffley JFX & Stillman B (1989a). Similarity between the transcriptional silencer binding proteins ABF1 and RAP1. Science 246: 1034–1-38

Diffley JFX & Stillman B (1989b). Transcriptional silencing and lamins. Nature 342: 24

Diffley JFX & Stillman B (1990). The initiation of chromosomal DNA replication in eucaryotes. Trends in Genetics 6: 427–432

Dolan JW & Fields S (1991). Cell-type-specific transcription in yeast. Biochim. Biophys. Acta 1088: 155–169

Dorsman JC, Gozdzicka-Jozefiak A, van Heeswijk WC & Grivell LA (1991). Multi-functional DNA proteins in yeast: the factors GF1 and GFII are identical to the ARS-binding factor ABFI and the centromere-binding factor CPF1 respectively. Yeast 7: 401–412

Dubois E, Bercy J & Messenguy F (1987). Characterization of two genes, *ARGRI* and *ARGRIII* required for specific regulation of arginine metabolism in yeast. Mol. Gen. Genet. 207: 142–148

Eisenberg S, Civalier C & Tye B-K (1988). Specific interaction between a *Saccharomyces cerevisiae* protein and a DNA element associated with certain autonomously replicating sequences. Proc. Natl. Acad. Sci. USA 85: 743–746

Fedor MJ, Lue NF & Kornberg RD (1988). Statistical positioning of nucleosomes by specific protein-binding to an upstream activating sequence in yeast. J. Mol. Biol. 204: 109–127

Francesconi SC & Eisenberg S (1991). The multifunctional protein OBF1 is phosphorylated at serine and threonine residues in *Saccharomyces cerevisiae*. Proc. Natl. Acad. Sci. USA 88: 4089–4093

Giesman D, Best L & Tatchell K (1991). The role of RAP1 in the regulation of the *MATα* locus. Mol. Cell. Biol. 11: 1069–1079

Gottschling DE, Aparicio OM, Billington BL & Zakian VA (1990). Position effect at *S. cerevisiae* telomeres: Reversible repression of polII transcription. Cell 63: 751–762

Halfter H, Kavety B, Vandekerckhove, Kiefer F & Gallwitz D (1989). Sequence, expression and mutational analysis of BAF1, a transcriptional activator and ARS1-binding protein of the yeast *Saccharomyces cerevisiae*. EMBO J. 8: 4265–4272

Halfter H, Muller U, Winnacker E-L & Gallwitz D (1989). Isolation and DNA-binding characteristics of a protein involved in transcription activation of two divergently transcribed, essential yeast genes. EMBO J. 8: 3029–3037

Hamil KG, Nam HG & Fried HM (1988). Constituitive transcription of yeast ribosomal protein gene *TCM1* is promoted by uncommon *cis-* and *trans*-acting elements. Mol. Cell. Biol. 8: 4328–4341

Henry YAL, Chambers A, Tsang JSH, Kingsman AJ & Kingsman SM (1990). Characterization of the DNA binding domain of the yeast RAP1 protein. Nucleic Acids Res. 18: 2617–2623

Hofmann JF-X & Gasser SM (1991). Identification and purification of a protein that binds the yeast ARS consensus sequence. Cell 64: 951–960

Holland JP, Brindle PK & Holland MJ (1990). Sequences within an upstream activation site in the yeast enolase gene *ENO2* modulate repression of *ENO2* expression in strains carrying a null mutation in the positive regulatory gene *GCR1*. Mol. Cell. Biol. 10: 4863–4871

Holland MJ, Yokoi T, Holland JP, Myambo K & Innis MA (1987). The GCR1 gene encodes a positive transcriptional regulator of the enolase and glyceraldehyde-3-phosphate dehydrogenase gene families in *Saccharomyces cerevisiae*. Mol. Cell. Biol. 7: 813–820

Huet J & Sentenac A (1987). TUF, the yeast DNA-binding factor specific for UAS_{rpg} upstream activating sequences: identification of the protein and its DNA-binding domain. Proc. Natl. Acad. Sci. USA 84: 3648–3652

Huet J, Cottrelle P, Cool M, Vignais M-L, Thiele D, Marck C, Buhler J-M, Sentenac A & Fromageot P (1985). A general upstream binding factor for genes of the yeast translational apparatus. EMBO J. 4: 3539–3547

Jarvis EE, Clark KL & Sprague Jr GF (1989). The yeast transcription activator PRTF, a homolog of the mammalian serum response factor, is encoded by the *MCM1* gene. Genes & Dev. 7: 936–945

Ju Q, Morrow BE & Warner JR (1990). REB1, a yeast DNA-binding protein with many targets, is essential for growth and bears some resemblance to the oncogene *myb*. Mol. Cell. Biol. 10: 5226–5234

Kimmerly W & Rine J (1987). Replication and segregation of plasmids containing *cis*-acting regulatory sites of silent mating-type genes in *Saccharomyces cerevisiae* are controlled by the *SIR* genes. Mol. Cell. Biol. 7: 4225–4237

Kimmerly W, Buchman A, Kornberg R & Rine J (1988). Roles of two DNA-binding factors in replication, segregation and transcriptional repression mediated by a yeast silencer. EMBO J. 7: 2241–2253

Kuno K, Murakami S & Kuno S (1990). Single-strand-binding factor(s) which interact with *ARS1* of *Saccharomyces cerevisiae*. Gene 95: 73–77

Kurtz S & Shore, D. (1991). RAP1 protein activates and silences transcription of mating-type genes in yeast. Genes & Dev. 5: 616–628

Lustig AJ, Kurtz S & Shore D (1990). Involvement of the silencer and UAS binding protein RAP1 in regulation of telomere length. Science 250: 549–553

Mahoney DJ, Marquardt R, Shei G-J, Rose AB & Broach JR (1991). Mutations in the *HML* E silencer of *Saccharomyces cerevisiae* yield metastable inheritance of transcriptional repression. Genes & Dev. 5: 605–615

Maine GT, Sinha P & Tye B-K (1984). Mutants of *S. cerevisiae* defective in the maintenance of minichromosomes. Genetics 106: 365–385

Marshall M, Mahoney D, Rose A, Hicks JB & Broach JR (1987). Functional domains of SIR4, a gene required for position effect regulation in *Saccharomyces cerevisiae*. Mol. Cell. Biol. 7: 4441–4452

Mellor J, Jiang W, Funk M, Rathjen J, Barnes CA, Hinz T, Hegemann JH & Philippsen P (1990). CPF1, a yeast protein which functions in centromeres and promoters. EMBO J. 9: 4017–4026

Mellor J, Rathjen J, Jiang W & Dowell SJ (1991). DNA binding of CPF1 is required for optimal centromere function but not for maintaining methionine prototrophy in yeast. Nucleic Acids Res. 19: 2961–2969

Miller AM & Nasmyth KA (1984). Role of DNA replication in the repression of silent mating loci in yeast. Nature 312: 247–251

Moehle CM & Hinnebusch AG (1991). Association of RAP1 binding sites with stringent control of ribosomal protein gene transcription in *Saccharomyces cerevisiae*. Mol. Cell. Biol. 11: 2723–2735

Mohr IJ, Clark R, Sun S, Androphy EJ, MacPherson P & Botchan MR (1990). Targeting the E1 replication protein to the papillomavirus origin of replication by complex formation with the E2 transactivator. Science 250: 1694–1699

Morrow BE, Ju Q & Warner JR (1990). Purification and characterization of the yeast rDNA binding protein REB1. J. Biol. Chem. 265: 20778–20783

Nasmyth K & Shore D (1987). Transcriptional regulation in the yeast life cycle. Science 237: 1162–1170

Norman C, Runswick M, Pollock R & Treisman R (1988). Isolation and properties of cDNA clones encoding SRF, a transcription factor that binds to the *c-fos* serum response element. Cell 55: 989–1003

Passmore S, Maine GT, Elble R, Christ C & Tye B-K (1988). *Saccharomyces cerevisiae* protein involved in plasmid maintenance is necessary for mating of MATa cells. J. Mol. Biol. 204: 593–606

Passmore S, Elble R & Tye B-K (1989). A protein involved in minichromosome maintenance in yeast binds a transcriptional enhancer conserved in eucaryotes. Genes & Dev. 3: 921–935

Pillus L & Rine J (1989). Epigenetic inheritance of transcriptional states in *S. cerevisiae*. Cell 59: 637–647

Rhode PR, Sweder KS, Oegama KF & Campbell JL (1989). The gene encoding ARS-binding factor I is essential for the viability of yeast. Genes & Dev. 3: 1926–1939

Santangelo GM & Tornow J (1990). Efficient transcription of the glycolytic gene *ADH1* and three translational component genes requires the *GCR1* product, which can act through TUF/GRF/RAP binding sites. Mol. Cell. Biol. 10: 859–862

Schmidt AMA, Herterich SU & Krauss G (1991). A single-stranded DNA binding protein from *S. cerevisiae* recognizes the T-rich strand of the core sequence of ARS elements and discriminates against mutant sequences. EMBO J. 10: 981–985

Shore D & Nasmyth K (1987). Purification and cloning of a DNA binding protein from yeast that binds to both silencer and activator elements. Cell 51: 721–732

Shore D, Stillman DJ, Brand AH & Nasmyth KA (1987). Identification of silencer binding proteins from yeast: possible roles in SIR control and DNA replication. EMBO J. 6: 461–467

Stanway C, Mellor J, Ogden JE, Kingsman AJ & Kingsman SM (1987). The UAS of the yeast PGK gene contains functionally distinct domains. Nucleic Acids Res. 15: 6855–6873

Sussel L & Shore D (1991). Separation of transcriptional activation and silencing functions of the *RAP1*-encoded repressor/activator protein 1: isolation of viable mutants affecting both silencing and telomere length. Proc. Natl. Acad. Sci. USA 88: 7749–7753

Sweder KS, Rhode PR & Campbell JL (1988). Purification and characterization of proteins that bind to yeast *ARSs*. J. Biol. Chem. 263: 17270–17277

Thomas D, Cherest H & Surdin-Kerjan Y (1989). Elements involved in S-adenosylmethionine-mediated regulation of the *Saccharomyces cerevisiae* MET25 gene. Mol. Cell. Biol. 9: 3292–3298.

Tsang JS, Henry YA, Chambers A, Kingsman AJ & Kingsman SM (1990). Phosphorylation influences the binding of the yeast RAP1 protein to the upstream activating sequence of the PGK gene. Nucleic Acids Res. 18: 7331–7337

Vignais M-L, Woudt LP, Wassenaar GM, Mager WH, Sentenac A & Planta RJ (1987). Specific binding of TUF factor to upstream activation sites of yeast ribosomal protein genes. EMBO J. 6: 1451–1457

Walker SS, Francesconi SC, Tye B-K & Eisenberg S (1989). The OBF1 protein and its DNA-binding site are important for the function of an autonomously replicating sequence in *Saccharomyces cerevisiae*. Mol. Cell. Biol.: 2914–2921

Walker SS, Francesconi SC & Eisenberg S (1990). A DNA replication enhancer in *Saccharomyces cerevisiae*. Proc. Natl. Acad. Sci. USA 87: 4665–4669

Wang H, Nicholson PR & Stillman DJ (1990). Identification of a *Saccharomyces cerevisiae* DNA-binding protein involved in transcriptional regulation. Mol. Cell. Biol. 10: 1743–1753

Workman JL, Taylor IC & Kingston RE (1991). Activation domains of stably bound GAL4 derivatives alleviate repression of promoters by nucleosomes. Cell 64: 533–544

Wu L-C, Fisher PA & Broach JR (1986). A yeast plasmid partitioning protein is a karyoskeletal component. J. Biol. Chem. 262: 883–891

Antonie van Leeuwenhoek **62**: 35–46, 1992.

The genetics of nuclear pre-mRNA splicing: a complex story

Jeremy D. Brown, Mary Plumpton & Jean D. Beggs
Institute of Cell and Molecular Biology, University of Edinburgh, King's Buildings, Mayfield Road, Edinburgh EH9 3JR, Scotland

Key words: introns, pre-mRNA splicing, RNA processing, Saccharomyces cerevisiae, yeast genetics

Abstract

The occurrence of introns in nuclear precursor RNAs (pre-mRNAs) is widespread in eukaryotes, and the splicing process that removes them is basically the same in yeasts as it is in higher eukaryotes. Splicing takes place in a very large, multi-component complex, the spliceosome, and biochemical studies have been complicated by the large number of splicing factors involved. This review describes how genetic approaches used to study RNA splicing in *Saccharomyces cerevisiae* have complemented the biochemical studies and led to rapid advances in the field.

1. Introduction

Pre-mRNA splicing, the precise removal of introns from nascent transcripts, is an essential requirement of eukaryotic gene expression, and an important regulatory step in gene expression in higher eukaryotes. The splicing reaction pathway (Fig. 1) and the major components of the spliceosome complex in which the splicing reaction takes place have been established for several years. Recent efforts have been aimed mainly at understanding the dynamic composition and structure of the spliceosome during the splicing reaction, and the functions of the various factors involved. A vast amount of information has been generated since the discovery of splicing 15 years ago, and much of this has come from studying the process in *Saccharomyces cerevisiae* using the powerful genetic approaches available in yeast allied to biochemical studies and the manipulation of cloned genes *in vitro*.

In yeast pre-mRNA splicing is very similar to but lacks some of the flexibility seen in the process in higher eukaryotes. *S. cerevisiae* has relatively few intron-containing genes and only one example is known of a gene with more than one intron (*MATa1*; Miller, 1984). Intron definition sequences at the 5′ and 3′ splice sites and at the branch point are highly conserved and there are no known examples of alternatively spliced introns in *S. cerevisiae*. In higher eukaryotes most genes have multiple introns, the removal of which may be regulated, with alternative splice sites often being used. Compared to yeast, intron definition sequences are much less conserved, and mutations in them often promote the use of cryptic sites, something rarely seen in yeast. Thus the detailed analysis of splicing in yeast may yield a more defined view of the basic process.

Splicing is a dynamic process involving interactions between many proteins and RNAs in a cycle of events in which the spliceosome is presumed to be 'built' and 'dismantled' each time a transcript is spliced. Genetic approaches are ideal for analysing a complex system such as this, isolating mutations in components acting at various points in the cycle, the defects of which can then be analysed biochemically. As splicing is an essential process, conditional (such as heat or cold sensitive) mutations are sought, otherwise the mutants could not be propagated. The development of a yeast cell-free system for studying splicing *in vitro* (Lin et al. 1985) has

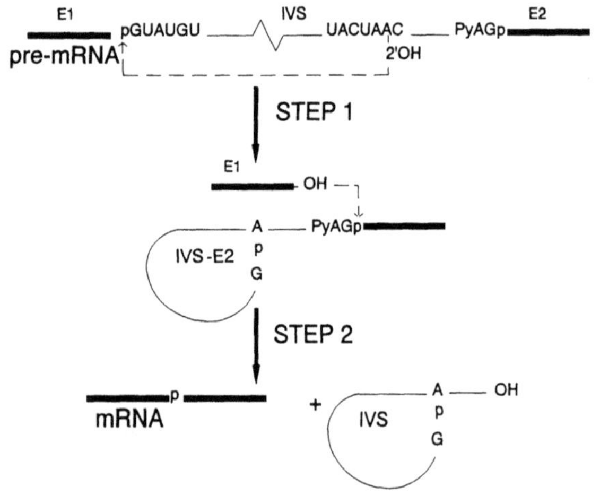

Fig. 1. The mechanism of pre-mRNA splicing involves two transesterification reactions; reactive phosphate and hydroxyl groups are shown, dotted arrows indicate nucleophilic attack. The conserved intron sequences of *S. cerevisiae* at the 5′ and 3′ splice sites and at the branch point are shown. E1 and E2 represent the 5′ and 3′ exons respectively. IVS indicates the intron.

greatly facilitated the analysis of such mutants, as temperature-sensitive forms of many splicing factors can be heat-inactivated *in vitro* (Lustig et al. 1986; Abovitch et al. 1990). Once mutants have been identified, gene cloning and characterisation can be achieved with relative ease in yeast, and attempts made to determine interactions with other spliceosomal factors.

Other recent reviews of splicing in yeast include Vijayraghavan and Abelson (1990), Guthrie (1991) and Ruby and Abelson (1991). This review emphasises how genetic approaches have furthered our understanding of splicing, yielding insights – sometimes unexpected – which build up a picture of the process of intron recognition and excision.

2. snRNAs and snRNPs

Nuclear pre-mRNA introns are thought to be processed via the same reaction pathway as group II introns found in fungal mitochondria and chloroplasts (Sharp 1987; Jacquier 1990), and the main difference between these introns is their information content. Group II introns have conserved secondary structural elements that form the catalytic site and bring the splice sites and the branch point together. Nuclear pre-mRNA introns do not have these secondary structural features and thus pre-mRNA splicing must be governed by the *trans*-acting factors of the spliceosome. Some group II introns self-splice *in vitro* in the absence of proteins and ATP, and the observed requirement for ATP in pre-mRNA splicing may be solely for assembly of the spliceosome and for guiding and regulating the reaction.

Best characterised of the *trans*-acting factors are the small nuclear ribonucleoprotein particles (snRNPs), considered to be major determinants of intron definition. Each snRNP is named after the small nuclear RNA (snRNA) U1, U2, U4, U5 or U6 it contains. The yeast genes (*SNR19, 20, 14, 7* and *6*) encoding the splicing-associated snRNAs (U1, U2, U4, U5 and U6 respectively) were identified by virtue of having limited primary sequence and secondary structural similarity to the equivalent metazoan genes. The *SNR* genes are single copy and essential. The snRNAs are precipitable by so called anti-Sm antibodies which recognise proteins found in splicing-associated snRNPs and have characteristic trimethylguanosine caps (except for U6 which is only Sm-precipitable by virtue of its base-pairing with U4, and has a different cap structure). For detailed reviews see Guthrie (1988), Guthrie and Patterson (1988) and Lührmann et al.(1990).

The U1 and U2 snRNAs have sequences at their 5′ ends which, as in their metazoan equivalents, are complementary to the 5′ splice site and branch point consensus sequences respectively (Fig. 2). That these sequences are important for splicing was shown by mutating the *SNR* genes *in vitro* and transforming them into cells carrying reporter genes with mutated intron sequences. In each case, when the *SNR* gene was mutated such that complementarity of the snRNA to the mutant reporter gene was restored, a significant increase in the efficiency of splicing of the reporter gene transcript was observed (Parker et al. 1987; Seraphin et al. 1988; Siliciano & Guthrie 1988). The allele-specificity of this effect confirmed the existence of the proposed Watson-Crick base-pairing between these snRNAs and the intron, and that these inter-

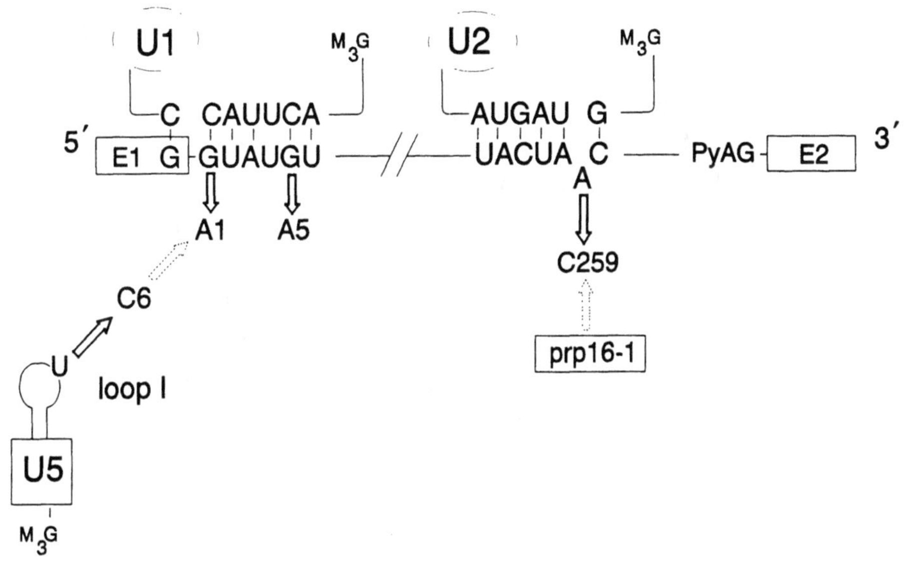

Fig. 2. The U1 and U2 snRNAs interact directly with the pre-mRNA. Base-pairing interactions U1:5′ splice site and U2:branch point are shown. G-A1 and A-C259 are point mutations used in selections for *trans*-acting extragenic suppressors. Interactions implied by the suppressor analyses are shown by dashed arrows. G-A5 indicates a point mutation which causes aberrant cleavages *in vivo*.

actions are functionally important. However, splicing was not returned to wild type levels in these experiments, implying that other factors also interact with these regions of the intron. In particular, a G to A mutation at position 5 of the 5′ splice site (Fig. 2) in several transcripts resulted in the use of aberrant cleavage sites (Jacquier et al. 1985; Parker & Guthrie 1985; Fouser & Frieson 1986), and a compensatory change in the U1 snRNA led to increased use of both genuine and aberrant cleavage sites (Seraphin et al. 1988; Siliciano & Guthrie 1988). Thus while U1:5′ splice site base-pairing is obviously important in defining a region in which cleavage can take place, the exact location of the cleavage may be determined by (an)other factor (s). Recent evidence points to the U5 snRNP being a candidate 5′ cleavage site definition factor (Newman & Norman 1991, and see section 4.2).

The U1 and U2 snRNPs are the first to bind to pre-mRNA *in vitro* (for review see Rosbash & Seraphin 1991). Two methods have been used to show this; targeted RNase H cleavage *in vitro* (Ruby & Abelson 1988; Legrain et al. 1988) and depletion *in vivo* (Seraphin & Rosbash 1989). RNase H cleavage is carried out by adding to cell extracts a DNA oligonucleotide complementary to an snRNA. The RNA component of DNA-RNA duplexes formed is then cleaved by endogenous RNase H. Efficient depletion *in vivo* has been achieved by fusing the *GAL1/10* transcription control sequences onto *SNR* genes which can then be repressed by growth on glucose. Cells with a chromosomal disruption of an *SNR* locus and carrying the *GAL-SNR* fusion construct on a plasmid can thus be depleted of that snRNA *in vivo*. In U2-depleted extracts prepared by either method, the U1 snRNP bound to pre-mRNA substrates in the absence of ATP. This U1:pre-mRNA complex was stable to subsequent competition with excess unlabelled substrate and could be chased into functional spliceosomes by the addition of U1-depleted extract. Interestingly, branch point as well as 5′ splice site mutations affected U1 snRNP binding, implying an interaction of the U1 snRNP either directly or indirectly (via a factor other than the U2 snRNP) with the branch point. A more recent study by Goguel et al. (1991) using genes containing duplicated branch point and 3′ splice site sequences showed that, *in vivo*, certain U1 snRNA mutations increased the usage of a normally non-preferred distal branch point and 3′ splice site. This suggests a more direct role for the U1 snRNP in 3′ splice site definition.

The *in vivo* depletion strategy (Patterson &

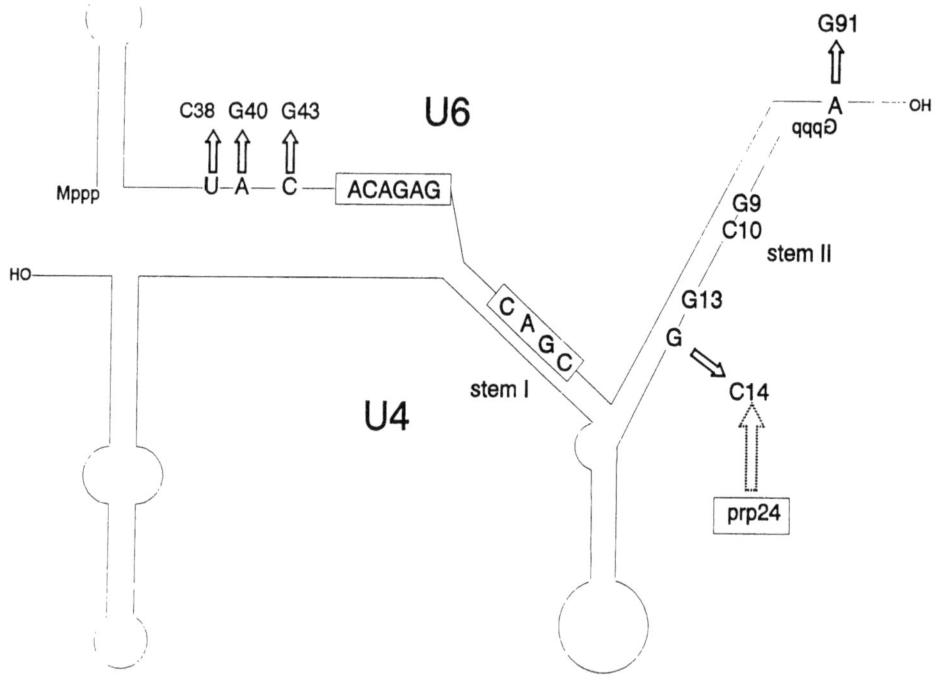

Fig. 3. The U4:U6 complex and genetic interactions. The snRNAs are shown base-paired together at stem I and stem II as in the U4/U6 snRNP particle. For simplicity, only those residues referred to in the text are shown. Most base changes at U4 positions 9, 10, 13 and 14 cause cold sensitive phenotypes *in vivo*. Suppression of U4 G14C by a mutation in *PRP24* is indicated by a dotted arrow. C38, G40, G43 and G91 are recessive mutations in U6 which suppress U4 G14C, and have been postulated to define binding sites for the PRP24 protein (Shannon & Guthrie 1991). Mutations affecting boxed residues in U6 result in conditional lethal phenotypes *in vivo* (Madhani et al. 1990) and block *in vitro* splicing reactions at either the first or second step (Fabrizio and Abelson 1990).

Guthrie 1987; Seraphin et al. 1991), is particularly useful for preparation of U5-depleted extracts as this snRNA is extremely resistant to RNase H cleavage in intact snRNPs. Extracts depleted for U5 snRNA accumulate an ATP-dependent precursor complex containing U1 and U2 snRNPs but no U4 or U6 snRNAs (Seraphin et al. 1991) which are believed to join the pre-spliceosome along with U5 in a tri-snRNP complex.

Unlike U1, U2 and U5 snRNAs, which in yeast are much larger than their metazoan counterparts, U4 and U6 are conserved in length. U6 is also highly conserved in primary sequence, particularly in the central portion which includes sequences that base pair to U4 (Fig. 3). Conservation of primary sequences involved in base-pairing is unusual – often complementarity is conserved, but individual bases are not, compensatory changes being found on opposite sides of the helix. This led Brow and Guthrie (1988) to propose that U6 has a fundamental role in splicing, and the conservation of length and sequence in U6 was suggested to imply steric constraints imposed by its central position in the spliceosome and by important contacts with other spliceosomal factors.

Extensive mutagenesis has been carried out on U6 snRNA to identify important bases. Mutant U6 genes were constructed by doped synthesis or site directed mutagenesis followed by screening for mutations affecting function either *in vivo* (Madhani et al. 1990) or by reconstitution of mutant U6 snRNAs into snRNPs *in vitro* (Fabrizio & Abelson 1990). Two regions of U6 were defined in which mutation or deletion (bases 47–52 and 58–61; Fig. 3) had a large effect on splicing – either a lethal or conditional lethal defect *in vivo*, or a block of the first or second step of splicing *in vitro*. These results, and the finding of introns in the U6 genes of *Schizosaccharomyces pombe* and *Rhodotorula dacryoidum* within the first and second of these regions respectively, and several introns within the conserved region of U6 in *Rhodosporidium hasega-*

wae (Tani & Ohshima 1989, Tani & Ohshima 1991), support the suggestion by Brow and Guthrie (1988) that U6 snRNA comes into very close proximity with the intron during the splicing reaction. U6 has been proposed by Brow and Guthrie (1989) to be catalytic. They suggested that an intron could have become inserted into the U6 catalytic site by a reverse splicing event, the sequence being transferred to the genome by reverse transcription and homologous recombination. In this model, U4 may act as an anti-sense regulator of U6, sequestering the active residues in the U6 snRNA until the appropriate point in splicing has been reached. The observation that the U4:U6 interaction is destabilised just prior to the splicing reaction (U4 is lost from spliceosomes upon fractionation in non-denaturing gels (Pikielny et al. 1986; Cheng & Abelson 1987)) is consistent with this role for U4. Tani & Ohshima (1991) proposed that the structure of the catalytic site and the mechanism of action of U6 may be similar to that of the negative strand of the tobacco ring spot virus [(-)sTRSV]. The (-)sTRSV RNA undergoes self-catalysed cleavage during replication, and the catalytic domain has sequence similarities to the region of U6 proposed by Brow and Guthrie to be catalytic. It also forms a structure almost identical to one that U6 has been predicted to adopt in the absence of U4 snRNA (see Tani & Ohshima 1991).

More limited mutagenesis has been carried out with U4 snRNA. Certain single base changes in the stem II region of the U6 interaction domain (Fig. 3) were found to cause a cold sensitive phenotype *in vivo*. Restoration of base-pairing by complementary base changes in U6 alleviated the phenotype, and Shannon and Guthrie (1991) suggested that the cold sensitivity was due to destabilisation of the U4/U6 base-paired complex to favour an alternative complex in which the snRNAs are at least partially unwound, this complex being hyperstabilised in the mutants at low temperatures. Isolation of suppressors of one of these cold-sensitive U4 mutants led to the identification of PRP24 as a putative U4:U6 annealing factor (Shannon & Guthrie 1991, and see section 4.1).

3. Isolation and characterisation of splicing factors

Brute force screening of temperature-sensitive strains has led to the identification of many genes encoding splicing factors. The first mutants isolated were thought to affect ribosome biosynthesis (Hartwell et al. 1970), but were subsequently shown to be primarily defective in pre-mRNA splicing (Rosbash et al. 1981, Lustig et al. 1986). Introns occur in only a small proportion of genes in *S. cerevisiae*, but, as many of the ribosomal protein genes do contain an intron, splicing defects result in a secondary defect in ribosome formation. More recent screens to specifically identify splicing mutants were based on the identification of conditional lethal mutants that accumulated pre-mRNA, intermediates or intron product under non-permissive conditions (Vijayraghavan & Abelson 1989; Strauss & Guthrie 1991; Maddock, Roy & Woolford pers. comm.). Mutations affecting various points in the splicing pathway have been identified, attesting to the efficacy of the approach, however, screens that have identified new complementation groups have not isolated new alleles of all those previously known. This suggests that these searches were not saturating and that other genes remain to be identified. Approximately 35 complementation groups (termed *PRP* for *p*recursor *R*NA *p*rocessing) are currently known, and with more to be found, the complexity of the splicing apparatus probably equates to that of a ribosomal subunit. Interestingly, a fairly high proportion of temperature sensitive-mutants – up to about 5% – have turned out to be splicing mutants. This may be related to the fact that splicing takes place in a large complex with many transient interactions and as such may be inherently sensitive to subtle changes in protein stability and conformation.

Wild type *PRP* genes are cloned by complementation of the mutant phenotype using yeast genomic DNA plasmid libraries. All the *PRP* genes that have been cloned and characterised have been found to be essential, single copy and encode proteins. Functional characterisation of a PRP protein often begins with heat-inactivated extracts from temperature sensitive *prp* strains and with the aid of antibodies generated against the

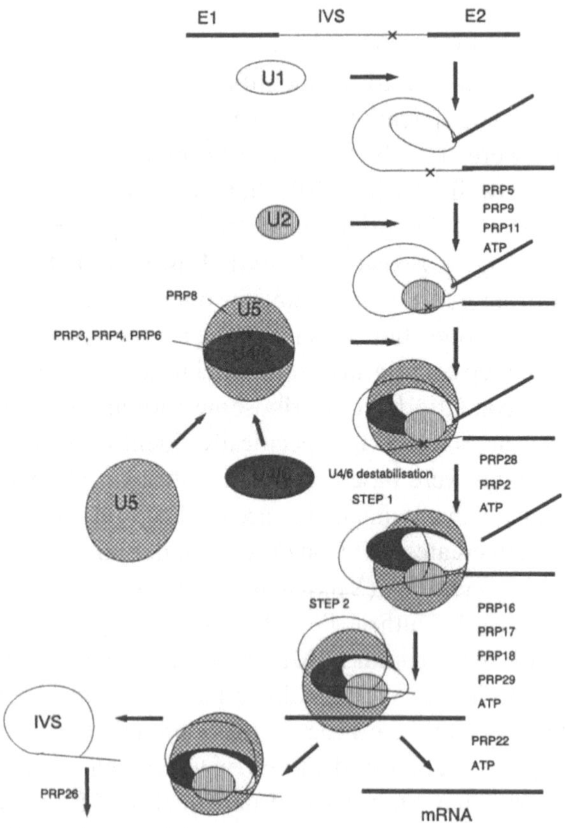

Fig. 4. A model illustrating spliceosome assembly and splicing. The stepwise interaction of snRNPs and the proposed action points for many of the PRP proteins are shown.

protein. The *in vitro* and *in vivo* data together provide knowledge of where the protein is acting in the splicing pathway, whether it is a component of a snRNP and perhaps an insight into the function of the protein from primary sequence analysis.

A number of the *PRP* gene products have been assigned to particular temporal action points in the splicing pathway and/or localised to particular snRNPs. PRP8 protein is a U5 snRNP component (Lossky et al. 1987) and PRP3, PRP4 and PRP6 are all U4/U6 associated (Anthony & Woolford pers. comm.; Banroques & Abelson 1989; Peterson-Bjorn et al. 1989; Abovitch et al. 1990). PRP2 and PRP16 are required both *in vitro* and *in vivo* for the first and second step in splicing respectively (Rosbash et al. 1981; Lustig et al. 1986; Schwer & Guthrie 1991) and have been shown by immunoprecipitation studies to interact transiently with spliceosomes at these stages (King & Beggs 1990; Schwer & Guthrie 1991). The products of the *PRP17*, *PRP18*, and *PRP29* genes are also required for the second step of the splicing reaction (Vijayraghavan et al. 1989; Vijayraghavan & Abelson 1990; Maddock, Roy & Woolford pers. comm.), as indicated by *in vivo* accumulation of intron-exon 2 intermediates in temperature sensitive mutants at the non-permissive temperature. Another class of mutants (*prp22*, *prp26* and *prp27*) corresponds to activities required late in the splicing pathway and accumulate excised lariat-intron. PRP22 is required for release of the spliced message from the spliceosome (Company et al. 1991), and PRP26 has been shown to be the lariat debranching enzyme (Chapman & Boeke 1991). A more complete assignation of where splicing factors act is shown in Fig. 4.

Not unexpectedly, many PRP proteins have putative nucleic acid binding motifs. PRP2, PRP6, PRP9 and PRP11 all have zinc finger-like motifs (Chen & Lin 1990; Legrain & Choulika 1990), PRP24 contains three copies of the RNP-motif found in many RNA binding proteins (Shannon & Guthrie 1991) while PRP5 and PRP28 proteins (Dalbadie-McFarland & Abelson 1990; Strauss & Guthrie 1991) have motifs that characterise the ATP-dependent RNA helicase DEAD-box family (D-E-A-D being single letter code for amino acids) whose founder member eIF-4A has been shown to have ATP-dependent RNA helicase activity (Ray et al. 1985; Rozen et al. 1990). PRP2, PRP16 and PRP22 contain related motifs that define a subclass of the DEAD-box family, the DEAH-box proteins. Purified PRP16 protein has been shown to have an ATPase activity *in vitro* that is stimulated by RNA (Schwer & Guthrie 1991). This is the first purification and biochemical characterisation of a yeast splicing factor. Activities that have been assigned to other splicing factors on the basis of sequence analysis will also have to be verified biochemically.

4. Linking up the components of the spliceosome

Characterised splicing mutants have been used in various genetic selections to identify components which physically or functionally interact. The most widely used method is the isolation of *trans*-acting extragenic suppressor mutations. Others include the isolation of high copy-number or dosage suppressors, and the identification of synergistic mutations. These genetic approaches offer alternatives to biochemical studies which may not easily identify low abundance factors, especially where interactions are only transient. Suppressors of mutations in *cis*-acting elements (conserved intron sequences), or in genes encoding *trans*-acting factors (*SNR* and *PRP* genes) have provided exciting new insights into the mechanism of splice site recognition, and spliceosome composition and assembly.

4.1. Suppressors of defects in trans*-acting factors*

The classical approach of isolating extragenic suppressor mutations has been particularly productive, not only in analysing interactions between components, but also in identifying novel splicing factors. The general approach is as follows: under non-permissive conditions, a mutation results in a conformational change in factor X preventing an essential association with a second component, factor Y; suppression of this defect may be achieved by a mutation in the gene encoding factor Y which alters its conformation such that the interaction is restored. In practice, conditional-lethal (usually temperature sensitive) splicing mutants are incubated under non-permissive conditions to select for growth of pseudorevertants. Cloning and further analysis of suppressor genes is facilitated if they confer an additional phenotype that is linked to suppression. Subsequent analysis must distinguish between the sought-after interactive suppressors, and other classes of suppressors which could be present in a pool of pseudorevertants. These include translational suppressors, suppressors which act by modulating the level of expression of the mutant protein, and by-pass suppressors i.e. unrelated factors that substitute for the mutant activity.

As described in section 2, in order to explain the cold-sensitivity of certain mutations in the region of U4 snRNA that base-pairs with U6, Shannon and Guthrie (1991) proposed the existence of an alternative complex that would be hyperstabilised in the cold-sensitive mutants. According to this model, mutations which destabilise the competing complex might suppress the cold sensitivity of these U4 defects. Such mutations were sought by selecting for recessive suppressors of the U4 mutation G14C. One class of suppressor mutations mapped to the U6 gene outwith the U4:U6 interaction domain (Fig. 3). This suggested that U6 snRNA might form part of the competing complex in association with other *trans*-acting factors. They therefore sought mutations in these putative factors, and to prevent isolation of U6 mutants, suppressors were selected in U4-G14C haploids that contained an extra copy of *SNR6* on a plasmid. Three recessive suppressors were isolated that mapped to the same gene, and one of these, by virtue of it having a temperature-sensitive phenotype, permitted complementation analysis. The factor was characterised as the previously identified PRP24 protein, and the recessive mutations in U6 were proposed to define possible binding sites for this protein.

Consistent with the existence of RNP-motifs in PRP24, immunoprecipitation experiments using antibodies raised against the protein demonstrated a strong association between this factor and the U6 snRNA in wild type extracts (Shannon & Guthrie 1990). However, in extracts prepared from U4-G14C mutants, PRP24 was found to be associated with both U4 and U6, whereas it was not associated with either snRNA in extracts from cells carrying *snr6* or *prp24* recessive suppressor mutations. Thus molecular evidence was obtained for a model in which PRP24 protein binds strongly to U6 RNA, stabilising the unwound form, and binds transiently to the destabilised form of U4/U6 in an alternative complex involved perhaps in reformation of the U4/U6 snRNP. This study demonstrates the immense power of suppressor analysis in identifying interacting factors.

The interaction of the U5 snRNP protein PRP8 with a putative RNA helicase has been suggested by a suppressor approach; *spp81* is one of several cold-sensitive suppressors of a mutation in the *PRP8* gene (Jamieson et al. 1991). This *spp81* mutation maps to the *DED1* gene which encodes a member of the DEAD-box family of putative RNA helicases. A direct interaction between the spp81 and PRP8 proteins was suggested by the observation that increased copy-number of the wild type *PRP8* gene impairs the growth of *spp81* cells, indicating that excess PRP8 protein may sequester the essential DED1 protein, thus limiting growth. Strauss and Guthrie (1991) have genetic evidence for PRP8 interacting with another RNA helicase protein PRP28; a mutant allele of the *PRP8* gene was isolated as a suppressor of a cold-sensitive mutation in *PRP28*. It is known that the PRP28 protein is essential for splicing, however it remains to be seen whether DED1/SPP81 is normally involved in splicing, or is recruited to this role under special circumstances.

During attempts to clone the wild type *SPP81/DED1* gene by complementation of the cold sensitive *spp81* mutation, another gene was cloned by virtue of its ability, when present on high copy-number plasmids, to suppress the *spp81* defect (Jamieson & Beggs 1991). This gene encodes yet another member of the DEAD box family, and was designated *DBP1* (*D*ead *B*ox *P*rotein). The predicted DBP1 protein sequence is very closely related to that of DED1 (82% similarity, 72% identity) and so presumably suppresses *spp81* mutations by substituting for the activity that is defective at low temperatures. This is an example of dosage suppression, models for which are discussed by Last et al.(1987).

Dosage suppression has also identified a novel factor, SPP2, which may interact with the PRP2 protein (Last et al. 1987). When present at elevated copy-number, the *SPP2* gene complements *prp2* mutations. This does not occur simply as a bypass of PRP2 function, because elevated levels of SPP2 will not suppress a null allele of *PRP2*. Importantly, temperature-sensitive alleles of *SPP2*, created by *in vitro* mutagenesis, show that SPP2 protein is itself essential for splicing. The dosage suppression is reciprocal; overexpression of the *PRP2* gene can suppress temperature sensitive alleles of *SPP2*, providing strong evidence for a direct interaction between these splicing factors (Kim & Woolford pers. comm.).

Multiple copies of the *PRP3* gene have been shown to suppress the temperature sensitivity of several *prp4* alleles, but not a *prp4* null allele (Last et al. 1987). In this case, dosage suppression is not reciprocal; extra copies of *PRP4* do not suppress *prp3* mutations. However, certain cold-sensitive *prp4* mutants specifically suppress temperature-sensitive *prp3* mutants (Woolford 1989). These observations, which on a genetic level suggest a specific interaction between the PRP3 and PRP4 proteins, are consistent with molecular evidence for the association of both proteins with U4/U6 snRNPs (Banroques & Abelson 1989; Peterson-Bjorn et al. 1989; Anthony & Woolford pers. comm.).

4.2. Suppressors of defects in cis*-acting sequences*

Mutations in the conserved yeast intron sequences generally cause severe splicing defects, providing an ideal system for the genetic identification of splicing factors that interact with these *cis*-acting elements of the pre-mRNA. A strong selection for suppressors of intron mutations can be achieved by the use of a reporter construct that contains a defective intron.

This approach was adopted by Couto et al. (1987) using constructs containing an A to C branch point mutation (designated C259, see Fig. 2) which blocks the first step in the splicing reaction. They isolated a dominant suppressor by its ability to overcome the splicing defect conferred by the *C259* mutation in an actin-*HIS4* gene fusion. Growth on the histidine precursor histidinol was not possible unless the intron was accurately spliced, so pseudo-revertants could be selected by growth on histidinol. The screen was elegantly designed, employing diploids to allow the recovery of recessive-lethal mutations, and also because this results in the isolation of dominant suppressors which are likely to represent a gain of function. To establish that the suppressors were indeed extragenic and *trans*-acting, they were tested for the ability to suppress a second, independent C259 mutation present in another reporter gene. Only one of the dominant suppressors satisfied these criteria and this mapped to the *PRP16* gene. The mutant protein encoded by this allele, *prp16–1*, was thus identified as being able to influence the fidelity of branch point recognition, however, as *prp16–1* haploids are viable, the mutant protein must still fulfil its normal role. Although *prp16–1* seems to influence the recognition of the branchpoint in the pre-mRNA (i.e. before step 1), PRP16 depleted extracts carry out the first cleavage-ligation step (Schwer & Guthrie 1991). Thus the *prp16–1* mutation may confer an extra property on the PRP16 protein. The *prp16–1* mutation causes a single amino acid change in the putative ATP binding domain of the protein (Burgess et al. 1990), supporting a role for PRP16 at one of the ATP-requiring steps in splicing. A kinetic proof-reading model was proposed in which the aberrant function of the prp16–1 protein was due to a relaxation in the fidelity of recognition of its target, caused by a change in the rate of ATP binding and/or hydrolysis (Burgess et al. 1990).

A similar approach was used by Newman and Norman (1991) to identify factors involved in 5′ splice site recognition. They isolated suppressors of a 5′ splice site mutation in which the invariant G residue at intron position 1 was substituted by an A (Fig. 2). This mutation inhibits step 2 of splicing and the products of step 1 accumulate. A dominant *trans*-acting suppressor mutant was isolated that activated a cryptic splice site 12 nucleotides upstream of the authentic 5′ splice junction. This resulted in the production of an in-frame mRNA with a short deletion. Cloning and sequencing identified the suppressor as a mutant form of the U5 snRNA with a single U to C base change in a phylogenetically conserved loop (loop I U6 to C6; Fig. 2). Subsequently, from a collection of U5 snRNA mutants generated by *in vitro* mutagenesis, several more suppressors that activate cryptic splice sites were identified. All but one suppressor carried the U6 to C6 mutation, suggesting a specific effect of this base change. The only suppressor without this change contained a single deletion of an A residue within the U5 loop, and caused the activation of another cryptic splice site 4 nucleotides 3′ to the normal 5′ splice junction. In a further screen in which the major cryptic splice site at position − 12 was removed, a U5 loop mutant was isolated which could process the defective branch point lariat intermediate (that contains an A-A 2′-5′phosphodiester bond), permitting step 2 of the splicing reaction. These results provide evidence for a role of the U5 loop I element in 5′ splice site recognition, and also perhaps in the fidelity of branchpoint recognition during step 2 of splicing.

4.3. Synergistic mutations

In a multi-subunit complex, the activity of one component may be dependent on the integrity/stability of another. Thus interacting factors may be identified by screening combinations of known mutations for those that are lethal when present together in a haploid cell under conditions which are permissive for the individual mutations. This phenomenon, termed synthetic lethality, has been demonstrated for mutations in *PRP3* and *PRP4* (Woolford 1989). Such a genetic interaction between *prp3* and *prp4* mutants *in vivo* agrees with *in vitro* complementation studies in which prp3 and prp4 heat-inactivated splicing extracts complement only weakly compared to many other combinations of mutant extracts (Lustig et al. 1986), and with the discovery that the PRP3 and PRP4 proteins are both associated with the U4/U6 snRNP. Synthetic lethality has also been shown for certain alleles of *PRP24* and *PRP28* (Strauss & Guthrie 1991). As PRP28 is a putative RNA helicase, and PRP24 is a U6 binding protein, Strauss and Guthrie (1991) proposed that these proteins both act in the U4/U6 cycle of unwinding and reannealing.

4.4. Dominant negative mutations

Another genetic approach well suited to the analysis of multi-component splicing complexes is the generation of dominant negative mutations. Usually a cloned gene is mutated *in vitro*, and alleles are selected *in vivo* which, when overexpressed from an inducible promoter, compete with and disrupt the activity of the corresponding wild type factor (either RNA or protein). Several models have been proposed for the way in which these mutations may cause their phenotype (Herskowitz

1987). Changes in a specific, overexpressed factor may permit its entry into complexes in preference to the wild type version, but prevent subsequent activity and/or turnover of the complex. Production of these 'frozen', dead-end complexes could facilitate biochemical analysis of normally transient events.

In a study to test the function of variant U2 RNAs, the first example was isolated of a dominant-negative mutation affecting a yeast splicing factor (Miraglia et al. 1991). A chimeric yeast U2 gene was constructed in which the branch point interaction region (Fig. 2) was replaced by the corresponding sequence from a trypanosome U2 snRNA gene. Overexpression of this gene caused a severe splicing defect, and a model was proposed in which the mutant snRNA was not released from complexes, thus sequestering essential limiting factors.

Production of dominant-negative mutations may also provide an excellent basis for the isolation of dosage suppressors, as overexpression of an interacting factor may titrate out the toxic mutant RNA or protein.

5. Conclusions

The use of genetic approaches is, as we hope to have demonstrated here, resolving some of the complexities of splicing. This process involves many *trans*-acting factors, whose roles are becoming clear. Considering the small proportion of genes in *S. cerevisiae* that contain introns, it is curious that this extremely elaborate, energy-consuming process has been maintained. Fink (1987) has proposed that many introns may have been lost from the genome of *S. cerevisiae* as a consequence of the replacement of genomic sequences by pseudogenes (endogenous cDNAs) through homologous recombination. If this is the case, the most likely reason for splicing being maintained would be that splicing in yeast, as in higher eukaryotes, plays an important role in the regulation of gene expression.

The recent finding that certain proteins are specifically involved in the splicing of particular yeast transcripts in several important areas of cell biology – meiosis (Engelbrecht et al. 1991), mating type determination (Nakazawa et al. 1991) and ribosome formation (Eng & Warner 1991) – opens up new avenues of research. Although the extent of regulated splicing in yeast may be small compared to that seen in higher eukaryotes, it should be amenable to the genetic approaches that have already provided such rapid advances and have made yeast a most attractive organism in which to study complex processes.

Acknowledgements

The authors acknowledge The Wellcome Trust and the Science and Engineering Research Council for studentships and The Royal Society for a senior fellowship. Nik Somia, Peter Hodges and Stefan Teigelkamp are thanked for generous help in the preparation of this manuscript.

References

Abovich N, Legrain P & Rosbash M (1990) The yeast *PRP6* gene encodes a U4/U6 small nuclear ribonucleoprotein particle (snRNP) protein, and the *PRP9* gene encodes a protein required for U2 snRNP binding. Mol. Cell. Biol. 10: 6417–6425

Banroques J & Abelson JN (1989) PRP4: a protein of the yeast U4/U6 small nuclear ribonucleoprotein particle. Mol. Cell. Biol. 9: 3710–3719

Brow DA & Guthrie C (1988) Spliceosomal RNA U6 is remarkably conserved from yeast to mammals. Nature 334: 213–218

Brow DA & Guthrie C (1989) Splicing a spliceosomal RNA. Nature 337: 14–15

Burgess S, Couto JR & Guthrie C (1990) A putative ATP binding protein influences the fidelity of branchpoint recognition in yeast splicing. Cell 60: 705–717

Chapman KB & Boeke JD (1991) Isolation and characterisation of the gene encoding yeast debranching enzyme. Cell 65: 483–492

Chen J-H & Lin R-J (1990) The yeast PRP2 protein, a putative RNA-dependent ATP-ase, shares extensive sequence homology with two other pre-mRNA splicing factors. Nucleic Acids Res. 18: 6447

Cheng S-C & Abelson J (1987) Spliceosomal assembly in yeast. Genes & Dev. 1: 1014–1027

Company M, Arenas J & Abelson J (1991) Requirement of the

RNA helicase-like protein PRP22 for release of messenger RNA from spliceosomes. Nature 349: 487–493

Couto JR, Tamm J, Parker R & Guthrie C (1987) A *trans*-acting suppressor restores splicing of a yeast intron with a branch point mutation. Genes & Dev. 1: 445–455

Dalbadie-McFarland G & Abelson J (1990) PRP5: a helicase-like protein required for mRNA splicing in yeast. PNAS 87: 4236–4240

Eng FJ & Warner JR (1991) Structural basis for the regulation of splicing of a yeast messenger RNA. Cell 64: 797–804

Engerbrecht J, Voelkel-Meiman K & Roeder GS (1991) Meiosis-specific splicing in yeast. Cell 66: 1257–1268

Fabrizio P & Abelson J (1990) Two domains of yeast U6 small nuclear RNA required for both steps of nuclear precursor messenger RNA splicing. Science 250: 404–409

Fink GR (1987) Pseudogenes in yeast? Cell 49: 5–6.

Fouser LA & Frieson JD (1986) Mutations in a yeast intron demonstrates the importance of specific conserved nucleotides for the two stages of nuclear mRNA splicing. Cell 45: 81–93

Goguel V, Liao X, Rymond BC & Rosbash M (1991) U1 snRNP can influence 3′-splice site selection as well as 5′-splice site selection. Genes & Dev. 5: 1430–1438

Guthrie C (1988) Genetic analysis of yeast snRNAs. In: Birnstiel M (Ed) Structure and Function of Major and Minor Small Nuclear Ribonucleoprotein Particles (pp 196–211). Springer-Verlag, Berlin, Heidelberg, New York

Guthrie C (1991) Messenger RNA splicing in yeast: Clues to why the spliceosome is a ribonuceoprotein. Science 253: 157–163

Guthrie C & Patterson B (1988) Spliceosomal snRNAs. Ann. Rev. Genet. 22: 387–419

Hartwell LH, McLaughlin CS & Warner JR (1970) Identification of ten genes that control ribosome formation in yeast. Mol. Gen. Genet. 109: 42–56

Herskowitz I (1987) Functional inactivation of genes by dominant negative mutations. Nature 329: 219–222

Jacquier A (1990) Self-splicing group II and nuclear introns: how similar are they? TIBS 15: 351–354

Jacquier A, Rodriguez JR & Rosbash M (1985) A quantitative analysis of the effects of 5′ junction and TACTAAC box mutants and mutant combinations on yeast mRNA splicing. Cell 43: 423–430

Jamieson DJ & Beggs JD (1991) A suppressor of yeast *spp81/ ded1* mutations encodes a very similar putative ATP-dependent RNA helicase. Mol. Microb. 5: 805–812

Jamieson DJ, Rahe B, Pringle J & Beggs JD (1991) A suppressor of a yeast splicing mutation (*prp8–1*) encodes a putative ATP-dependent RNA helicase. Nature 349: 715–717

King DS & Beggs JD (1990) Interactions of PRP2 protein with pre-mRNA splicing complexes in *Saccharomyces cerevisiae*. Nucleic Acids Res. 18: 6559–6564

Last RL, Maddock JR & Woolford Jr JL (1987) Evidence for related functions of the *RNA* genes of *Saccharomyces cerevisiae*. Genetics 117: 619–631

Legrain P & Choulika A (1990) The molecular characterisation of *PRP6* and *PRP9* yeast genes reveals a new cysteine/histidine motif common to several splicing factors. EMBO J. 9: 2775–2781

Legrain P, Seraphin B & Rosbash M (1988) Early commitment of yeast pre-mRNA to the spliceosome pathway. Mol. Cell. Biol. 8: 3755–3760

Lin R-J, Newman AJ, Cheng S-C & Abelson J (1985) Yeast mRNA splicing in vitro. J. Biol. Chem. 260: 14780–14792

Lossky M, Anderson GJ, Jackson SP & Beggs JD (1987) Identification of a yeast snRNP protein and detection of snRNP-snRNP interactions. Cell 51: 1019–1026

Lührmann R, Kastner B & Bach M (1990) Structure of spliceosomal snRNPs and their role in pre-mRNA splicing. Biochem. Biophys. Acta 1087: 265–292

Lustig AJ, Lin R-J & Abelson J (1986) The yeast *RNA* gene products are essential for mRNAsplicing in vitro. Cell 47: 953–963

Madhani HD, Bordonne R & Guthrie C (1990) Multiple roles for U6 snRNA in the splicing pathway. Genes & Dev. 4: 2264–2277

Miller AM (1984) The yeast *MATa1* gene contains two introns. EMBO J. 3: 1061–1065

Miraglia L, Seiwert S, Igel AH & Ares M (1991) Limited functional equivalence of phylogenetic variation in small nuclear RNA: Yeast U2 RNA with altered branchpoint complementarity inhibits splicing and produces a dominant lethal phenotype. PNAS 88: 7061–7065

Nakazawa N, Harashima S & Oshima Y (1991) *AAR2*, a gene for splicing pre-mRNA of the *MATa1* cistron in cell type control of *Saccharomyces cerevisiae*. Mol. Cell. Biol. 11: 5693–5700

Newman A & Norman C (1991) Mutations in yeast U5 snRNA alter the specificity of 5′splice site cleavage. Cell 65: 115–123

Parker R & Guthrie C (1985) A point mutation in the conserved hexanucleotide at a yeast 5′ splice site uncouples recognition, cleavage, and ligation. Cell 41: 107–118

Parker R, Siliciano PG & Guthrie C (1987) Recognition of the TACTAAC box during mRNA splicing in yeast involved base pairing to the U2-like snRNA. Cell 49: 229–239

Patterson B & Guthrie C (1987) An essential yeast snRNA with a U5-like domain is required for splicing *in vivo*. Cell 49: 613–624

Petersen-Bjorn S, Soltyk A, Beggs JD & Friesen JD (1989) *PRP4 (RNA4)* from *Saccharomyces cerevisiae*: its gene product is associated with the U4/U6 small nuclear ribonucleoprotein particle. Mol. Cell. Biol. 9: 3698–3709

Pikielny CW, Rymond BC & Rosbash M (1986) Electrophoresis of ribonucleoproteins reveals an ordered assembly pathway of yeast splicing complexes. Nature 324: 341–345

Ray BK, Lawson TG, Kramer JC, Cladras MH, Grifo JA, Abramson RD, Merrick WC & Thach RE (1985) ATP-dependent unwinding of messenger RNA structure by eukaryotic initiation factors. J. Biol. Chem. 260: 7651–7658

Rosbash M & Seraphin B (1991) Who's on first? The U1 snRNP-5′splice site interaction and splicing. TIBS 16: 187–190

Rosbash M, Harris PKW, Woolford JL & Teem JL (1981) The effect of temperature-sensitive RNA mutants on the transcription products from cloned ribosomal protein genes of yeast. Cell 24: 679–686

Rozen F, Edery I, Meerovitch K, Dever TE, Merrick WC & Sonenberg N (1990) Bidirectional RNA helicase activity of eukaryotic translation initiation fasctors 4A and 4F. Mol. Cell. Biol. 10: 1134–1144

Ruby SW & Abelson J (1988) An early heirarchic role of U1 small nuclear ribonucleoprotein in spliceosome assembly. Science 242: 1028–1035

Ruby SW & Abelson J (1991) Pre-mRNA splicing in yeast. TIG 7: 79–85

Schwer B & Guthrie C (1991) PRP16 is an RNA-dependent ATPase that interacts transiently with the spliceosome. Nature 349: 494–499

Seraphin B & Rosbash M (1989) Identification of functional U1 snRNA-pre-mRNA complexes committed to spliceosome assembly and splicing. Cell 59: 349–358

Seraphin B, Kretzner L & Rosbash M (1988) A U1 snRNA:pre-mRNA base pairing interaction is required early in yeast spliceosome assembly but does not uniquely define the 5′ cleavage site. EMBO J. 7: 2533–2538

Seraphin B, Abovich N & Rosbash M (1991) Genetic depletion indicates a late role for U5 snRNP during in vitro spliceosome assembly. Nucleic Acids Res. 19: 3857–3860

Sharp PA (1987) Splicing of Messenger RNA precursors. Science 235: 766–771

Shannon KW & Guthrie C (1991) Suppressors of a U4 snRNA mutation define a novel U6 snRNP protein with RNA-binding motifs. Genes & Dev. 5: 773–785

Siliciano PG & Guthrie C (1988) 5′ splice site selection in yeast: genetic alterations in base-pairing with U1 reveal additional requirements. Genes & Dev. 2: 1258–1267

Strauss EJ & Guthrie C (1991) A cold-sensitive mRNA splicing mutant is a member of the RNA helicase gene family. Genes & Dev. 5: 629–641

Tani T & Ohshima Y (1989) The gene for the U6 small nuclear RNA in fission yeast has an intron. Nature 337: 87–90

Tani T & Ohshima Y (1991) mRNA-type introns in U6 small nuclear RNA genes: implications for the catalysis in pre-mRNA splicing. Genes & Dev. 5: 1022–1031

Vijayraghavan U & Abelson J (1989) Pre-mRNA Splicing in yeast. In: Eckstein F and Lilley DMJ (Eds) Nucleic Acids and Molecular Biology, Vol. 3 (pp 197–215). Springer-Verlag, Berlin, Heidelberg

Vijayraghavan U & Abelson J (1990) PRP18, a protein required for the second reaction in pre-mRNA splicing. Mol. Cell. Biol. 10: 324–332

Vijayraghavan U, Company M & Abelson J (1989) Isolation and characterisation of pre-mRNA splicing mutants of *Saccharomyces cerevisiae*. Genes & Dev. 3: 1206–1216

Woolford JL (1989) Nuclear pre-mRNA splicing in yeast. Yeast 5: 439–458

Antonie van Leeuwenhoek **62**: 47–62, 1992.

Molecular biology of translation in yeast

Patrick Linder
Department of Microbiology, Biozentrum, 70 Klingelbergstrasse, CH-4056 Basel, Switzerland

Key words: Saccharomyces cerevisiae, genetics, translation regulation

Abstract

The combination of genetic, molecular and biochemical approaches have made the yeast *Saccharomyces cerevisiae* a convenient organism to study translation. The sequence similarity of translation factors from yeast and other organisms suggests a high degree of conservation in the translational machineries. This view is also strengthened by a functional analogy of some proteins implicated in translation. Beautiful genetic experiments have confirmed existing models and added new insights in the mechanism of translation. This review summarizes recent experiments using yeast as a model system for the analysis of this complex process.

Introduction

The study of translation is attractive for different reasons. Translation is an important step in gene expression and many examples have been found in which translation initiation plays a crucial role in the regulation of gene expression. Initiation involves a large number of factors that interact with each other, with RNA and with ribosomes. One of the main goals in the study of translation is to understand how protein-protein, protein-RNA and RNA-RNA interactions occur. Although mutants in protein synthesis in yeast have been isolated a long time ago (Hartwell & McLaughlin 1968b), the advantages of this organism have been reconsidered only recently. The reader is referred to several reviews dealing with different aspects of translation in eukaryotes (Linder & Prat 1990; Moldave 1985; Rhoads 1988; Riis et al. 1990; Sonenberg 1988; Trachsel 1991; Tuite 1989; Warner 1989). In this short overview I will summarize how the use of molecular genetic methods can lead to a better understanding of translation in the yeast *Saccharomyces cerevisiae*.

Overview on translation in eukaryotes

Translation can be divided into three steps: i) Initiation, the association of the ribosomal subunits on a mRNA, ii) elongation, the movement of the ribosomes along the mRNA and the formation of peptide bonds and iii) termination, the release and dissociation of the ribosomes at the end of an open reading frame.

Most of the knowledge on translation comes from *in vitro* studies of higher eukaryotes. Translation initiation in eukaryotes is a complex process which requires approximately 10 initiation factors composed of about 25 polypeptides (Moldave 1985; Tuite 1989). Initiation can be subdivided into the preparation of the mRNA, preparation of the small ribosomal subunit (43S preinitiation complex) and scanning of the small ribosomal subunit until the initiator codon is reached. At this point the large ribosomal subunit joins the preinitiation complex and translation can start (Fig. 1).

In the first step, the cap structure at the 5′ end of the mRNA is recognized by the cap-binding protein, eIF-4E (eIF = eukaryotic Initiation Factor). Subsequently a high molecular weight protein (220kDa, 150kDa in yeast) and eIF-4A form to-

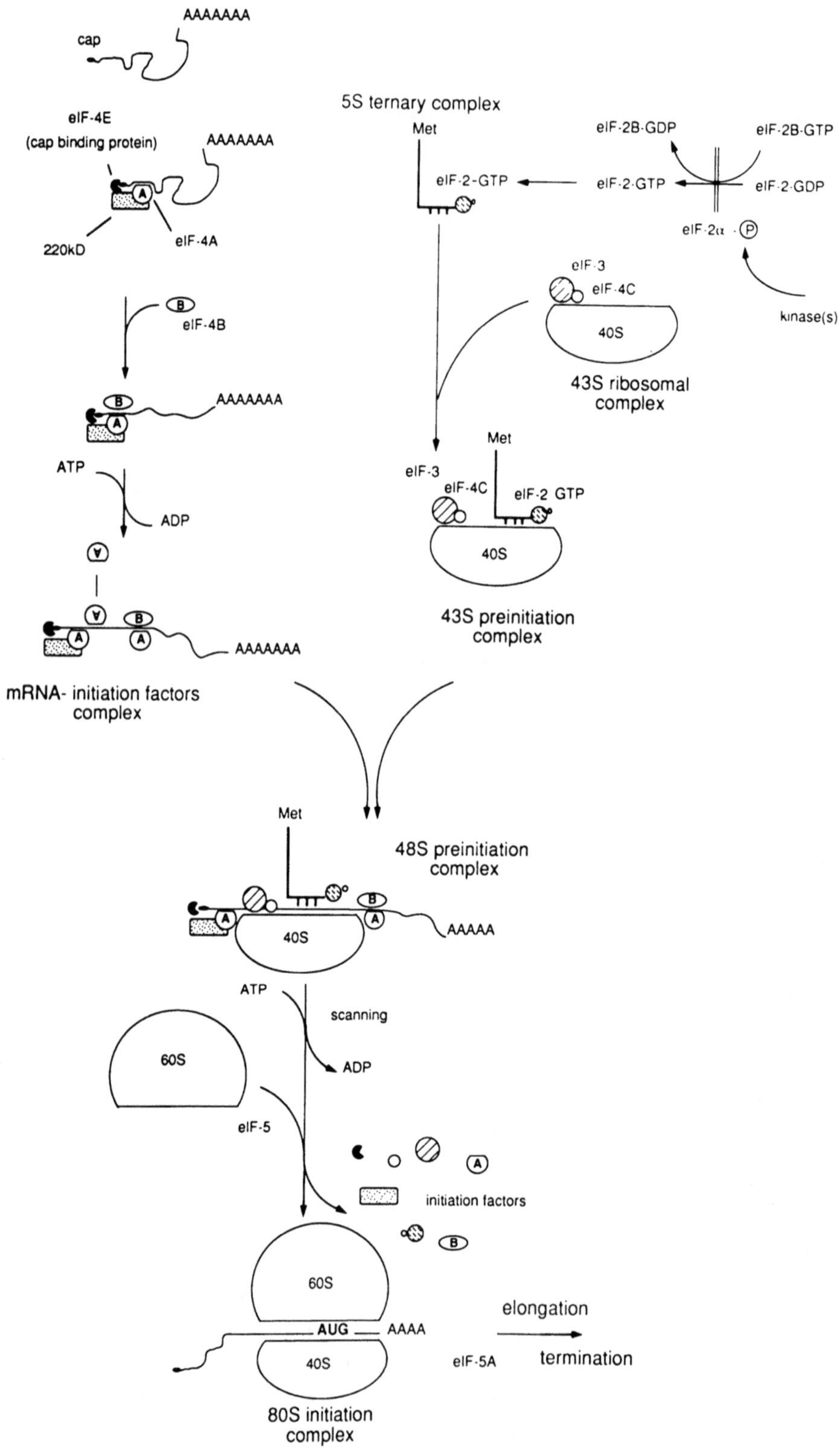

Fig. 1. The different steps of translation initiation. The data leading to the proposed scheme are mainly obtained from *in vitro* systems of higher eukaryotic cells. The individual factors are described in the text. This figure is reprinted in a slightly modified form from (Linder & Prat 1990).

gether with eIF-4E the cap binding protein complex, also referred to as eIF-4F. Alternatively it is possible that this complex could associate prior to cap binding. In fact, eIF-4F or the CBP (cap-binding-protein) complex can be isolated by affinity chromatography, in which the complex binds to a cap analogue (Edery et al. 1983). Subsequent to the binding of eIF-4F to the cap, more eIF-4A, together with eIF-4B, are thought to bind to the mRNA and remove potential secondary structures or other RNA-RNA interactions, and thus stimulate the binding of the small ribosomal subunit to the mRNA (Sonenberg 1988).

Before binding the mRNA, the small ribosomal subunit interacts with the 5S ternary complex consisting of the initiator-tRNA^{Met}, eIF-2 and GTP to form the 43S preinitiation complex. The initiation factor eIF-2 consists of three subunits and GTP, which is hydrolyzed upon initiation. The bound GDP is then regenerated by eIF-2B. This step is blocked in reticulocytes due to phosphorylation of the α-subunit of eIF-2 by two protein kinases, HRI (heme-regulated inhibitor) and DAI (dsRNA-activated inhibitor) (Fig. 1).

In the third step, the 43S preinitiation complex binds to the mRNA and begins to scan for the initiator codon occurs. In contrast to the prokaryotic system, no Shine-Dalgarno sequence is present to guide the ribosome to the initiator codon. In most cases, the small ribosomal subunit scans until it reaches the first AUG codon present on the mRNA (Kozak 1978; Kozak 1989). The initiator-tRNA and initiation factor eIF-2 seem to play a crucial role in this scanning process. Cigan and co-workers (Cigan et al. 1988) changed the AUG codon of the yeast *HIS4* gene to a non-functional AGG triplet and made the corresponding change in one of the four initiator-tRNA^{Met} genes to yield a complementary 3′-UCC-5′ anticodon. A strain bearing both of these mutations (his4-AGG and tRNA^{Met}-UCC) does grow on medium lacking histidine provided that the methionyl-tRNA synthetase (*MES1*) can charge the modified tRNA either due to a mutation or an elevated gene dosage. Mutations in the initiator codon can also be overcome by suppressor mutation in the α- or β-subunits of eIF-2, pointing to a role of this factor in the scanning process (see below; Cigan et al. 1989; Donahue et al. 1988).

The scanning process requires hydrolysis of ATP, an activity which has been attributed to eIF-4A. It is generally assumed that this factor together with eIF-4B uses this energy to remove RNA-RNA pairing (Jackson 1991). This is in accordance with the requirement of ATP hydrolysis for the *in vitro* RNA helicase activity described for the eIF-4A/eIF-4B system (Rozen et al. 1990). At the AUG, initiation factor eIF-5 catalyzes the association of the two ribosomal subunits and the bound initiation factors dissociate. Initiation factor eIF-5A (previously called eIF-4D) seems to be involved in the formation of the first peptide bond and therefore holds a special place between initiation and elongation.

In the elongation process the ribosome moves relative to the mRNA and peptidyl bonds arc formed (Fig. 2). Two elongation factors, EF-1 and EF-2, are required (for a recent review see Riis et al. 1990). The α subunit of factor EF-1 binds GTP and the aminoacyl-tRNA and places the the tRNA in the A-site of the ribosome. After release of EF-1α-GDP the β subunit of EF-1 is required for the nucleotide exchange on EF-1α. After the transfer of the nascent peptide chain onto the aminoacyl-tRNA located in the A-site (a process which requires the activity of the peptidyl transferase of the ribosome) the ribosome is translocated with respect to the mRNA and the tRNA is ejected from the P-site. This step requires EF-2 and GTP hydrolysis. Elongation factor EF-2 is present in two conformations, bound to GTP or GDP. It is generally assumed that the conformational change in EF-2 drives the translocation step. After translocation, the A-site is again free to accept a new aminoacyl-tRNA. A peculiarity of the yeast system is the presence of a third elongation factor, EF-3, which stimulates the positioning of the aminoacyl-tRNA by EF-1α.

The model for translation termination at one of the termination codons implies a release factor that is specific for the termination codon. Binding of the release factor to the A-site is followed by hydrolysis of the peptidyl-tRNA ester bond at the P-site and hydrolysis of GTP. The peptide is released and the

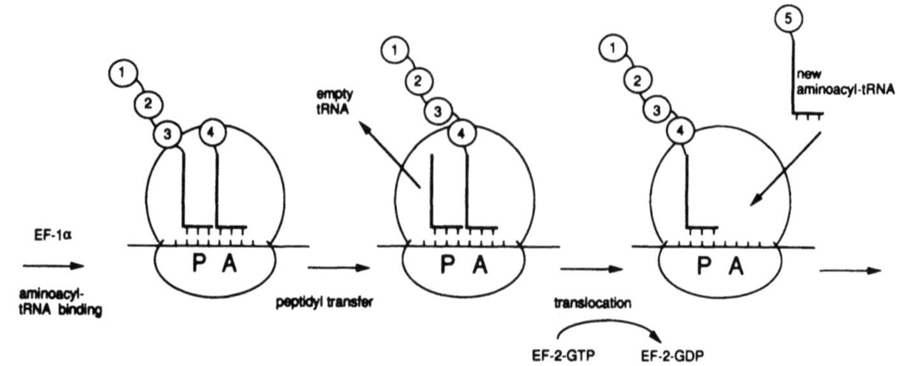

Fig. 2. Schematic representation of one cycle of the elongation process. The elongation factor EF-3 which stimulates EF-1 activity is not indicated.

ribosome dissociates from the mRNA. At present, no molecular data on eukaryotic cytoplasmic release factors are available.

Yeast as a 'test tube' for translation analysis

Yeast offers the advantage of permitting a combination of both biochemical and molecular genetic approaches to investigate complex processes such as translation. A simplified overview of such approaches is given here. The pioneering genetic work of Hartwell's group in the sixties led to the isolation of several conditional lethal mutations which are defective in translation (Hartwell & McLaughlin 1968a; Hartwell & McLaughlin 1968b). A well characterized example is the mutation in the *PRT1* gene (previously called ts187), a mutation which also causes a cell cycle arrest at the nonpermissive temperature.

Mutational analysis and molecular genetics gave insights into the general control of amino acid biosynthesis (for review see Hinnebusch 1988). In this system the expression of a transcriptional activator (*GCN4*) is regulated at the translational level. Mutations have been isolated which influence its expression, some being constitutively induced for *GCN4* expression (GCD=general amino acid control constitutively derepressed), others abolishing the induction (GCN=general amino acid control nonderepressible). The molecular analysis of the 5′ untranslated region of the *GCN4* mRNA revealed that four short upstream open-reading frames are required for translational control (Mueller & Hinnebusch 1986). Due to the combination of biochemical, molecular genetic and immunological methods in this system, models which explain the unusual control of *GCN4* expression have emerged. The study of translational control of the *GCN4* gene is not only of interest *per se*, but of general use since many factors involved in the *GCN4* expression are likely to be general translation initiation factors.

A useful approach is the isolation of suppressors of already existing mutations. The aim in this approach is to characterize factors or elements which interact in some way with the product of a mutant gene (protein or RNA). An exciting example is the isolation of mutations in the α and β subunits of the initiation factor eIF-2 (see below). These mutations can suppress the effect of initiation codon alterations. This result, together with others, clearly demonstrated the importance of eIF-2 in the scanning process (Cigan et al. 1989; Donahue et al. 1988).

In many cases the effect of a suppressor mutation is indirect. An indication for direct interaction of two components is the creation of synthetic lethals. A combination of conditional mutations in *GCN3* and *SUI2*, encoding the a subunit of eIF-2, is lethal to the cell and thus suggests an interaction of the two gene products (Williams et al. 1989). This is also supported by biochemical analysis of GCD1-GCD2-GCN3 complexes associated with eIF-2 (Cigan et al. 1991).

The yeast system also benefits from the fact that genes can easily be introduced in the cell, either on a plasmid or by homologous recombination in the genome. Using this approach, predictions made on the function of a gene can be tested *in vivo*. A family of putative RNA helicases has recently been defined with the translation initiation factor eIF-4A as a member (Linder et al. 1989). Mutational analysis of some of the conserved residues of translation initiation factor eIF-4A, a member of that family, was carried out and the mutant genes tested for *in vivo* activity (Schmid & Linder 1991). This *in vivo* study is the first mutational analysis of a D-E-A-D box family member.

The availability of a cell-free translation system from yeast (Gasior et al. 1979) has permitted the use of this organism to study both translation initiation and elongation. The knowledge acquired from the *in vivo* analysis of mutants can be extended by studying the effects of these mutations in a yeast cell free translation system. In that fashion, mutants in translation initiation factor 4E have been tested in *in vitro* translation assays (Altmann et al. 1989c). In cases of newly isolated genes, potentially encoding translation initiation factors, the analysis of mutant cell extracts helps to determine whether this particular gene product is involved in translation (Blum et al. 1989).

Translation initiation factors from yeast

Different approaches have been used to isolate translation initiation factors from yeast (Table 1). As described above, mutational analysis led to the characterization of *prtl* (formerly called ts187). At present, it is still not known if *PRT1* encodes a known translation initiation factor. Cell extracts from mutant *prtl* strains are deficient in the association of the 5S ternary complex with the small ribosomal subunit to form the 43S preinitiation complex (Feinberg et al. 1982).

To find genes involved in translation initiation a much more specific approach is the use of antibodies directed against *in vitro* characterized initiation factors. Using such antibodies it is possible to

Table 1. Translation factors from yeast, their (putative) functions and methods of isolation of their genes.

Factor	Gene(s)	Yeast/mam.	Mass	Function	Isolation procedure
eIF-4E	*CDC33* (*TIF45*)	33%	24 kDa	Cap binding	λgt11 library
eIF-4A	*TIF1* *TIF2*	65%	46 kDa	RNA helicase together with 4B	Suppressor of mitochondrial mutation
eIF-2α	*sui2*	58%	34 kDa	Formation of ternary complex	Suppressor of AUG mutant
eIF-2β	*SUI3*	42%	31 kDa	Formation of ternary complex	Suppressor of AUG mutant
eIF-5A	*TIF51A* *TIF51B*	63%	17 kDa	First ribosome translocation	Screening with eIF-5A cDNA from human 5A
150 kDa	*TIF463A* *TIF463B*	–	107 kDa	Part of the cap-binding-protein complex	Cap binding λgt11 library
ts-187	*prtl*	–	88 kDa	Subunit of eIF-3(?)	ts-mutant from the Hartwell collection
EF-1α	*TEF1* *TEF2*	81%	5 kDa	Positioning of aminoacyl-tRNA in A-site	Screening for EF-1α antigen screening by hybridization
EF-2	*EFT1* *EFT2*	66%	93 kDa	Translocation of ribosome	Hybridization with oligonucleotide corresponding to the diphtamide domain
EF-3	*YEF3*	–	116 kDa	Stimulation of EF-1α activity	

Yeast translation factors, their functions and their genes. The yeast/mammalian homology represents the percentage of identical amino acids between the two proteins. In analogy with the genes *TIF51* the gene designation *TIF45* is proposed for the gene encoding eIF-4E (M. Altmann, pers. communication). The designations *sui2* and *SUI3* refer to the mutants isolated originally. The genes encoding the 150 kDa protein has tentatively been designated *TIF463A* and *TIF463B*. The indicated masses of the proteins are deduced from sequence data. The genes involved in the expression of *GCN4* are not listed, although some of them may be *bona fide* translation factors.

screen a λgt11 library. The translation initiation factor eIF-4E has been isolated by using a polyclonal antibody directed against proteins which bind to a cap-analogue (m^7-GDP)-agarose column (Altmann et al. 1985; Altmann et al. 1987). The initiation factor eIF-4E is encoded by a single copy gene, which has also been isolated as the cell cycle arrest mutant cdc33 (Brenner et al. 1988). Comparison of eIF-4E proteins from different species revealed a relatively high proportion of tryptophan residues which are all conserved. Mutational analysis of these tryptophan residues shows that at least two of them are required for cap binding activity *in vitro* (Altmann et al. 1988). Other mutations in the eIF-4E protein affect *in vitro* translation and cell growth (Altmann et al. 1989c; Altmann & Trachsel 1989). Conditional mutants in the gene encoding the cap-binding protein have been used to set up a 4E-dependent cell-free translation system (Altmann et al. 1989c). This system is particularly useful in the analysis of internal, 4E-independent translation initiation (Altmann et al. 1990). Finally, the cap-binding protein from mouse was the first mammalian translation initiation factor which has been shown to be able to functionally replace its yeast counterpart *in vivo* (Altmann et al. 1989b).

A gene encoding a 20kDa protein has also been isolated using the same polyclonal antibody alluded to previously (Altmann et al. 1989a). However, its definitive role in translation initiation has not been established. The characterization of a gene possibly encoding the large molecular weight subunit (220kDa in mammals, 150kDa in yeast, Goyer et al. 1989) of eIF-4F is also based on this method (C. Goyer pers. communication). Growth analysis of strains having this gene disrupted indicates that it is not essential, although it affects growth. The nonessentiality of this gene, tentatively termed *TIF463A* (eIF-4F, γ-subunit) is somewhat surprising in the current view of translation initiation. This could be explained by the presence of a second gene (*TIF463B*) which is 54% identical to the first one, the homology being even much higher in the C-terminal half of the gene (approx. 80%). This second gene has been isolated by low stringency hybridization using a region encoding a Cys-rich area of the *TIF463A*. It is interesting to note that in both cases the mRNAs are much longer than the length required for the protein. Preliminary experiments indicate the presence of six AUGs in the 5′UTR of *TIF463A* but none in that of *TIF463B*. Further experiments have to show whether these genes can substitute for each other *in vivo* and *in vitro* and whether they encode translation initiation factors (C. Goyer, pers. communication).

Initiation factor 4A (eIF-4A) has been isolated as a gene dosage dependent suppressor of a mitochondrial mutation (Linder & Slonimski 1989). The mechanism of suppression is not yet known. Since eIF-4A acts as an RNA helicase, one could argue that overexpression of the genes encoding eIF-4A alters the translation efficiency of some mRNAs, thereby permitting some mitochondrial proteins to be made in an excess. This could lead to an unbalance in the level of proteins involved in mitochondrial functions and thus cause the missense suppression. The translation initiation factor eIF-4A is encoded by two genes, *TIF1* and *TIF2*, which differ only by five silent substitutions in the open reading frames. Under normal laboratory conditions the two genes are redundant, although the level of expression of the *TIF2* gene is several fold higher than that of the *TIF1* gene. However, under extreme conditions, such as low temperature, the strains containing only *TIF1* grow drastically slower than the corresponding strain having only *TIF2* (A. Prat and P.L., unpublished).

Comparison of the eIF-4A sequence with other protein sequences allowed the definition of a new family of proteins (D-E-A-D-protein family, named after one of the conserved motifs (Linder et al. 1989). These proteins are involved in different processes such as splicing, ribosome assembly, translation, growth and development, and are found in organisms ranging from bacteria to human (for review see Schmid & Linder, in press).

A conditional expression system was developed to study the effect of low levels of eIF-4A on cell growth. In this system the *TIF1* gene is under the control of a galactose promoter and synthesis of the Tif protein can be switched off by transferring the cells to glucose containing medium. After depletion of eIF-4A, although the cell free translation extract is no longer active, it can be reactivated upon addition of purified protein (Blum et al. 1989). This clearly showed that the *TIF1* and *TIF2* genes code for a translation factor. The 4A-dependent translation system is currently used to analyze of mutations which have been made in the *TIF1/TIF2* genes. Surprisingly, despite its high homology to the mammalian eIF-4A protein, the yeast protein cannot be replaced by its mouse counterpart, neither *in vivo* nor *in vitro* (Prat et al. 1990). In an *in vitro* helicase assay, however, the mammalian 4A protein can be replaced by the yeast protein, indicating that the mammalian eIF-4B and the yeast eIF-4A proteins can function together (Jaramillo et al. 1990).

Interestingly the identification and characterization of yeast cap-binding proteins revealed that in this system the initiation factor 4A is absent from the cap-binding complex (Goyer et al. 1989). Thus it seems that the interaction between eIF-4A and the other subunits of the cap-binding complex is different between the yeast and the mammalian system This could explain why the mouse eIF-4A fails to functionally replace its yeast counterpart.

To better understand the role of the conserved domains in the D-E-A-D-protein family of putative RNA helicases, an extensive mutational analysis of the *TIF1/TIF2* genes has been carried out (Schmid & Linder 1991). It was shown that three highly conserved domains of the D-E-A-D protein family are essential for protein function. Only in rare cases could a conservative substitution be made without affecting cell growth. The isolated mutants show a clear correlation between their growth and *in vivo* translation rates. One mutation that results in a temperature-sensitive phenotype, reveals an immediate decrease in translation activity following a shift to the nonpermissive temperature. These *in vivo* results confirm the conclusion drawn from *in vitro* data showing that the *TIF1* and *TIF2* gene products are absolutely required for translation.

Another method to isolate translation initiation factors is based on the high conservation of the translation machinery between yeast and higher eukaryotes. By this approach, the mammalian gene is used as a probe to screen a yeast library. This method was successfully used to clone two yeast genes, *TIF51A* and *TIF51B*, encoding eIF-5A (Schnier et al. 1991). The eIF-5A protein purified from yeast can functionally replace its mammalian counterpart in an *in vitro* methionyl-puromycin synthesis assay. One of the lysine residues of eIF-5A is post-translationally modified by spermidine to form hypusine. This particular modification has been shown to be essential for the *in vitro* function of the mammalian enzyme (Hershey et al. 1990). Mutational analysis of this residue in the yeast gene shows that it is also essential *in vivo* (Schnier et al. 1991).

The *TIF51A* and *TIF51B* genes had been initially identified as two loci, trl and tr2 (*ANB1*), that respond differently to oxygen (Mehta et al. 1990). Gene disruption experiments show that (under normal laboratory conditions) one gene (*TIF51B*) is completely dispensable, whereas inactivation of the second gene (*TIF51A*) impairs growth considerably and disruption of both genes is lethal. In contrast to the initiation factor 4A, encoded by *TIF1* and *TIF2*, the two eIF-5A proteins share only 90% identity.

Based on the finding that translation of most mRNAs is initiated at the first AUG, Donahue and Cigan set up a screen to identify factors involved in the initiator codon selection. They searched for suppressor mutations that would allow yeast ribosomes to initiate translation in the *his4* gene at other codons than the AUG (Cigan et al. 1989; Donahue et al. 1988). Three suppressor mutations, *sui1*, *sui2* and *SUI3* (sui for *su*ppressor of *i*nitiator codon), were isolated by a double screen for a histidine prototrophy and blue color phenotype in a strain in which the AUG codons of both the *HIS4* gene and a *HIS4-lacZ* fusion gene were altered. Molecular analysis of the dominant *SUI3* suppressor and comparison of the mutant Sui3 protein with its wild type counterpart revealed that this gene encodes one of the three subunits (the β-subunit) of initiation factor 2 (Donahue et al. 1988). Wild type eIF-2 binds initiator tRNA and GTP to form the 5S ternary complex (Fig. 1). Interestingly, the mutations in three *SUI3* alleles lie close to a putative Zn-finger motif that may be involved in RNA binding. A similar analysis of the recessive *sui2* suppressor mutation (Cigan et al. 1989) showed that this gene encodes the α-subunit of initiation factor 2

(Table 1). It has been demonstrated that phosphorylation of eIF-2α in mammals correlates with translation initiation inhibition (for a review see Hershey 1989). In rabbit reticulocytes, two protein kinases, HRI (heme-regulated inhibitor) and DAI (dsRNA-activated inhibitor), phosphorylate Ser-51 on eIF-2α to inhibit the GDP to GTP exchange by eIF-2B (also called GEF for guanine nucleotide exchange factor) thereby compromising the next round of initiation (Fig 1). In the mammalian protein, replacing Ser-51 with Ala partially relieves inhibition by the activated DAI kinase. In yeast, the α-subunit of initiation factor 2 is also phosphorylated, but the responsible protein kinase(s) has not been experimentally defined. A putative candidate for the kinase is the protein encoded by the *GCN2* gene. In fact, a recent report revealed a high similarity of the Gcn2 protein with the mammalian heme-regulated eIF-2α kinase (Chen et al. 1991). The exact mechanism of suppression by *SUI3* and *sui2* is not known but the analysis of proteins expressed from the *his4-lacZ* fusion revealed that in these mutant strains translation initiated at an UUG codon.

Translation elongation factors from yeast

Elongation factor EF-1 is one of the most abundant proteins in eukaryotic cells (Slobin 1980). This fact was used to clone the two copies of *Saccharomyces cerevisiae* EF-1α genes, *TEF1* and *TEF2* (Cottrelle et al. 1985a; Cottrelle et al. 1985b; Nagata et al. 1984) by using cDNA clones made from a mRNA fraction containing EF-1α transcripts. One group (Nagata et al. 1984) used in a second step a mixture of the *E. coli tufB* and the yeast *tufM* genes (coding for bacterial and mitochondrial elongation factors, respectively) as a probe for further screening, whereas Cottrelle and co-workers searched for clones able to produce the EF-1α antigen. Schirmaier and Philippsen have also cloned the *TEF1* gene by virtue of its high expression level (Schirmaier & Philippsen 1984). The proteins encoded by the two genes *TEF1* and *TEF2* are identical. The genes only differ by two nucleotides in the coding region, whereas the flanking sequences are divergent. Inactivation of either gene has no effect on cell growth under normal conditions, but a double disruption is lethal. As for other proteins involved in translation, the EF-1α protein from yeast is highly similar to the one of other eukaryotes (81% to human). The mammalian EF-1α factor is post-translationally modified to carry dimethyl- and trimethyl-lysine, as well as glycerylphosphoryl-ethanolamine (Dever et al. 1989). In yeast it seems that four modified lysines are present but that only one of them matches the modifications of the other EF-1α proteins (Riis et al. 1990).

It has been found that EF-1α plays a role in translational fidelity. Mutant *TEF2* genes were able to suppress frameshift and nonsense mutations in amino acid biosynthesis genes (Sandbaken & Culbertson 1988). During the analysis of an anti-suppressor, Song and co-workers found that an additional copy of the *TEF1* gene allows suppression of nonsense mutations in the *LEU2* and *MET8* genes (Song et al. 1989). Further genetic analysis has shown that deletion of the *TEF2* gene cosegregates with a decreased level of misreading induced by paromomycin. These results suggest that EF-1α plays a role in translation fidelity.

Two genes, *EFT1* and *EFT2*, coding for translation elongation factor EF-2 are present in yeast (Perentesis et al. 1991). As in the case of *TEF1*/*TEF2* and *TIF1*/*TIF2* the two genes encode exactly the same protein and differ by only 4 substitutions at the nucleotide level. The *EFT1* and *EFT2* genes have been isolated by screening a *S. cerevisiae* genomic library with an oligonucleotide corresponding to the diphtamide domain of EF-2 (see below). The EF-2 protein is post-translationally modified by changing a histidine to a diphtamide. In yeast, five complementation groups are involved in the formation of the diphtamide (Chen et al. 1985). The diphtamide residue in EF-2 is the target site of the diphteria toxin and mutants lacking the diphtamide modification are viable and resistant to the toxin (Chen & Bodley 1988). A mutant EF-2 protein, however, which has one of its histidine residues mutated is inactive (Omura et al. 1989). EF-2 is also phosphorylated at a threonine in the N-terminal region of the protein by the EF-2 kinase (or calmodulin-dependent protein kinase

III). Since the phosphorylated EF-2 protein is inactive in translation it might be a target for translational control (Ryazanov et al. 1988).

In yeast, a third elongation factor is present. The EF-3 protein is encoded by a single copy gene (*YEF3*) and is essential for cell viability (Qin et al. 1987; Qin et al. 1990; Sandbaken et al. 1990). The EF-3 factor stimulates EF-1α mediated binding of the aminoacyl-tRNA by the ribosome. *In vitro* studies using ribosomal subunits from heterologous sources have shown an effect of EF-3 only in conjunction with the yeast 40S subunit, but not with the 60S subunit. Overexpression of the EF-3 encoding gene leads to an increase in sensitivity to hygromycin and paromomycin, antibiotics which increase the level of translational errors.

An intriguing finding is that many translation factor genes are duplicated (*TIF1/2*, *TIF463A/B*, *TIF51A/B*, *TEF1/2* and *EFT1/2*), a situation also typical for ribosomal protein genes. The reason for these gene duplications is unknown but it will be interesting to see whether duplicated translation factor genes respond differently to changes in environmental conditions.

Translation termination

In *E. coli* the two release factors RF1 and RF2 recognize the termination codons UAA/UAG and UAA/UGA, respectively (Sharp & Bulmer 1988). Analysis of the termination codons in *S. cerevisiae* shows a preference for UAA, followed by UGA and UAG. The preference is more pronounced in genes with a high codon bias. As not all stop codons respond in the same way to suppressors the nucleotide sequences around the stop codon were analyzed to explain such differences (Brown et al. 1990). Brown and co-workers found that in *S. cerevisiae* 50% of highly expressed genes terminate with *UAAG* and 31% with *UAAA*, suggesting that a tetranucleotide sequence is recognized by the postulated release factor.

Translational fidelity

The low frequency of mistakes in protein synthesis requires an accurate positioning of the aminoacyl-tRNA as well as a proofreading of the codon:anticodon interaction. The accuracy of this process can be disturbed by antibiotics like hygromycin or paromomycin or by the presence of mutations in genes involved in translation. These errors can lead to the suppression of nonsense mutations. It is obvious that an elevated error rate caused by the combination of antibiotics and suppressor mutations is lethal to the cell.

Different types of mutations affecting the translational fidelity are known (for reviews see Sherman 1982; Surguchov 1988). Omnipotent suppressors can suppress frameshift mutations and all three types of nonsense codons. Such suppressors have been isolated by different groups and have received different names (*sup1* = *sup45*; *sup2* = *sup35* = *suf12*). It is interesting that the *SUP2* gene has also been isolated in an analysis of G1->S transition mutants (*GST1*, (Kikuchi et al. 1988)). This is in agreement with other protein synthesis mutants which show a cell cycle deficient phenotype (*cdc33*, *prt1*) and with the finding that protein synthesis is required for completion of the cell cycle (Burke & Church 1991).

Molecular analysis of the SUP2 gene revealed that it encodes a protein with 62% identity of the encoded protein to EF-1α (Kikuchi et al. 1988; Kushnirov et al. 1988; Wilson & Culbertson 1988). Since null mutants of either gene are lethal, it is clear that the two proteins can not replace each other. From the characteristics of the *sup* mutants, one could imagine that these genes encode ribosome-associated proteins involved in positioning of the aminoacyl-tRNA and thus participate in proofreading (for review see Surguchov 1988).

Other mutants enhance (allo-suppressors, *SAL*) or decrease (anti-suppressor, *ASU*) the efficiency of an already existing suppressor. The *sal* mutants restore suppression of the tRNA *SUQ5* suppressor in a [ψ^-] background. The [ψ^+]-factor is probably an extrachromosomal element which enhances the effect of altered tRNA suppressors. A combination of the [ψ^+] factor and the *sal* mutants as well as the *sal3*/*sal4* mutants are lethal. The *sal3* mutants confer hypersensitivity to paromomycin, which is consistent with an increase in misreading. The *SAL3* gene encodes a 80kDa protein, suggesting that it is not part of the ribosome itself (Crouzet & Tuite 1987). The *SAL4* gene is identical to the *SUP1* gene (Crouzet et al. 1988). Further genetic and biochemical analysis of such suppressor genes will help to elucidate the mechanisms involved in maintenance of translation accuracy and termination.

Involvement of the 5′ end in translation initiation

The analysis of many yeast genes has shown that in accordance with the scanning hypothesis in over 95% of mRNAs, the 5′ proximal AUG is used as the initiator codon (Cigan & Donahue 1987; Kozak 1978; Kozak 1989). The average length of 5′ untranslated regions in yeast mRNAs is 52 nucleotides. Furthermore, it is devoid of strong secondary structures (Cigan & Donahue 1987). Deletion analysis of the 5′-untranslated leader of the *S. cerevisiae tcm1* gene (encoding a mutant form of the L3 ribosomal protein conferring trichodermin resistance) indicates that no 5′-leader is required for translation initiation (Maicas et al. 1990). The efficiency of translation initiation in these mutant constructs has been verified *in vivo* by resistance to trichodermin. When placed on a high copy number vector, constructs in which the first nucleotide of the mRNA is the A of the initiator codon still confer resistance, albeit to a reduced level. This is in accordance with results obtained from a deletion analysis of the 5′-noncoding region of the *PGK* gene (van den Heuvel et al. 1989). Reduction of the leader length from 45 to 27 nucleotides does not affect translation efficiency while shortening to 7 nucleotides decreases the rate of *PGK* synthesis by a factor of two.

Although the length of the 5′ noncoding sequence seems not to have a major influence on the translation efficiency, the composition of this leader is important. The natural sequences are generally rich in A's and T's and insertion of sequences rich in G's (>40%) lowers the translation efficiency considerably (van den Heuvel et al. 1990).

The comparison of the sequences surrounding the AUG codon indicates the existence of a preferred context specific for yeast (5′-A [A/Y] A [A/U] A A U G U C U-3′). However, experiments to establish the requirement for such a preference showed that deviations from this sequence context are not detrimental for efficient translation (Cigan et al. 1988).

According to the current model of translation initiation (Fig. 1), the 40S ribosomal subunit scans along the mRNA until it reaches the initiator AUG. Upstream AUGs and secondary structures are thus believed to hinder this process. Indeed, translation initiation of a *cyc1* mutant gene, bearing a secondary structure element 5′ to the AUG, is severely reduced and can only be restored through i) extensive sequence changes that prevent formation of this element; ii) substitutions in the three bases 5′ to the AUG codon that alter the sequence context of the initiator codon and the stability of the stem-loop structure; iii) acquisition of a new AUG codon 5′ to the secondary structure that allows initiation from a region devoid of inhibitory secondary structures. In this last class, the secondary structure is now 3′ of the new AUG and appears not to be an obstacle for the speeding ribosome (Baim & Sherman 1988).

Nevertheless, a highly expressed gene bearing a strong potential secondary structure in its 233-nucleotide-long 5′ UTR (untranslated region) has been described (Capieaux et al. 1989). The secondary structure in the 5′-UTR of the PMA1 gene (encoding a H^+-ATPase), comprises almost the entire non-translated region and makes it difficult to envisage how ribosomes could bind to the mRNA. Despite this unorthodox initiation region, the *PMA1* gene is expressed at a high level. Interestingly, a sequence complementary to the 16S rRNA lies near the AUG codon of the *PMA1* mRNA suggesting that, similar to translation initiation in prokaryotes, the ribosome may bind directly to the mRNA near the AUG codon rather than scanning from the 5′-cap. It will be interesting to see if the translation initiation of this mRNA is eIF-4E independent.

Translational control by open reading frames in the 5′ untranslated region (UTR)

A few genes have small upstream open reading frames (uORF) in the 5′ untranslated region (UTR) of their mRNA (CBP1 (Dieckmann et al. 1984), *CPA1* (Werner et al. 1987), *DCD1* (McIntosh & Haynes 1986), *GCN4* (Hinnebusch 1988), *PPR1* (Losson et al. 1983), *PET111* (Strick & Fox 1987) and *MOD5* (Najarian et al. 1987)). Two genes, *CPA1* and *GCN4*, have been analyzed in detail in this respect. The mRNA of *CPA1*, a gene involved in arginine biosynthesis, carries a 25 codon long uORF which was shown by mutational analysis to be required for translational repression of the *CPA1* open reading frame when arginine is present in the medium (Werner et al. 1987). At present, it is not clear how this small peptide is involved in regulation of *CPA1* expression.

In the case of *GCN4*, which encodes a protein involved in general control of amino acid biosynthesis, four uORFs present in a long 5′ region are required for translational regulation of the *GCN4* open reading frame (Hinnebusch 1988). A large number of independent genes, found by genetic methods, are involved in this regulation. The products of the *GCD* genes act negatively on the expression of *GCN4* and are thought to be factors involved in general protein synthesis. In accordance with this, it has been found that mutations in the genes encoding the α and β subunits of eIF-2 have a gcd phenotype (Williams et al. 1989). The negative regulatory effect of the *GCD* genes is antagonized by *GCN* genes, *GCN2* and *GCN3*. The uORFs are essential for the regulation of *GCN4* expression, i.e. repression under nonstarvation conditions and derepression under starvation conditions (Mueller & Hinnebusch 1986). The fourth uORF4 (most closely located to the AUG of the main open reading frame) exerts a negative effect on translation of the *GCN4* mRNA whereas uORF1 is required to overcome the inhibitory effect of uORF4 under starvation conditions. From the molecular analysis it became clear that both the distance between the uORF4 and the GCN4 reading frame and the uORF4 stop codon with the following 10 nucleotides immediately downstream play an important role in this negative regulation (Miller & Hinnebusch 1989). The presence of these uORFs raised the possibility that during starvation conditions, the scanning might be affected (leading to initiation at a downstream initiator codon) or alternatively that internal initiation could take place as it does in certain viral systems of higher eukaryotes (for review see Sonenberg & Pelletier 1989). A recent report by Abastado et al. presents considerable evidence for a scanning mechanism in which the ribosomes recover more or less rapidly from initiation at uORF1 depending on the growth conditions (Abastado et al. 1991). In non-starvation conditions the ribosomes recover rapidly and can reinitiate at uORF4, while they fail to do so at the following AUG, the *GCN4* initiator AUG. Under starvation conditions however, the ribosomes and associated initiation factors need more time to recover from the initiation event at uORF1 and are therefore unable to reinitiate at uORF4, but can do so at the AUG of the *GCN4* reading frame. In accordance with this model is the observation that depletion of the ribosomal protein L16 (which leads to the formation of half-mers polyribosomes) induces the expression of *GCN4* under nonstarvation conditions (similar to a defect in *GCD2*) and can no longer be induced to elevated levels by starvation (Foiani et al. 1991). This suggests that a higher proportion of small ribosomal subunits can continue scanning and eventually pick-up a 60S subunit at the *GCN4* initiator codon.

Analysis of general protein synthesis by measuring incorporation of radiolabeled amino acids and by comparing polysome gradients showed that the *GCD2* gene most likely encodes a general translation factor (Foiani et al. 1991). In *gcd2* mutants, the 43S or 48S complexes are enriched for the eIF-2α subunit indicating a block after the assembly of the small ribosomal subunit with the 5S ternary complex (initiator-tRNA^{Met}-eIF-2-GTP; Foiani et al. 1991). Moreover, part of the eIF-2 pool, present in the cell, has been found to be associated with Gcd1-Gcd2-Gcn3 which sediments as a 600kDa complex (Cigan et al. 1991). An interesting speculation is that this complex has an eIF-2B like activity (Fig. 1). Furthermore, the Gcn2 protein could function as a kinase involved in phosphorylation of eIF-2α and it shares similarities with the catalytic domains of other protein kinases and histidinyl tRNA synthetase (Wek et al. 1989) and more specifically with the heme-regulated eIF-2α kinase (Chen et al. 1991). Thus, the presence of uncharged tRNA molecules, due to amino acid starvation, could provoke phosphorylation of eIF-2α and modify the scanning process on the *GCN4* mRNA.

The codon usage is biased

In 1982, Bennetzen and Hall reported that the use of codons is different for highly versus poorly expressed genes (Bennetzen & Hall 1982). This was confirmed on the basis of codon usage data from 100 yeast genes (Sharp et al. 1986). It is assumed that the codons used in highly expressed genes represent triplets with multiple corresponding tRNAs whereas poorly expressed genes do not need such abundant tRNAs. Indeed, the substitution of frequently used codons by rare codons in the highly expressed gene *PGK1* leads to a reduction in expression (Hoekema et al. 1987). In a very elegant approach it has been demonstrated that this can also be used for regulation of gene expression (Belcourt & Farabaugh 1990). The two proteins TyA and TyB are made by processing of a larger protein. The TyB protein, however, is far less abundant than TyA and the two reading frames are not in frame. In fact it has been proposed that the TyB protein is made by translation of the TyA reading frame and a frameshift to yield a TyAB protein with proteolytic cleavage of the fusion protein. It has been suggested that the ribosome stalls at two consecutive leucine codons (decoded by a leucine isoacceptor which recognizes all six leucine codons) caused by the following triplet which has to be decoded by a rare $tRNA^{Arg}$. Thus the stalling of the ribosome would permit the ribosome to change the reading frame and allow translation of TyB. Belcourt and Farabaugh have found support for this hypothesis by overproduction of the rare $tRNA^{Arg}$ gene thereby abolishing the frameshift.

Is the 3′-end of the mRNA involved in translational control?

Prior to transport into the cytoplasm, the mRNA is processed by the addition of a 5′-cap-structure and a 3′-poly(A)-tail and, if necessary, or the removal of introns. A temperature-sensitive mutation in the poly(A)-binding protein results in shortening of poly(A)-tails as well as an inhibition of translation initiation as evidenced by a reduction of polysomes and accumulation of salt-sensitive monosomes (Sachs & Davis 1989). It has therefore been proposed that the poly(A)-binding protein somehow interacts with the translational machinery at the 5′ end of the mRNA. For the cell this would be an efficient means to sense the state of a mRNA and avoid translation of partially degraded messages.

To get more evidence for this, several suppressors of the temperature-sensitive poly(A)-binding protein mutant (*pab1-F364*) have been isolated. One of the suppressors, *spb2* (spb = suppressors of *pab1*), has been found to be allelic to the *RPL46* gene, encoding the L46 protein of the large ribosomal subunit. Another suppressor of this temperature-sensitive mutant encodes a putative RNA helicase, *spb4* (Sachs & Davis 1990) (for a review see Schmid & Linder, in press). As in *spb2*, the amount of 60S ribosomal subunit in a *spb4* mutant strain is reduced. Analysis of rRNA has shown that the amount of 25S rRNA is reduced relative to the 18S rRNA. This indicates that the *spb4* mutation alters the production of 25S rRNA. Interestingly, more than 75 independent complementation groups were found in which lethal mutations can be suppressed by the *spb4–1* allele (Sachs 1990). It is intriguing that 75% of these mutations can also be suppressed by deletion of the *spb2* gene encoding the large ribosomal L46 protein. These alterations do not lead to a general suppressor activity of amber- or ochre-codons, since such mutations in synthetic b-galactosidase genes or highly suppressible auxotrophic markers are not suppressed. Protein analysis from wild-type *SPB4* and mutant *spb4–1* strains revealed considerable differences in the amount of individual proteins, which could indicate different efficiencies of translation initiation or RNA stability. It will be interesting to identify these proteins and their corresponding genes in order to learn more about this type of regulation.

Conclusions and perspectives

In conclusion, it has become apparent that the process of translation in yeast is highly similar to that of higher eukaryotes. Although studies of translation in *S. cerevisiae* have permitted the accumulation of a great deal of information during the past few years, considerable work remains in order to understand the role of all the factors involved. The genetic and biochemical analysis will help to improve existing models. It is envisageable that new, unknown initiation factors will be discovered. At present, no natural system of internal initiation as described for virus infected mammalian cells has been described for yeast. It will be interesting to see whether such sequences are also present in yeast genes.

Acknowledgements

I would like to thank C. Goyer for communication of the unpublished results on the 150kDa protein and J. W. Bodley for communication of results on EF-2 prior to publication. I am grateful to R. Coppolecchia, R. Gosh, C. Goyer, P. Müller and S. Schmid for critical reading of the manuscript. Work in the author's lab is supported by grants from Swiss National Science Foundation.

References

Abastado JP, Miller PF, Jackson BM & Hinnebusch AG (1991) Suppression of ribosomal reinitiation at upstream open reading frames in amino acid-starved cells forms the basis for *GCN4* translational control. Mol. Cell. Biol. 11: 486–496

Altmann M & Trachsel H (1989) Altered mRNA cap recognition activity of initiation factor 4E in the yeast cell cycle division mutant cdc33. Nucleic Acids Res. 17: 5923–5931

Altmann M, Edery I, Sonenberg N & Trachsel H (1985) Purification and characterization of protein synthesis initiation factor eIF-4E from the yeast *Saccharomyces cerevisae*. Biochem. 24: 6085–6089

Altmann M, Handschin C & Trachsel H (1987) mRNA cap-binding protein: cloning of the gene encoding protein synthesis initiation factor eIF-4E from *Saccharomyces cerevisae*. Mol. Cell. Biol. 7: 998–1003

Altmann M, Edery I, Trachsel H & Sonenberg N (1988) Site-directed mutagenesis of the tryptophan residues in yeast eukaryotic initiation factor 4E. Effects on cap binding activity. J. Biol. Chem. 263: 17229–17232

Altmann M, Krieger M & Trachsel H (1989a) Nucleotide sequence of the gene encoding a 20 kDa protein associated with the cap binding protein eIF-4E from *Saccharomyces cerevisae*. Nucleic Acids Res. 17: 7520

Altmann M, Müller PP, Pelletier J, Sonenberg N & Trachsel H (1989b) A mammalian translation initiation factor can substitute for its yeast homologue *in vivo*. J. Biol. Chem. 264: 12145–12147

Altmann M, Sonenberg N & Trachsel H (1989c) Translation in *Saccharomyces cerevisae*: initiation factor 4E-dependent cell-free system. Mol. Cell. Biol. 9: 4467–4472

Altmann M, Blum S, Wilson TMA & Trachsel H (1990) The 5′-leader sequence of tobacco mosaic virus RNA mediates initiation-factor-4E-independent, but still initiation-factor-4A-dependent translation in yeast extracts. Gene 91: 127–129

Baim SB & Sherman F (1988) mRNA structures influencing translation in the yeast *Saccharomyces cerevisae*. Mol. Cell. Biol. 8: 1591–1601

Belcourt MF & Farabaugh PJ (1990) Ribosomal frameshifting in the yeast retrotransposon Ty: tRNAs induce slippage on a 7 nucleotide minimal site. Cell 62: 339–352

Bennetzen JL & Hall BD (1982) Codon selection in yeast. J. Biol. Chem. 257: 3026–3031

Blum S, Mueller M, Schmid SR, Linder P & Trachsel H (1989) Translation in *Saccharomyces cerevisae*: initiation factor 4A-dependent cell-free system. Proc. Natl. Acad. Sci. USA 86: 6043–6046

Brenner C, Nakayama N, Goebl M, Tanaka K, Toh-e A & Matsumoto K (1988) CDC33 encodes mRNA cap-binding protein eIF-4E of *Saccharomyces cerevisae*. Mol. Cell. Biol. 8: 3556–3559

Brown CM, Stockwell PA, Trotman CN & Tate WP (1990) Sequence analysis suggests that tetra-nucleotides signal the termination of protein synthesis in eukaryotes. Nucleic Acids Res. 18: 6339-6345

Burke DJ & Church D (1991) Protein synthesis requirements for nuclear division, cytokinesis, and cell separation in *Saccharomyces cerevisae*. Mol. Cell. Biol. 11: 3691–3698

Capieaux E, Vignais ML, Sentenac A & Goffeau A (1989) The yeast H^+-ATPase gene is controlled by the promoter binding factor TUF. J. Biol. Chem. 264: 7437–7446

Chen JJ, Throop MS, Gehrke L, Kuo I, Pal JK, Brodsky M & London IM (1991) Cloning of the cDNA of the heme-regulated eukaryotic initiation factor 2a (eIF-2a) kinase of rabbit reticulocytes: homology to yeast GCN2 protein kinase and human double-stranded-RNA-dependent eIF-2 alpha kinase. Proc. Natl. Acad. Sci. USA 88: 7729–7733

Chen JY & Bodley JW (1988) Biosynthesis of diphthamide in

Saccharomyces cerevisiae. Partial purification and characterization of a specific S-adenosylmethionine: elongation factor 2 methyltransferase. J. Biol. Chem. 263: 11692–11696

Chen JY, Bodley JW & Livingston DM (1985) Diphtheria toxin-resistant mutants of *Saccharomyces cerevisae*. Mol. Cell. Biol. 5: 3357–3360

Cigan AM & Donahue TF (1987) Sequence and structural features associated with translational initiator regions in yeast – a review. Gene 59: 1–18

Cigan AM, Feng L & Donahue TF (1988) tRNA$_i$(met) functions in directing the scanning ribosome to the start site of translation. Science 242: 93–97

Cigan AM, Pabich EK & Donahue TF (1988) Mutational analysis of the *HIS4* translational initiator region in *Saccharomyces cerevisae*. Mol. Cell. Biol. 8: 2964–2975

Cigan AM, Pabich EK, Feng L & Donahue TF (1989) Yeast translation initiation suppressor *sui2* encodes the a subunit of eukaryotic initiation factor 2 and shares sequence identity with the human a subunit. Proc. Natl. Acad. Sci. USA 86: 2784–2788

Cigan AM, Foiani M, Hannig EM & Hinnebusch AG (1991) Complex formation by positive and negative translational regulators of *GCN4*. Mol. Cell. Biol. 11: 3217–3228

Cottrelle P, Cool M, Thuriaux P, Price VL, Thiele D, Buhler JM & Fromageot P (1985a) Either one of the two yeast EF-1 alpha genes is required for cell viability. Curr. Genet. 9: 693–697

Cottrelle P, Thiele D, Price VL, Memet S, Micouin JY, Marck C, Buhler JM, Sentenac A & Fromageot P (1985b) Cloning, nucleotide sequence, and expression of one of two genes coding for yeast elongation factor 1 alpha. J. Biol. Chem. 260: 3090–3096

Crouzet M & Tuite MF (1987) Genetic control of translational fidelity in yeast: molecular cloning and analysis of the allosuppressor gene *SAL3*. Mol. Gen. Genet. 210: 581–583

Crouzet M, Izgu F, Grant CM & Tuite MF (1988) The allosuppressor gene *SAL4* encodes a protein important for maintaining translational fidelity in *Saccharomyces cerevisae*. Curr. Genet. 14: 537–43

Dever TE, Costello CE, Owens CL, Rosenberry TL & Merrick WC (1989) Location of seven post-translational modifications in rabbit elongation factor 1 alpha including dimethyllysine, trimethyllysine, and glycerylphosphorylethanolamine. J. Biol. Chem. 264: 20518–20525

Dieckmann CL, Homison G & Tzagoloff A (1984) Assembly of the mitochondrial membrane system. Nucleotide sequence a yeast nuclear gene (*CBP1*) involved in 5′ end processing cytochrome b pre-mRNA. J. Biol. Chem. 259: 4732–4738

Donahue TF, Cigan AM, Pabich EK & Castilho Valavicius B (1988) Mutations at a Zn(II) finger motif in the yeast eIF-2β gene alter ribosomal start-site selection during the scanning process. Cell 54: 621–632

Edery I, Humbelin M, Darveau A, Lee KAW, Milburn S, Hershey JWB, Trachsel H & Sonenberg N (1983) Involvement of eukaryotic initiation factor 4A in the cap recognition process. J. Biol. Chem. 258: 11398–11403

Feinberg B, McLaughlin CS & Moldave K (1982) Analysis of temperature-sensitive mutant ts 187 of *Saccharomyces cerevisae* altered in a component required for the initiation of protein synthesis. J. Biol. Chem. 257: 10846–10851

Foiani M, Cigan AM, Paddon CJ, Harashima S & Hinnebusch AG (1991) *GCD2*, a translational repressor of the GCN4 gene, has a general function in the initiation of protein synthesis in *Saccharomyces cerevisae*. Mol. Cell. Biol. 11: 3203–3216

Gasior E, Herrera F, Sadnik I, McLaughlin CS & Moldave K (1979) The preparation and characterization of a cell-free system from *Saccharomyces cerevisae* that translates natural messenger ribonucleic acid. J. Biol. Chem. 254: 3965–3969

Goyer C, Altmann M, Trachsel H & Sonenberg N (1989) Identification and characterization of cap-binding proteins from yeast. J. Biol. Chem. 264: 7603–7610

Hartwell BLH & McLaughlin CS (1968a) A mutant of yeast apparently defective in the initiation of protein synthesis. Biochem. 62: 468–474

Hartwell LH & McLaughlin CS (1968b) Temperature-sensitive mutants of yeast exhibiting a rapid inhibition of protein synthesis. J. Bacteriol. 96: 1664–1671

Hershey JWB (1989) Protein phosphorylation controls translation rates. J. Biol. Chem. 264: 20823–20826

Hershey JWB, Smit-McBride Z & Schnier J (1990) The role of mammalian initiation factor eIF-4D and its hypusine modification in translation. Biochim. Biophys. Acta 1050: 160–162

Hinnebusch AG (1988) Mechanisms of gene regulation in the general control of amino acid biosynthesis in *Saccharomyces cerevisae*. Microbiol. Rev. 52: 248–273

Hoekema A, Kastelein RA, Vasser M & de Boer HA (1987) Codon replacement in the PGK1 gene of *Saccharomyces cerevisae*: experimental approach to study the role of biased codon usage in gene expression. Mol. Cell. Biol. 7: 2914–2924

Jackson R (1991) The ATP requirement for initiation of eukaryotic translation varies according to the mRNA species. Eur. J. Biochem. 200: 285–294

Jaramillo M, Browning K, Dever TE, Blum S, Trachsel H, Merrick WC, Ravel JM & Sonenberg N (1990) Translation initiation factors that function as RNA helicases from mammals, plants and yeast. Biochim. Biophys. Acta 1050: 134–139

Kikuchi Y, Shimatake H & Kikuchi A (1988) A yeast gene required for the G1-to-S transition encodes a protein containing an A-kinase target site and GTPase domain. EMBO J. 7: 1175–1182

Kozak M (1978) How do eucaryotic ribosomes select initiation regions in messenger RNA? Cell 15: 1109–1123

Kozak M (1989) The scanning model for translation: an update. J. Cell Biol. 108: 229–241

Kushnirov VV, Ter AMD, Telckov MV, Surguchov AP, Smirnov VN & Inge VSG (1988) Nucleotide sequence of the *SUP2* (*SUP35*) gene of *Saccharomyces cerevisae*. Gene 66: 45–54

Linder P & Prat A (1990) Baker's yeast, the new work horse in protein synthesis studies: analyzing eukaryotic translation initiation. BioEssays 12: 519–526

Linder P & Slonimski PP (1989) An essential yeast protein, encoded by duplicated genes *TIF1* and *TIF2* and homologous to the mammalian translation initiation factor eIF-4A, can suppress a mitochondrial missense mutation. Proc. Natl. Acad. Sci. USA 86: 2286–2290

Linder P, Lasko PF, Ashburner M, Leroy P, Nielsen PJ, Nishi K, Schnier J & Slonimski PP (1989) Birth of the D-E-A-D box. Nature 337: 121–122

Losson R, Fuchs RPP & Lacroute F (1983) *In vivo* transcription of eukaryotic regulatory gene. EMBO J. 2: 2179–2184

Maicas E, Shago M & Friesen JD (1990) Translation of the *Saccharomyces cerevisae tcml* gene in the absence of a 5′-untranslated leader. Nucleic Acids Res. 18: 5823–5828

McIntosh EM & Haynes RH (1986) Sequence and expression of the dCMP deaminase gene (DCD1) of *Saccharomyces cerevisae*. Mol. Cell. Biol. 6: 1711–1721

Mehta KD, Leung D, Lefebvre L & Smith M (1990) The ANB1 locus of *Saccharomyces cerevisae* encodes the protein synthesis initiation factor eIF-4D. J. Biol. Chem. 265: 8802–8807

Miller PF & Hinnebusch AG (1989) Sequences that surround the stop codons of upstream open reading frames in *GCN4* mRNA determine their distinct functions in translational control. Genes Dev. 3: 1217–1225

Moldave K (1985) Eukaryotic protein synthesis. Ann. Rev. Biochem. 54: 1109–1149

Mueller PP & Hinnebusch AG (1986) Multiple upstream AUG codons mediate translational control of *GCN4*. Cell 45: 201–207

Nagata S, Nagashiman K, Tsunetsugu-Yokota Y, Fujimura K, Miyazaki M & Kaziro Y (1984) Polypeptide chain elongation factor 1α (EF-1α from yeast: nucleotide sequence of one of the two genes for EF-1α from *Saccharomyces cerevisiae*. EMBO J. 3: 1825–1830

Najarian D, Dihanich ME, Martin NC & Hopper AK (1987) DNA sequence and transcript mapping of MOD5: features of the 5′ region which suggest two translational starts. Mol. Cell. Biol. 7: 185–191

Omura F, Kohno K & Uchida T (1989) The histidine residue of codon 715 is essential for function of elongation factor 2. Eur. J. Biochem. 180: 1–8

Perentesis JP, Phan LD, Gleason WB, LaPorte DC, Livingston DM & Bodley JW (1991) *Saccharomyces cerevisae* elongation factor 2: genetic cloning, characterization of expression, and G-domain modeling. J. Biol. Chem. 267: 1190–1197

Prat A, Schmid SR, Buser P, Blum S, Trachsel H, Nielsen PJ & Linder P (1990) Expression of translation initiation factor 4A from yeast and mouse in *Saccharomyces cerevisae*. Biochim. Biophys. Acta 1050: 140–145

Qin SL, Moldave K & McLaughlin CS (1987) Isolation of the yeast gene encoding elongation factor 3 for protein synthesis. J. Biol. Chem. 262: 7802–7807

Qin SL, Xie AG, Bonato MC & McLaughlin CS (1990) Sequence analysis of the translational elongation factor 3 from *Saccharomyces cerevisae*. J. Biol. Chem. 265: 1903–1912

Rhoads RE (1988) Cap recognition and the entry of mRNA into the protein synthesis initiation cycle. Trends Biochem. Sci. 13: 52–56

Riis B, Rattan SIS, Clark BFC & Merrick WC (1990) Eukaryotic protein elongation factors. Trends Biochem. Sci. 15: 420–424

Rozen F, Edery I, Meerovitch K, Dever TE, Merrick WC & Sonenberg N (1990) Bidirectional RNA helicase activity of eucaryotic translation initiation factors 4A and 4F. Mol. Cell. Biol. 10: 1134–1144

Ryazanov AG, Shestakova EA & Natapov PG (1988) Phosphorylation of elongation factor 2 by EF-2 kinase affects rate of translation. Nature 334: 170–173

Sachs AB (1990) Isolation of the *lsd* mutations in *Saccharomyces cerevisae*. In: McCarthy JEG & Tuite MF (Eds) Post-transcriptional Control of Gene Expression (pp 549–555). NATO ASI Series, Springer-Verlag, Berlin

Sachs AB & Davis RW (1989) The poly(A) binding protein is required for poly(A) shortening and 60S ribosomal subunit-dependent translation initiation. Cell 58: 857–867

Sachs AB & Davis RW (1990) Translation initiation and ribosomal biogenesis: involvement of a putative rRNA helicase and *RPL46*. Science 247: 1077–1079

Sandbaken MG & Culbertson MR (1988) Mutations in elongation factor EF-1 alpha affect the frequency of frameshifting and amino acid misincorporation in *Saccharomyces cerevisae*. Genetics 120: 923–934

Sandbaken MG, Lupisella JA, DiDomenico B & Charkraburtty K (1990) Protein synthesis in yeast. J. Biol. Chem. 265: 15838–15844

Schirmaier F & Philippsen P (1984) Identification of two genes coding for the translation elongation factor EF-1α of *S. cerevisiae*. EMBO J. 3: 3311–3315

Schmid SR & Linder P (1991) Translation initiation factor 4A from *Saccharomyces cerevisae*: analysis of residues conserved in the D-E-A-D family of RNA helicases. Mol. Cell. Biol. 11: 3463–3471

Schmid SR & Linder P (1992) D-E-A-D protein family of putative RNA helicases. Mol. Microbiol. 6: 283–292

Schnier J, Schwelberger HG, Smit-McBride Z, Ah Kang H & Hershey JWB (1991) Translation initiation factor 5A and its hypusine modification are essential for cell viability in the yeast *Saccharomyces cerevisae*. Mol. Cell. Biol. 11: 3105–3114

Sharp PM & Bulmer M (1988) Selective differences among translation termination codons. Gene 63: 141–145

Sharp PM, Tuohy TM & Mosurski KR (1986) Codon usage in yeast: cluster analysis clearly differentiates highly and lowly expressed genes. Nucleic Acids Res. 14: 5125–5143

Sherman F (1982) Suppression in the yeast *Saccharomyces cerevisae*. In: Strathern JN, Jones EW & Broach JR (Eds) The Molecular Biology of the Yeast Saccharomyces. Metabolism and Gene Expression (pp 463–486). Cold Spring Harbor Laboratory, Cold Spring Harbor, NY

Slobin LI (1980) The role of eucaryotic factor Tu in protein synthesis. The measurement of the elongation factor Tu content of rabbit reticulocytes and other mammalian cells by a sensitive radioimmunoassay. Eur. J. Biochem. 110: 555–563

Sonenberg N (1988) Cap-binding proteins of eukaryotic messenger RNA: functions in initiation and control of translation. Prog. Nucleic Acid Res. Mol. Biol. 35: 173–207

Sonenberg N & Pelletier J (1989) Poliovirus translation: a paradigm for a novel initiation mechanism. BioEssays 11: 128–132

Song JM, Picologlou S, Grant CM, Firoozan M, Tuite MF & Liebman S (1989) Elongation factor EF-1 alpha gene dosage alters translational fidelity in *Saccharomyces cerevisae*. Mol. Cell. Biol. 9: 4571–4575

Strick CA & Fox TD (1987) *Saccharomyces cerevisae* positive regulatory gene *PET111* encodes a mitochondrial protein that is translated from an mRNA with a long 5′ leader. Mol. Cell. Biol. 7: 2728–2734

Surguchov AP (1988) 'Omnipotent' nonsense suppressors: new clues to an old puzzle. TIBS 13: 120–123

Trachsel H (1991). Translation in Eukaryotes. CRC Press, Boca Raton, Ann Arbor, Boston and London

Tuite MF (1989) Protein synthesis. In: Rose AH & Harrison JS (Eds) The Yeasts (pp 161–204). Academic Press, NY

van den Heuvel JJ, Bergkamp RJ, Planta RJ & Raue HA (1989) Effect of deletions in the 5′-noncoding region on the translational efficiency of phosphoglycerate kinase mRNA in yeast. Gene 79: 83–95

van den Heuvel JJ, Planta RJ & Raue HA (1990) Effect of leader primary structure on the translational efficiency of phosphoglycerate kinase mRNA in yeast. Yeast 6: 473–482

Warner JR (1989) Synthesis of ribosomes in *Saccharomyces cerevisae*. Microbiol. Rev. 53: 256–271

Wek RC, Jackson BM & Hinnebusch AG (1989) Juxtaposition of domains homologous to protein kinases and histidyl-t RNA synthetases in GCN2 protein suggests a mechanism for coupling GCN4 expression to amino acid availability. Proc. Natl. Acad. Sci. USA 86: 4579–4583

Werner M, Feller A, Messenguy F & Piérard A (1987) The leader peptide of yeast gene CPA1 is essential for the translational repression of its expression. Cell 49: 805–813

Williams NP, Hinnebusch AG & Donahue TF (1989) Mutations in the structural genes for eukaryotic initiation factors 2 alpha and 2 beta of *Saccharomyces cerevisae* disrupt translational control of GCN4 mRNA. Proc. Natl. Acad. Sci. USA 86: 7515–7519

Wilson PG & Culbertson MR (1988) SUF12 suppressor protein of yeast. A fusion protein related to the EF-1 family of elongation factors. J. Mol. Biol. 199: 559–573

Antonie van Leeuwenhoek **62**: 63–78, 1992.

Peroxisome biogenesis in *Saccharomyces cerevisiae*

Wolf-H. Kunau[1] & Andreas Hartig[2]
[1] *Abteilung Zellbiochemie, Institut für Physiologische Chemie, Medizinische Fakultät der Ruhr-Universität, D-4630 Bochum, Germany;* [2] *Institut für Allgemeine Biochemie der Universität und Ludwig Boltzmann-Forschungsstelle für Biochemie, Währinger Strasse 38, A-1090 Wien, Austria*

Key words: biogenesis, gene expression, mutants, peroxisomes, yeast

Abstract

The observation that peroxisomes of *Saccharomyces cerevisiae* can be induced by oleic acid has opened the possibility to investigate the biogenesis of these organelles in a biochemically and genetically well characterized organism. Only few enzymes have been identified as peroxisomal proteins in *Saccharomyces cerevisiae* so far; the three enzymes involved in β-oxidation of fatty acids, enzymes of the glyoxylate cycle, catalase A and the PAS3 gene product have been unequivocally assigned to the peroxisomal compartment. However, more proteins are expected to be constituents of the peroxisomes in *Saccharomyces cerevisiae.*

Mutagenesis of *Saccharomyces cerevisiae* cells gave rise to mutants unable to use oleic acid as sole carbon source. These mutants could be divided in two groups: those with defects in structural genes of β-oxidation enzymes (fox-mutants) and those with defects in peroxisomal assembly (pas-mutants). All fox-mutants possess morphologically normal peroxisomes and can be assigned to one of three complementation groups (FOX1, 2, 3). All three FOX genes have been cloned and characterized. The pas-mutants isolated are distributed among 13 complementation groups and represent 3 different classes: peroxisomes are either morphologically not detectable (type I) or present but non-proliferating (type II). Mislocalization concerns all peroxisomal proteins in cells of these two classes. The third class of mutants contains peroxisomes normal in size and number, however, distinct peroxisomal matrix proteins are mislocalized (type III). Five additional complementation groups were found in the laboratory of H.F. Tabak. Not all PAS genes have been cloned and characterized so far, and only for few of them the function could be deduced from sequence comparisons.

Proliferation of microbodies is repressed by glucose, derepressed by non-fermentable carbon sources and fully induced by oleic acid. The regulation of four genes encoding peroxisomal proteins (PAS1, CTA1, FOX2, FOX3) occurs on the transcriptional level and reflects the morphological observations: repression by glucose and induction by oleic acid. Moreover, trans-acting factors like ADR1, SNF1 and SNF4, all involved in derepression of various cellular processes, have been demonstrated to affect transcriptional regulation of genes encoding peroxisomal proteins.

The peroxisomal import machinery seems to be conserved between different organisms as indicated by import of heterologous proteins into microbodies of different host cells. In addition, many peroxisomal proteins contain C-terminal targeting signals. However, more than one import route into peroxisomes does exist. Dissection of the import mechanism in a genetically well suited organism like *Saccharomyces cerevisiae* together with further characterization and functional assignment of the PAS gene products will provide insight into the biogenesis of peroxisomes. Moreover, these studies will lead to a good model system for elucidation of the mechanisms underlying human peroxisomal disorders.

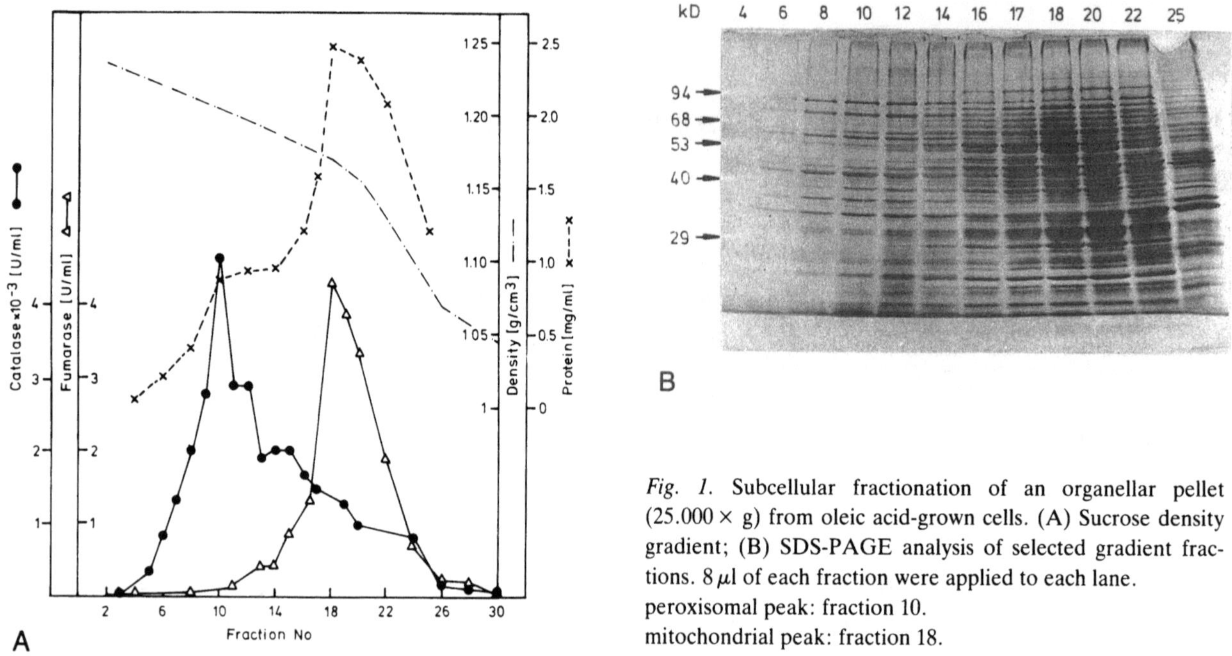

Fig. 1. Subcellular fractionation of an organellar pellet (25.000 × g) from oleic acid-grown cells. (A) Sucrose density gradient; (B) SDS-PAGE analysis of selected gradient fractions. 8 μl of each fraction were applied to each lane.
peroxisomal peak: fraction 10.
mitochondrial peak: fraction 18.

Introduction

The budding yeast *Saccharomyces cerevisiae* is well suited for investigating fundamental questions of cell biology because of the ease with which it can be manipulated and because of the wealth of information available concerning its physiology, biochemistry, and genetics (Rose & Harrison 1987–1991). In addition, in recent years recombinant DNA procedures have opened in an unprecedented way new avenues to analyze and manipulate the yeast genome (Botstein & Fink 1988). Combined genetic, biochemical and molecular genetic approaches have been used very successfully to investigate the biogenesis of subcellular structures, including mitochondria (Glick et al. 1991; Pfanner et al. 1991), endoplasmic reticulum (Newman & Ferro-Novick 1990; Hicke & Schekman 1990), vacuoles (Klionsky et al. 1990; Rothman et al. 1989), and nuclei (Silver 1991). Peroxisomes are late comers to this field.

The history of peroxisomes in yeasts began when Avers and Federman demonstrated the presence of these organelles in *S. cerevisiae* (Avers & Federman 1968). However, in contrast to other fungal species (Tanaka & Fukui 1989; Veenhuis & Harder 1991) thereafter very little was heard of peroxisomes in *S. cerevisiae*. Their presence in *S. cerevisiae* was even questioned when the particulate catalase (catalase A) was assigned to vacuoles (Susani et al. 1976). These initial observations can be explained in view of our current knowledge because peroxisomes are rare and small under conditions originally chosen for growth and isolation (Veenhuis et al. 1987).

Accumulated evidence has demonstrated that proliferation of peroxisomes in yeasts is under the control of glucose repression and derepression as well as induction by distinct carbon and nitrogen sources (Harder & Veenhuis 1989; Tanaka & Fukui 1989; Veenhuis & Harder 1991). This inducibility of peroxisome proliferation in yeasts and also to some extent in higher eukaryotes (animals and plants) reflects one specific feature of these organelles. Another striking property is their functional diversity. Their enzyme patterns vary markedly depending on organism, cell type, and environmental conditions (Fahimi & Sies 1987). Thus, media differing in carbon and nitrogen sources lead to induction of different peroxisomal metabolic pathways in cells of various species of yeasts and filamentous fungi. Especially thoroughly studied examples are *Hansenula polymorpha, Pichia pastoris* (Gleeson & Sudberry 1988), and *Neurospora crassa* (Wanner & Theimer 1982, Desel et al. 1982, Kunau et al. 1987). In contrast, for many years no

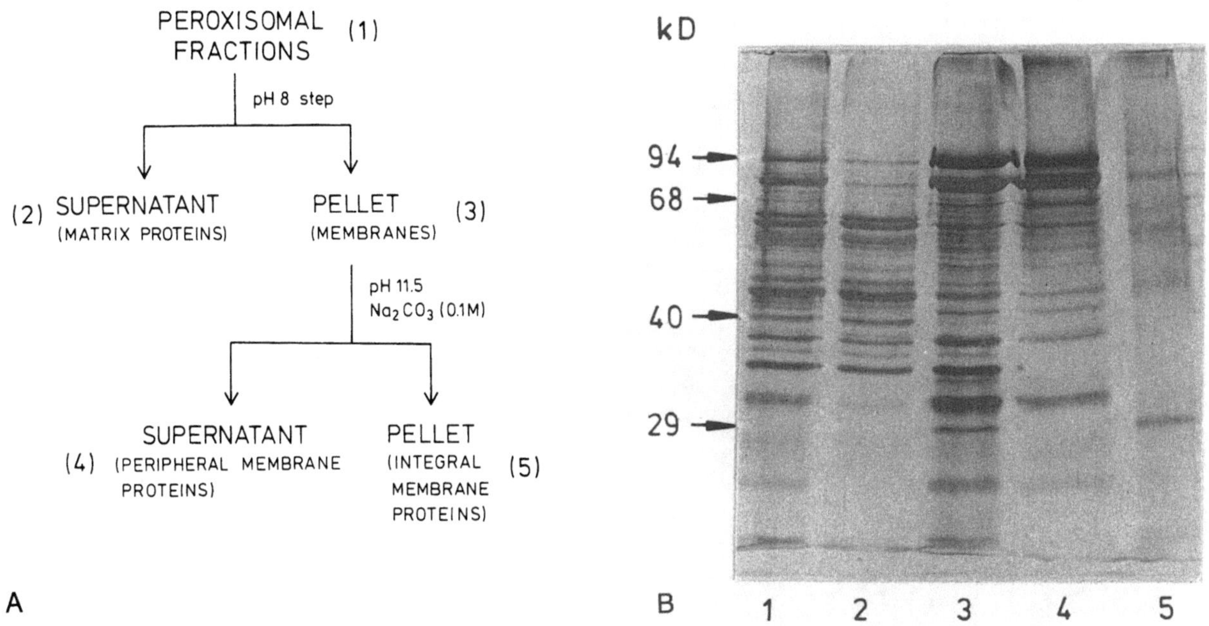

Fig. 2. (A) Subfractionation scheme of gradient fractions containing peroxisomes; (B) SDS-PAGE analysis of peroxisomal subfractions. Numbers correspond to numbers given in brackets in Fig. to A: (1) isolated peroxisomes (sucrose density gradient); (2) pH-8-step: supernatant (matrix proteins); (3) pH-8-step: pellet (membrane proteins); (4) pH-11.5-step: supernatant (peripheral membrane proteins); (5) pH-11.5-step: pellet (integral membrane proteins) about 15 μg of proteins were applied to each lane.

progress was made regarding biogenesis of peroxisomes in *S. cerevisiae* despite many efforts of various laboratories. In a very fruitful collaboration between the laboratory of one of us (Wolf-H. Kunau) and that of Marten Veenhuis (University of Groningen/The Netherlands) the carbon source oleic acid was identified as inducer of both the peroxisomal β-oxidation and the proliferation of peroxisomes in *S. cerevisiae* (Veenhuis et al. 1987). This observation started the renaissance of peroxisomal research in this yeast. Since oleic acid is still the only peroxisomal inducer known for *S. cerevisiae*, in this respect this organism lags behind other yeasts. Nevertheless, the extensive knowledge of its biochemistry, genetics and molecular biology together with the possibility to induce its peroxisomes makes *S. cerevisiae* a well suited organism to study peroxisome biogenesis.

The purpose of this article is to review the first attempts to dissect the mechanism of peroxisome biogenesis by a combined biochemical and genetic approach. To do this we will describe (i) proteins known or suggested to be constituents of peroxisomes in *S. cerevisiae*, (ii) the peroxisomal mutants (fox- and pas-mutants) recently isolated and the first PAS genes which have been characterized, (iii) what is known about the import of peroxisomal proteins and (iv) the regulation of genes encoding peroxisomal proteins. Although focussing on *S. cerevisiae* some results obtained with other yeasts will be included for comparison. However, we make no attempt for an exhaustive coverage of peroxisomal research in other yeasts.

Peroxisomal proteins

Compartmentation of metabolic pathways is essential for eukaryotic cells. Each subcellular structure has its own distinct set of proteins and these proteins have to be imported into their target organelle. The knowledge which proteins are constituents of the organelle under study is a prerequisite for an experimental approach towards organelle biogenesis. For the sake of simplicity it is convenient to classify peroxisomal proteins into mem-

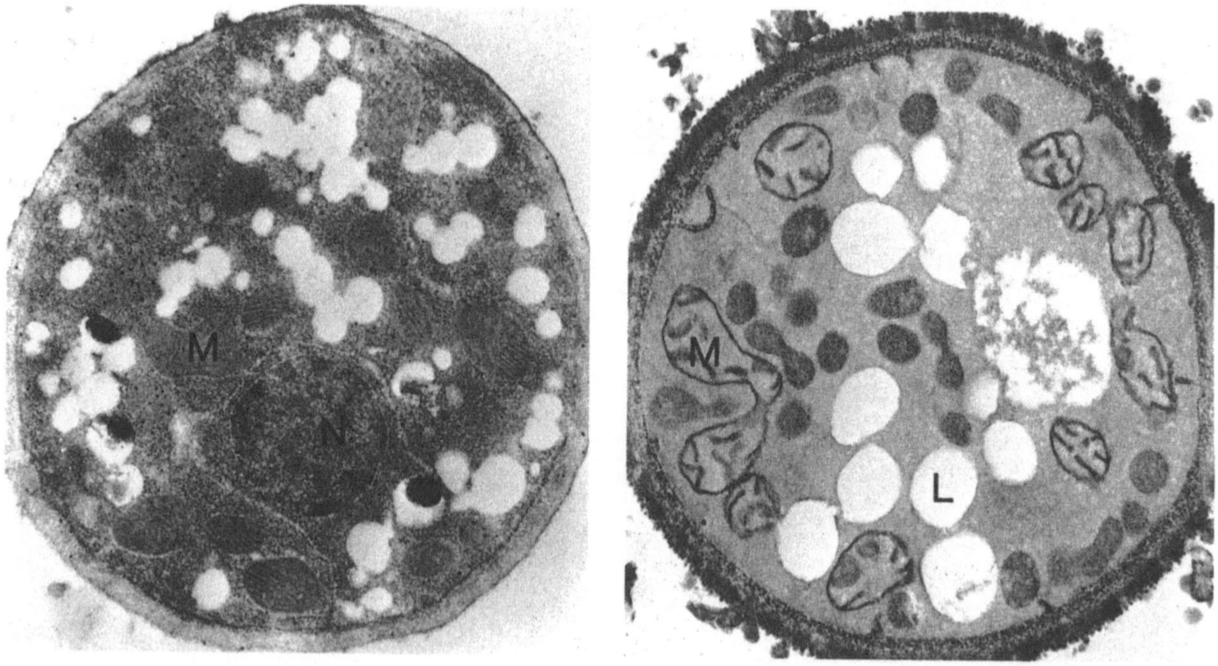

A

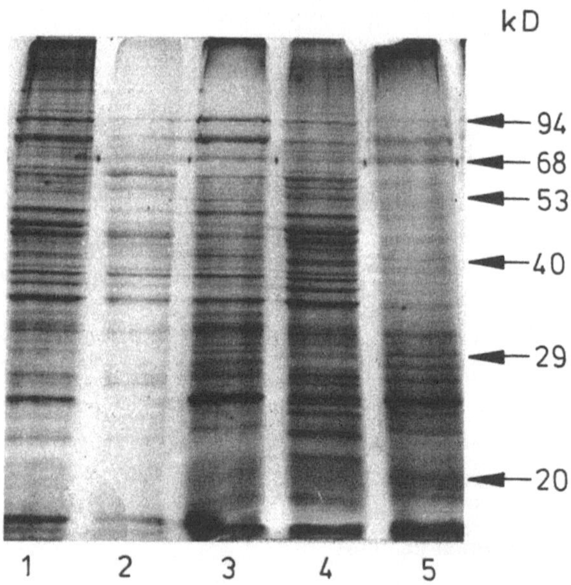

B

Fig. 3. Electron micrographs of *Saccharomyces cerevisiae* cells after growth on oleate as sole carbon source. (A) Immunocytochemical staining with anti-thiolase antibodies; (B) Cytochemical staining with diaminobenzidin (DAB) for catalase activity.

brane and matrix proteins. In addition, the former ones are frequently subdivided into integral and peripheral membrane proteins based entirely on an operational criterion (solubility in 0.1 M Na-carbonate pH 11,5, Fujiki et al. 1982). Analyses of this kind have been performed for mammalian tissues, especially liver (Lazarow 1984; Hashimoto et al. 1986; Hardeman et al. 1990) and some yeasts, such as *Candida boidinii* (Goodman et al. 1986, 1990) and *H. polymorpha* (Sulter et al. 1990). For *S. cerevisiae* much less is known (McCammon et al. 1990a, see Figs 1 and 2) reflecting the fact that peroxisomes of this yeast only recently became accessible for investigation.

Catalase A was the first and for some time the only peroxisomal matrix protein which was biochemically characterized. Mutants of catalase A and its cytosolic counterpart catalase T were isolated before the peroxisomal localization was firmly established (Spevak et al. 1983; Cohen et al. 1985). In the meantime, catalase A is besides 3-oxoacyl-CoA thiolase (see below) the most frequently used protein for biochemical or immunocytochemical characterization of peroxisomes in *S. cerevisiae* (see Fig. 3). Catalase A has been cloned and sequenced (Cohen et al. 1985, 1988) and the complex regulation of its expression is being studied (Simon et al. 1991).

With the demonstration of a peroxisomal β-oxidation system in *S. cerevisiae* (Veenhuis et al. 1987, Skoneczny et al. 1988) several new peroxisomal enzymes were identified. Peroxisomal β-oxidation systems consist of acyl-CoA oxidase, multifunctional protein and 3-oxoacyl-CoA thiolase. This was demonstrated in mammals (Hashimoto 1990), plants (Kirsch et al. 1986), various fungi (Kunau et al. 1988) and now also in *S. cerevisiae*. All three proteins have been purified and characterized (Kunau et al., in preparation) as well as cloned and sequenced (Dmochowska et al. 1990; Kunau et al., in preparation). These three proteins together catalyze the four reactions of the classical β-oxidation pathway.

While catalase, acyl-CoA oxidase and thiolase are matrix enzymes the multifunctional protein of *S. cerevisiae* is predominantly found in the peripheral membrane protein fraction. Whether peroxisomes of *S. cerevisiae* like other yeasts (Yamada et al. 1980; Numa 1981) also contain a peroxisomal acyl-CoA synthetase is not known.

A second major pathway frequently found in peroxisomes is the glyoxylate cycle. While plant glyoxysomes contain the enzymes of the entire cycle only the key enzymes, isocitrate lyase and malate synthase, have been demonstrated in peroxisomes of fungi (Kindl & Lazarow 1982). However, evidence for the presence of glyoxylate cycle enzymes in *S. cerevisiae* has been controversial. Early reports of the presence of enzymes of the glyoxylate cycle in peroxisomes (Szabo & Avers 1969) contrast with others assigning these enzymes to the cytosol (Duntze et al. 1969, Parish 1975). More recently, the possibility to induce peroxisome proliferation by growth on oleate prompted reinvestigation of this problem. Activities of malate synthase (McCammon et al. 1990a), citrate synthase (Lewin et al. 1990) and malate dehydrogenase (McCammon et al. 1990a; Minard & McAllister-Henn 1991) were demonstrated to co-migrate with peroxisomes in sucrose density gradients. The presence of the putative peroxisomal targeting signal S-K-L (Gould et al. 1989, see below) at the extreme C-termini of the citrate synthase 2 (Rosenkrantz et al. 1986) and malate synthase (A. Hartig et al., unpubl.) reinforced this conclusion. In addition, for malate synthase immunocytochemical evidence also indicated a peroxisomal location (McCammon et al. 1990a). For isocitrate lyase, however, the intracellular location is entirely enigmatic. While subcellular fractionation studies consistently showed more than 95% of the enzyme activity in the soluble fraction (Duntze et al. 1969; McCammon et al. 1990a) preliminary immunocytochemical evidence indicates a peroxisomal location (McCammon et al. 1990a). Although leakage of major fractions of activities have been reported also for other peroxisomal enzymes of *S. cerevisiae* such as malate dehydrogenase and malate synthase (McCammon et al. 1990a) it is not easy to envisage how an enzyme can be lost completely from peroxisomes even before or during the first sedimentation of these organelles from cell lysates. The difficulty to recover these enzymes predominantly in the peroxisomal fraction might reflect the lack of intactness of isolated peroxisomes much more than currently anticipated. Peroxisomes seem to be extremely fragile under conditions used for isolation. To date, it is not easy to assess whether the isolation of intact peroxisomes is a general problem or more pronounced for distinct organisms. The difficulties to set up a reliable *in vitro*-import system for peroxisomes (see below) seems to support the former possibility.

Very little is known about membrane proteins of peroxisomes from *S. cerevisiae* (McCammon et al. 1990a; V. Hines, pers. commun.; W.-H. Kunau, unpubl.). In some other yeasts membrane analyses have been reported (see below). Three peroxisomal membrane proteins of *C. boidinii* have been cloned and sequenced, PMP20 (Garrard & Goodman 1989), PMP31 (J.M. Goodman, pers. commun.) and PMP47 (McCammon et al. 1990b). Recently, the first primary structure of a peroxisomal integral membrane protein of *S. cerevisiae* (Pas3p) was reported (Höhfeld et al. 1991). As a protein encoded by a PAS gene this membrane protein is essential for peroxisome biogenesis. It has been partially characterized and shown to possess properties expected for a membrane bound receptor. The availability of Pas3p as membrane marker should now greatly facilitate identification of peroxisomal membranes in cells of *S. cerevisiae*. For higher eukaryotes a first integral membrane protein essential for peroxisome formation has also recently been identified (Tsukamoto et al. 1991).

Based on the number of polypeptide bands resolved by SDS-PAGE of peroxisomal fractions of *S. cerevisiae* (Fig. 1) it seems very likely that in the future more proteins will be identified as peroxisomal constituents of this yeast (see Table 1). On the basis of the assumption that peroxisomal import is conserved between different organisms (Gould et al. 1990), potential peroxisomal proteins of *S. cerevisiae* can be listed. A putative peroxisomal targeting signal (Ser-Lys-Leu at the extreme C-terminus, see below) has been identified for the gene products of PAS6 (A. Skaletz & W.-H. Kunau, unpubl.) and of DAL7 (Yoo & Cooper 1989, see Table 1B). Recently, the heterologous import of human peroxisomal Cu,Zn-superoxide dismutase into peroxisomes of *S. cerevisiae* has been reported (Keller et al. 1991). Since the yeast counterpart of this enzyme has a similar primary sequence (Bermingham-McDonogh et al. 1988), it is conceivable that Cu,Zn-superoxide dismutase of *S. cerevisiae* is a peroxisomal protein as well. Activation of fatty acids is a prerequisite for their degradation, and acyl-CoA synthetases have been demonstrated in peroxisomes of mammals (Alexson et al. 1985; Lageweg et al. 1991) and *Yarrowia lipolytica* (Numa 1981). Genetic and biochemical evidence for an acyl-CoA synthetase in *S. cerevisiae* has been reported (Kamiryo et al. 1976), however, the activity was measured only in glucose-grown cells. It seems likely that a second oleic acid-induced synthetase is present, resembling the situation in *Y. lipolytica.* Another candidate for a peroxisomal protein in *S. cerevisiae* is the nonspecific lipid-transfer protein, demonstrated in peroxisomes of mammals (Mori et al. 1991) and *C. tropicalis* (Szabo et al. 1989, Tan et al. 1990). Assuming that induction by acetate or ethanol is an indication for peroxisomal location (e.g. malate synthase, McCammon et al. 1990a) then the alanine-glyoxy-

Table 1. Proteins detected (A) or expected to be (B) in peroxisomes of *Saccharomyces cerevisiae.*

Proteins			Sequence	Localized	C-term.	Gene
A	catalase A	Sc	(1)	px (2)	-SKF	CTA1
	acylCoA-oxidase	Sc	(3)	px (2)	-INK	POX1
	multifunctional pr.	Sc	(4)	px (4)	-SKL	FOX2
	3-keto-thiolase	Sc	(5)	px (5)	-IKE	FOX3
	citrate synthase 2	Sc	(6)	px (7)	-SKL	CIT2
	malate DH 2	Sc	(8)	px (9)	-ASS	MDH2
	isocitrate lyase	Sc	(10)	? (9)	-VKK	ICL1
	malate synthase	Sc	(11)	px (9)	-SKL	MLS1
	PAS3 gene product	Sc	(12)	px (12)	-FKP	PAS3
B	PAS6 gene product	Sc	(13)	no	-SKL	PAS6
	DAL7 gene product (malate synthase?)	Sc	(14)	no	-SKL	DAL7

Abbreviations used: Cb – *Candida boidinii*; Ct – *Candida tropicalis*; Hp – *Hansenula polymorpha*; Sc – *Saccharomyces cerevisiae.* (1) = Cohen et al. 1988; (2) = Skoneczny et al. 1988; (3) = Dmochowska et al. 1990; (4) = K. Hiltunen et al. manuscript submitted; (5) = W.-H. Kunau et al. unpubl.; (6) Rosenkrantz et al. 1986; (7) = Lewin et al. 1990; (8) = Minard & McAllister-Henn 1991; (9) = McCammon et al. 1990a; (10) = E. Fernandez et al. unpubl., EMBL accession nr X61271; (11) = A. Hartig et al. unpubl.; (12) = Höhfeld et al. 1991; (13) = A. Skaletz & W.-H. Kunau, unpubl.; (14) = Yoo & Cooper 1989.

late amino transferase (Takada & Noguchi 1985) should be found in peroxisomes of *S. cerevisiae*, as was reported for other organisms (Sakuraba et al. 1991). If peroxisomal location in other organisms is a first hint for the same location of the protein under study HMG-CoA reductase can be considered a potential candidate, too. In mammalian cells distinct HMG-CoA reductase proteins are present in peroxisomes as well as in the ER (Rusnak et al. 1990). However, in *S. cerevisiae* the products of both genes identified so far are located in the ER (Sengstag et al. 1990) in glucose-grown cells. It would be interesting to look for still another HMG-CoA reductase in oleic acid-grown cells. The fact that mitochondrial import depends on the action of the heat shock proteins hsp70 and hsp60 inside the organelle (Neupert et al. 1990) suggests the existence of hsps also in peroxisomes. However, this hypothesis implies high similarity of the two import mechanisms, which remains to be shown. In other yeasts amine oxidase, urate oxidase and D-amino acid oxidase have been found in peroxisomes (Veenhuis & Harder 1991). It would be worthwhile to look for these activities in peroxisomes of *S. cerevisiae* grown under optimal induction conditions. Identification of more peroxisomal proteins will broaden the experimental basis to investigate peroxisome biogenesis.

Peroxisomal mutants of S. cerevisiae

The recently described induction of peroxisomal proliferation in *S. cerevisiae* (Veenhuis et al. 1987), using oleic acid as sole carbon source, opened the possibility to dissect peroxisome biogenesis genetically and thus to identify components essential for this process. As a first step a screening procedure for cells with peroxisomal defects (Erdmann et al. 1989) was devised which resulted in mutants defective in a broad array of peroxisomal gene products (Höhfeld et al. 1992). Cells able to grow on acetate or glycerol but unable to grow on plates containing oleic acid as sole carbon source (oleic acid non-utilizing phenotype, *onu*-strains) were isolated. This screening procedure discriminated between colonies which cannot utilize oleic acid and those with other metabolic defects. Like other yeasts (Kunau et al. 1988) *S. cerevisiae* does not contain a mitochondrial β-oxidation system (Veenhuis et al. 1987). Therefore, the inability to grow on oleic acid can be used as a criterion for an impaired peroxisomal fatty acid degradation. In order to identify mutants with a defective β-oxidation among the *onu*-strains biochemical methods were applied. Such mutant phenotypes can be due to either defects in structural genes of β-oxidation enzymes (fox-mutants) or defects in peroxisomal assembly (pas-mutants). The determination of β-oxidation activities in either whole-cell lysates or subcellular fractions obtained from lysates of spheroplasts allow detection of deficiencies of individual enzyme activities as well as detection of all or only distinct peroxisomal matrix enzymes in the cytosol rather than in peroxisomes. This mislocalization of peroxisomal enzymes was taken as an indication that peroxisomes are absent or import-incompetent. These conclusions were further verified by electron microscopy and immunocytochemistry.

The fox-mutants fall into three complementation groups (FOX1-3) in agreement with the biochemical results that in peroxisomes three different proteins catalyze the four reactions of the β-oxidation cycle. An acyl-CoA oxidase (FOX1 or POX1, Dmochowska et al. 1990) converts the activated fatty acid to a corresponding α, β-transunsaturated enoyl-CoA ester. The second and third reaction is catalyzed by a multifunctional protein (FOX2) converting this unsaturated acyl-CoA ester to the corresponding β-oxoacyl-CoA ester. This metabolite, in turn, is cleaved by a thiolase (FOX3) resulting in an activated fatty acid, which is two carbon atoms shorter than the starting compound and again serves as substrate for the first reaction. An acyl-CoA oxidase gene (FOX1 or POX1) was cloned searching for a completely unrelated gene as an unidentified open reading frame. It was identified on the basis of its sequence similarity to the acyl-CoA oxidase genes POX4 and POX5 of C. tropicalis (Dmochowska et al. 1990). Functional complementation of fox2 and fox3 mutants using a genomic library of *S. cerevisiae* led to the characterization of the FOX2 (K. Hiltunen, manuscript submitted) and FOX3 genes (W.-H. Kunau et al. unpubl.), respectively. All fox mutants possess morphologically normal peroxisomes.

There are other mutants with defects in genes encoding peroxisomal matrix enzymes: cta1, deficient in catalase A (Cohen et al. 1985), cit2, deficient in citrate synthase 2 (Rosenkrantz et al. 1986), and mdh2 deficient in malate dehydrogenase 2 (Minard & McAllister-Henn 1991). They were isolated using entirely different screening procedures specific for the individual enzymes and not with the intention of finding peroxisomal mutants. In all three cases the wild-type genes were cloned by functional complementation of the corresponding mutants.

Mutants with defects in structural genes of peroxisomal matrix proteins for which the phenotypical difference to wild-type strains allows the application of an assay system should be very valuable as *in vivo*-import systems. The possibility to transform them with plasmids carrying genes modified by *in vitro*-mutagenesis and to assay import by functional complementation make them at present an attractive alternative to *in vitro*-import systems (see below).

The pas-mutants are the second group of peroxisomal mutants of *S. cerevisiae* isolated besides the fox-mutants employing the above described screening procedure. This group comprises strains defective in peroxisome assembly and the mutants isolated so far fall into 13 complementation groups. Five additional complementation groups were found in the laboratory of H.F. Tabak (I. van der Leij et al., manuscript submitted). Mislocalization of peroxisomal matrix proteins to the cytosol was found for oleic acid-induced cells of all pas-mutants. The complexity of peroxisome assembly is not only reflected by the genetic diversity of the peroxisomal mutants but also by their different phenotypes representing three different classes. Mislocalization concerns either all (type I and type II) or only distinct peroxisomal matrix proteins (type III). Peroxisomes are normal in size and number (type III), present but non-proliferating (type II) or morphologically not detectable (type I) (for a detailed discussion see Höhfeld et al. 1992).

The pas-mutants of seven of our thirteen complementation groups were classified according to the three pas-types (Table 2). Absence or presence of few small peroxisomes can only be unambiguously decided by investigating tight phenotypes. For the remaining six complementation groups deletion mutants are not yet available. Therefore, it cannot be decided whether these mutants belong to type I or type II phenotype. Biochemical evidence excluded the type III phenotype.

Originally, we defined the pas-phenotype by the absence of morphologically detectable peroxisomes resulting in mislocalization of matrix pro-

Table 2. pas-mutants and PAS genes.

Complementation group	Number of alleles	Pastype	Corresponding wild type genes		
			Cloned	Sequenced	Features
PAS1	4	I	+	+	putative ATPase
PAS2	1	I	+	+	putative UBC[a]
PAS3	4	I	+	+	integral PMP[b]
PAS4	1	II	+	+	Zn-finger like motif required for proliferation
PAS5	1	I	+	+	Zn-finger like motif
PAS6	2	II(?)	+	+	C-terminal S-K-L
PAS7	2	III	+	+	required for thiolase import
PAS8	3	n.d.	−	−	
PAS9	1	n.d.	+	+	
PAS10	1	n.d.	−	−	
PAS11	1	n.d.	+	−	
PAS12	1	n.d.	+	−	
PAS13	1	n.d.	−	−	

[a] ubiquitin conjugating enzyme; [b] peroxisomal membrane protein.

teins and the inability to grow on oleic acid (type I) (Erdmann et al. 1989). However, detailed analysis of the peroxisomal mutants revealed that some of them possess morphologically recognizable peroxisomes (type II and III). These findings lead to a new definition of the pas-phenotype: common to all pas-mutants is the absence of functional peroxisomes resulting in (i) inability to grow on oleic acid and (ii) mislocalization of peroxisomal matrix enzymes. Their defects are not due to mutations in structural genes coding for metabolic enzymes (excluding e.g. fox-mutants, cta 1, cit2, and mdh2).

The isolation of pas-mutants allowed the use of functional complementation to identify genes encoding proteins essential for peroxisome biogenesis. Based on this approach the genes PAS1 to 7, 9, and 12 have been cloned in our laboratory (see Table 2). With the exception of PAS12 the complementing DNA-fragments have been sequenced and their open reading frames have been found to encode proteins not identical to any known protein. Four of the PAS gene products (Pas1p, Erdmann et al. 1991, Pas2p, F.F. Wiebel & W.-H. Kunau, in preparation, Pas4p, D. Mertens & W.-H. Kunau, in preparation, and Pas5p, A. Skaletz & W.-H. Kunau, unpubl.) exhibited at least partial sequence similarities to known proteins (see Table 2 and Höhfeld et al. 1992), one (Pas3p) contains features expected for a membrane bound receptor (Höhfeld et al. 1991) and one (Pas6p, A. Skaletz & W.-H. Kunau, unpubl.) contains the C-terminal putative peroxisomal targeting signal, Ser-Lys-Leu (see below). For all the other PAS gene products sequence analysis does not permit any conclusions regarding their function or intracellular location.

The further development of classical and molecular genetics in other yeasts, e.g. *P. pastoris* (Gleeson & Sudberry 1988) and *H. polymorpha* (Gleeson & Sudberry 1988, Cregg et al. 1990) increases the number of organisms suitable for genetic dissection of peroxisome biogenesis. This is emphasized by the isolation of peroxisomal mutants of *H. polymorpha* (Cregg et al. 1990; Didion & Roggenkamp 1990) and *P. pastoris* (J.M. Cregg & M. Veenhuis, pers. commun.). A genetic approach using appropriate mutants was also successful for the identification of a protein essential for peroxisome biogenesis in CHO-cells (Tsukamoto et al. 1991).

Import of peroxisomal proteins

Experimental approach

Peroxisomes, like mitochondria and chloroplasts, are thought to be formed from pre-existing organelles. Peroxisomal proteins are nuclear-encoded, synthesized on free polyribosomes and post-translationally imported into the organelles (for a review see Lazarow & Fujiki 1985). Most of them are synthesized at their final size, excluding major processing steps associated with import. Many peroxisomal functions as well as the import machinery seem to be conserved between organisms of different kind (Borst 1989). Heterologous peroxisomal proteins are imported into microbodies of different host cells including *S. cerevisiae* (Distel et al. 1987; Hansen & Roggenkamp 1989; Gödecke et al. 1989; Gould et al. 1990; McCammon et al. 1990b; Keller et al. 1991). However, peroxisomal targeting signals from proteins of *S. cerevisiae* have not yet been found experimentally. The main obstacle is the lack of an efficient *in vitro*-import system for peroxisomes, comparable to the system established for mitochondria (Hartl et al. 1987). With such a system the nature of the targeting signal(s) as well as the mechanism of import into peroxisomes could be investigated. The major advantages of the *in vitro*-import system for mitochondria are the existence of an assay for the quality of the organelle preparation (tight coupling of respiration and oxidative phosphorylation) and the fact that most mitochondrial proteins are made as larger precursors and are processed only inside the organelle, leaving no doubt whether the object of choice was at least partially translocated to the mitochondrial matrix or not. Compared to mitochondria, peroxisomes are very fragile organelles (Alexson et al. 1985). To date the most reliable peroxisomal *in vitro*-import system established is that for rat liver peroxisomes (Fujiki & Lazarow 1985). However, all *in vitro*-systems for peroxisomes are limited by the fact that there are no definite criteria for the functional intactness of isolated organelles (see e.g. Bellion & Goodman 1987; Imanaka et al. 1987) not for import itself, since peroxisomal proteins are usually not processed. The exceptions to that rule, malate dehydrogenase of watermelon (Gietl 1990) and rat liver thiolase (Hijikata et al. 1987), have not been tried in an *in vitro*-system from yeast. Therefore, the only criterion that a polypeptide has been imported is the protection of this protein against proteases in the presence of a very fragile and fairly protease sensitive organelle. The small percentage of membrane associated or imported protein which is usually observed (Imanaka et al. 1987; Small et al. 1987, 1988; Miyazawa et al. 1989; Thieringer et al. 1991) does not allow a reliable conclusion whether the labeled protein has been translocated or not. As long as clear evidence for true import into peroxisomes *in vitro* is missing, all results based on this kind of experiments require confirmation by an independent method. Additionally, microbodies of *S. cerevisiae* seem to be more fragile than peroxisomes from other organisms. This difference, however, might be due to the procedures employed for induction and isolation of these organelles from *S. cerevisiae*.

In yeasts, especially in *S. cerevisiae*, targeting experiments can also be carried out *in vivo*, taking advantage of the easy manipulation of a unicellular eukaryote. However, in most cases, particularly when an assay system is not yet available, overexpression of peroxisomal proteins from multicopy vectors and/or strong promoters seems to be essential to even recognize the gene product of interest in the cell with antibodies. As far as we know the only exception to that is thiolase. This might be due to antibodies of extreme high affinity to this protein. For *in vivo* experiments transformed cells are grown in the presence of an inducer of peroxisomes and processed for analysis. In most cases the determination of the protein location occurs via immunological methods after fixation (for electron microscopy) or spheroplasting and immobilization (for immunofluorescence) or spheroplasting followed by gentle lysis, differential centrifugation and separation of organelles on a density gradient. The last method, although in principle suited to be performed in every laboratory might result in artefacts due to inclusion body formation after overexpression of proteins (Mitraki & King 1989; Hartig et al. 1990). Conclusions reached on the basis of immunofluorescence in *S. cerevisiae* cells (Pringle et al. 1989) have to take into account that the cell is very small and subcellular structures are at the limit of the optical resolution of a microscope. While it is possible to define the cytosolic or the nuclear location of a protein, in most cases the organellar location of a protein can only be characterized by punctuate appearance of the fluorescent dye, which means association with particles. Whether these are small protein aggregates or mitochondria or peroxisomes can usually not be distinguished by immunofluorescence despite the possibility of double labelling. Improvements of currently employed techniques of optical resolution, e.g. by confocal laser scanning microscopy, might result in more specific data. Certainly the best way to define the location of a protein inside a *S. cerevisiae* cell is the use of an electron microscope and immunocytochemistry. Numerous reports have been published using this method, and it would be beyond the scope of this review to give any details (for an example see van der Klei 1991 and Fig. 3).

Targeting signals for peroxisomal proteins can also be found by functional complementation of corresponding yeast mutants taking advantage of all the knowledge of the genetics and biochemistry of *S. cerevisiae*. However, not all imported proteins are active probably due to lack of oligomerization (Distel et al. 1987). An interesting case may be the use of the fox-mutants. Functional complementation of these mutants with enzymes of the β-oxidation pathway from higher eukaryotes including their mitochondrial counterparts will provide new insights into the mechanisms of β-oxidation including oligomerization and stereochemical aspects. In *S. cerevisiae* an additional approach for mechanistic studies on transport of proteins is available: the use of semipermeable cells (Baker et al. 1988) might prove very valuable also for import into peroxisomes. However, the lack of a reliable test system for import might represent a major obstacle in this case, too (see above).

Targeting signals

In higher eukaryotes the best defined targeting signal for peroxisomal proteins consists of a C-terminal tripeptide (S/A/C – K/R/H – L), which is necessary and sufficient to direct the proteins themselves and cytosolic reporter proteins into peroxisomes (Gould et al. 1987, 1989; Osumi & Fujiki 1990). Peroxisomal proteins of various organisms can be imported into peroxisomes of *S. cerevisiae* demonstrating the conserved nature of peroxisomal targeting (see above). However, not all reporter proteins (e.g. DHFR) with the short tripeptide added to their C-termini enter peroxisomes of *S. cerevisiae in vivo* (B. Distel, pers. commun.; A. Hartig et al., unpubl.). They obviously lack an additional important feature necessary for import into peroxisomes of *S. cerevisiae* but present in the peroxisomal proteins themselves. Whether this feature is a defined sequence or a certain structure allowing the presentation of either S-K-L or any other signal to its recognition factor (receptor) is still unknown yet. From a number of import experiments with a series of fusion proteins no conclusive evidence regarding a peroxisomal targeting signal in *S. cerevisiae* could be deduced (A. Hartig et al. unpubl.). These hybrid proteins consisted of fusions between DHFR or subunit IV of cytochrome oxidase and parts of catalase A or C-termini of various peroxisomal proteins. In addition, gene deletion- and fusion-studies with alcohol oxidase of *H. polymorpha* in *S. cerevisiae* did not reveal any umambiguous signal for translocation either (Distel 1990). Therefore, the notion mainly developed for higher eukaryotes that peroxisomal targeting signals consist of only short tripeptides at the C-terminus could not be verified for *S. cerevisiae* in this simple form. Moreover, not all peroxisomal proteins do contain an S-K-L or any of its known degenerated versions at the C-terminus (see Table 1). The idea that this signal tripeptide can also act at internal locations as suggested (Gould et al. 1989) is not supported by experimental evidence yet. In contrast, there is growing evidence that such a sequence at internal locations is dispensable for import into peroxisomes (J.M. Goodman, pers. commun.).

Among the pas-mutants one complementation group was found whose defect was identified as the mislocalization of thiolase but not of other matrix proteins (see above). This finding already suggests the existence of at least two different import pathways for peroxisomal matrix proteins, one for thiolase and another one for the other enzymes tested. In addition, the observation that there is no competition for import between catalase A and thiolase (M. Marzioch & W.-H. Kunau, unpubl.; A. Hartig et al., unpubl.) supports this idea. Furthermore, the targeting signal of rat peroxisomal thiolase was identified to reside on the N-terminus of this peroxisomal protein (Swinkels et al. 1990). Moreover, for the glyoxysomal malate dehydrogenase of watermelon an NH_2-terminal targeting signal has been suggested (Gietl 1990). However, for the peroxisomal malate dehydrogenase MDH2 of *S. cerevisiae* (Minard & McAllister-Henn 1991) no evidence for an amino-terminal transit peptide exists. In the light of these recent findings an earlier observation becomes interesting: In addition to the C-terminal targeting signal the firefly luciferase obviously contains also in the N-terminal region sequences or structures necessary for import into peroxisomes (Gould et al. 1987). Altogether, in *S. cerevisiae* and probably in other organisms, too, for some proteins the C-terminal amino acids are necessary for translocation into peroxisomes, but they may not be sufficient. For others, the extreme C-terminus does not seem to be involved in targeting at all.

Peroxisomal membrane proteins presumably are targeted to peroxisomes by a separate pathway. Current knowledge about biogenesis of peroxisomes (Lazarow & Fujiki 1985) leads to the proposal that peroxisomal prestructures exist in pas-mutant cells (for a detailed discussion see Höhfeld et al. 1992), resembling peroxisomal membrane ghosts found in Zellweger syndrome fibroblasts (Santos et al. 1988). This hypothesis is strengthened by the experimental finding that proliferation of membranes and induction of membrane proteins precedes the induction of matrix proteins (Veenhuis & Goodman 1990; Lüers et al. 1990).

Regulation of gene expression

Proliferation of microbodies in yeasts is repressed by glucose and induced by various carbon or nitrogen sorces (Tanaka & Fukui 1989; Veenhuis & Harder 1991). Therefore, it would not be surprising if most genes encoding peroxisomal proteins were regulated in the same way. However, analysis of gene regulation with respect to peroxisome biogenesis has been performed exclusively in *S. cerevisiae*, simply because the corresponding methods are well established for this organism alone. The best studied example is the CTA1 gene encoding the peroxisomal catalase A. Glucose repression of catalase A activity has been known for some time (Cross & Ruis 1978; Rytka et al. 1978) and was demonstrated to occur at the transcriptional level (M. Simon et al., manuscript submitted). The same is true for the FOX3-gene encoding peroxisomal β-ketothiolase (Einerhand et al. 1991), the FOX2-gene encoding the multifunctional protein of peroxisomal β-oxidation (K. Hiltunen et al., 1992; M. Simon et al., 1992), the PAS1 gene (Erdmann et al. 1991) and the PAS3 gene (Höhfeld et al. 1991). Four genes (CTA1, FOX2, FOX3 and PAS1) were shown to be derepressed on a non-fermentable carbon source and fully induced on oleic acid (Erdmann et al. 1991; Einerhand et al. 1991; M. Simon et al., 1992). The two key enzymes of the glyoxylate cycle, isocitrate lyase and malate synthase, most probably also peroxisomal matrix enzymes (see above and Table 1) can be induced by oleic acid, too (McCammon et al. 1990a). Their activities can hardly be detected in cells grown on glucose (Duntze et al. 1969) and their regulation probably occurs on the transcriptional level.

In order to gain insight into the mechanisms regulating the expression levels of the genes mentioned above a molecular analysis of their promoter regions was started. Cis-acting regulatory DNA-elements and trans-acting factors were identified by comparing promoter sequences of genes encoding peroxisomal proteins with each other and with short sequences already known to be involved in regulation of gene expression. The promoter regions of the genes coding for the three β-oxidation enzymes, acylCoA-oxidase, multifunctional protein and thiolase, contain a short region of high similarity termed β-oxidation box and a similar sequence was identified in the promoter region of the CTA1 gene (Einerhand et al. 1991). It was demonstrated that this sequence of the thiolase promoter is responsible for mediating induced expression in the presence of oleic acid. A trans-acting factor binding to this newly identified cis-acting element has not been found yet. Upstream of that region in the FOX3-promoter a sequence was identified which was bound by the ARS binding factor *in vitro*. However, this element does not mediate a carbon source response *in vivo* (Einerhand et al. 1991). Similar experiments analyzing the CTA1-promoter resulted in the identification of an ADR1-binding site (Simon et al. 1991). ADR1, originally found to be necessary for derepression of the ADH2 gene (Denis & Young 1983), regulates the expression of the CTA1 gene as well as the expression of the genes FOX2, FOX3 and PAS1 (Simon et al. 1991). Moreover, using appropriate mutants two other genes, SNF1 and SNF4, known to control the derepression of glucose repressible genes (Carlson 1987), were identified to be involved in transcriptional regulation of the four genes CTA1, FOX2, FOX3 and PAS1 (M. Simon et al., 1992). These data suggest that defects in one of these trans-acting factors should lead to impaired peroxisome biogenesis. In fact, no peroxisomes could be detected in snf1- and snf4-mutant cells, and the peroxisomes observed in adr1-cells are small and not clustered like in wild type cells. Thus, it can be expected that genes encoding these and functionally similar proteins will be detected among the PAS genes. One example has already been found (PAS14 = SNF1, I. van der Leij et al., manuscript submitted) and with new mutant hunts using different screening procedures more such regulatory mutants might appear.

The results of these first few studies already suggest a very complex regulation of peroxisome biogenesis. The mode of action of the three trans-acting factors ADR1, SNF1 and SNF4 is different indicating the involvement of at least two separate signal pathways (M. Simon et al., 1992). It can be expected that more trans-acting factors will be identified which mediate specific and general metabolic regulations of genes encoding peroxisomal proteins. Moreover, the sequence of events leading to inheritance and proliferation of peroxisomes needs also a complicated regulatory network. The studies described in this section mark only the beginning of our knowledge concerning gene expression related to peroxisomes.

Concluding remarks

In the yeast *Saccharomyces cerevisiae* the application of two genetic selections (Erdmann et al. 1989; H.F. Tabak, pers. commun.) has resulted in the isolation of a large number of mutants with defects in peroxisomal assembly (pas-mutants) and/or peroxisomal functions (fox-mutants). Among the pas-mutants 18 different complementation groups have been identified at present. Their diverse biochemical and morphological phenotypes resemble many of the phenotypes reported for fibroblasts of patients with peroxisomal disorders. These disorders are a newly recognized class of human inborn errors with severe clinical symptoms (Lazarow & Moser 1989). In most cases they are lethal. It is tempting to assume that the phenotypical similarities between cells of fungal peroxisomal mutants and cells of patients with peroxisomal disorders reflect deficiencies in components conserved in yeast and man. The yeast mutants have led to the isolation of the first PAS-genes which encode proteins essential for the biogenesis of peroxisomes in *S. cerevisiae*. Experiments are under way to find their counterparts in higher eukaryotes.

Another important challenge in the study of peroxisome biogenesis will be to understand which step the distinct PAS-proteins catalyze. Elucidation of mechanisms underlying human diseases is often facilitated through model systems. All the studies summarized in this review suggest that *S. cerevisiae* is such an experimentally tractable system.

Acknowledgements

We are indebted to Jörg Höhfeld for fruitful discussions and helpful comments on the manuscript. We are also grateful to Marten Veenhuis for contributing the electron micrographs and to Jürgen Saidowsky for providing the analysis of peroxisomal fractions. We thank many colleagues for providing data prior to publication. The original research described in this review was supported by the Deutsche Forschungsgemeinschaft (grants Ku 329/14-4 and 11-4) and by the Austrian Fonds zur Förderung der wissenschaftlichen Forschung (grants P6711B and P7749B).

References

Alexson SEH, Fujiki Y, Shio H & Lazarow PB (1985) Partial disassembly of peroxisomes. J. Cell Biol. 101: 294–305

Avers CJ & Federman M (1968) The occurrence in yeast of cytoplasmic granules which resemble microbodies. J. Cell Biol. 37: 555–559

Baker D, Hicke L, Rexach M, Schleyer M & Schekman R (1988) Reconstitution of SEC gene product-dependent intercompartmental protein transport. Cell 54: 335–344

Bellion E & Goodman JM (1987) Proton ionophores prevent assembly of a peroxisomal protein. Cell 48: 165–173

Bermingham-McDonogh O, Gralla EB & Selverstone Valentine J (1988) The copper, zinc-superoxide dismutase gene of *Saccharomyces cerevisiae:* cloning, sequencing, and biological activity. Proc. Natl. Acad. Sci. USA 85: 4789–4793

Borst P (1989) Peroxisome biogenesis revisited. Biochim Biophys Acta 1008: 1–13

Botstein D & Fink GR (1988) Yeast: an experimental organism for modern biology. Science 240: 1439–1443

Carlson M (1987) Regulation of sugar utilization in Saccharomyces species. J. Bacteriol. 169: 4873–4877

Cohen G, Fessl F, Traczyk A, Rytka J & Ruis H (1985) Isolation of the catalase A gene of *Saccharomyces cerevisiae* by complementation of the cta1 mutation. Mol. Gen. Genet. 200: 74–79

Cohen G, Rapatz W & Ruis H (1988) Sequence of the *Saccharomyces cerevisiae* CTA1 gene and amino acid sequence of catalase A derived from it. Eur. J. Biochem. 176: 159–163

Cregg JM, van der Klei IJ, Sulter GJ, Veenhuis M & Harder W (1990) Peroxisome-deficient mutants of *Hansenula polymorpha*. Yeast 6: 87–97

Cross HS & Ruis H (1978) Regulation of catalase synthesis in *Saccharomyces cerevisiae* by carbon catabolite repression. Mol. Gen. Genet. 166: 37–43

Denis CL & Young ET (1983) Isolation and characterization of the positive regulatory gene ADR1 from *Saccharomyces cerevisiae* Mol. Cell. Biol. 3: 360–370

Desel H, Zimmermann R, Janes M, Miller F & Neupert W (1982) Biosynthesis of glyoxysomal enzymes in *Neurospora Crassa*. Ann. NY Acad. Sci. 386: 377–393

Didion T & Roggenkamr R (1990) Deficiency of peroxisome assembly in a mutant of the methylotrophic yeast *Hansenula polymorpha*. Curr. Genet. 17: 113–117

Distel B (1990) Import and assembly of peroxisomal proteins in yeast. PhD thesis, Amsterdam

Distel B, Veenhuis M & Tabak HF (1987) Import of alcohol oxidase into peroxisomes of *Saccharomyces cerevisiae*. EMBO J. 6: 3111–3116

Dmochowska A, Dignard D, Maleszka R & Thomas DY (1990) Structure and transcriptional control of the *Saccharomyces cerevisiae* POX1 gene encoding acyl-coenzyme A oxidase. Gene 88: 247–252

Duntze W, Neumann D, Gancedo JM, Atzpodien W & Holzer H (1969) Studies on the regulation and localization of the glyoxylate cycle enzymes in *Saccharomyces cerevisiae*. Eur. J. Biochem. 10: 83–89

Einerhand AW, Voorn-Brouwer TM, Erdmann R, Kunau WH & Tabak HF (1991) Regulation of transcription of the gene coding for peroxisomal 3-oxoacyl-CoA thiolase of *Saccharomyces cerevisiae*. Eur. J. Biochem. 200: 113–122

Erdmann R, Veenhuis M, Mertens D & Kunau WH (1989) Isolation of peroxisome-deficient mutants of *Saccharomyces cerevisiae*. Proc. Natl. Acad. Sci. USA 86: 5419–5423

Erdmann R, Wiebel FF, Flessau A, Rytka J, Beyer A, Fröhlich KU & Kunau WH (1991) PAS1, a yeast gene required for peroxisome biogenesis, encodes a member of a novel family of putative ATPases. Cell 64: 499–510

Fahimi HD & Sies H (1987) (Eds) Peroxisomes in Biology and Medicine. Springer-Verlag, Berlin

Fujiki Y & Lazarow PB (1985) Post-translational import of fatty acyl-CoA oxidase and catalase into peroxisomes of rat liver *in vitro*. J. Biol. Chem. 260: 5603–5609

Fujiki Y, Fowler S, Shio H, Hubbard AL & Lazarow PB (1982) Polypeptide and phospholipid composition of the membrane of rat liver peroxisomes: comparison with ER and mitochondrial membranes. J. Cell Biol. 93: 103–110

Garrard LJ & Goodman JM (1989) Two genes encode the major membrane-associated protein of methanol-induced peroxisomes from *Candida boidinii*. J. Biol. Chem. 264: 13929–13937

Gietl C (1990) Glyoxysomal malate dehydrogenase from watermelon is synthesized with an amino-terminal transit peptide. Proc. Natl. Acad. Sci. USA 87: 5773–5777

Gleeson MA & Sudbery E (1988) The methylotrophic yeasts. Yeast 4: 1–15

Glick B, Wachter C & Schatz G (1991) Protein import into mitochondria: two systems acting in tandem? Trends in Cell Biol. 1: 99–103

Gödecke A, Veenhuis M, Roggenkamp R, Janowicz ZA & Hollenberg CP (1989) Curr. Genet. 16: 13–20

Gödecke A, Veenhuis M, Roggenkamp R, Janowicz ZA & Hollenberg CP (1989) Biosynthesis of the peroxisomal dihydroxyacetone synthase from *Hansenula polymorpha* in *Saccharomyces cerevisiae* induces growth but not proliferation of peroxisomes. Curr. Genet. 16: 13–20

Goodman JM, Maher J, Silver PA, Pacifico A & Sanders D (1986) The membrane proteins of the methanol-induced peroxisome of *Candida boidinii:* initial characterization and generation of monoclonal antibodies. J. Biol. Chem. 261: 3464–3468

Goodman JM, Trapp SB, Hwang H & Veenhuis M (1990) Peroxisomes induced in *Candida boidinii* by methanol, oleic acid and D-alanine vary in metabolic function but share common integral membrane proteins. J. Cell Sci. 97: 193–204

Gould SJ, Keller GA & Subramani S (1987) Identification of a peroxisomal targeting signal at the carboxy terminus of firefly luciferase. J. Cell Biol. 105: 2923–2931

Gould SJ, Keller GA, Hosken N, Wilkinson J & Subramani S (1989) A conserved tripeptide sorts proteins to peroxisomes. J. Cell Biol. 108: 1657–1664

Gould SJ, Keller GA, Schneider M, Howell SH, Garrard LJ, Goodman JM, Distel B, Tabak HF & Subramani S (1990) Peroxisomal protein import is conserved between yeast, plants, insects and mammals. EMBO J. 9: 85–90

Hansen H & Roggenkamp R (1989) Functional complementation of catalase-defective peroxisomes in a methylotrophic yeast by import of catalase A of *Saccharomyces cerevisiae*. Eur. J. Biochem. 184: 173–179

Hardeman D, Versantvoort C, van den Brink JM & van den Bosch H (1990) Studies on peroxisomal membranes. Biochim. Biophys. Acta 1027: 149–154

Harder W & Veenhuis M (1989) Metabolism of one-carbon compounds. In: Rose AH & Harrison JS (Eds) The Yeasts, 2nd edition, Vol 3 (pp 289–316). Academic Press, London

Hartig A, Ogris M, Cohen G & Binder M (1990) Fate of highly expressed proteins destined to peroxisomes in *Saccharomyces cerevisiae*. Curr. Genet. 18: 23–27

Hartl FU, Ostermann J, Guiard B & Neupert W (1987) Successive translocation into and out of the mitochondrial matrix: targeting of proteins to the intermembrane space by a bipartite signal peptide. Cell 51: 1027–1037

Hashimoto T (1990) Purification, properties and biosynthesis of peroxisomal β-oxidation enzymes. In: Tanaka K & Coates PM (Eds) Fatty Acid Oxidation: Clinical, Biochemical and Molecular Aspects (pp 137–152). Alan R. Liss, Inc., New York

Hashimoto T, Kuwabara T, Usuda N & Nagata T (1986) Purification of membrane polypeptides of rat liver peroxisomes. J. Biochem. 100: 301–310

Hicke L & Schekman R (1990) Molecular machinery required for protein transport from the endoplasmic reticulum to the Golgi complex. BioEssays 12: 253–258

Hijikata M, Ishii N, Kagamiyama H, Osumi T & Hashimoto T (1987) Structural analysis of cDNA for rat peroxisomal 3-ketoacyl-CoA thiolase. J. Biol. Chem. 262: 8151–8158

Hiltunen JK, Wenzel B, Beyer A, Erdmann R, Fosså A & Kunau W-H (1992) Peroxisomal multifunctional β-oxidation protein of *Saccharomyces cerevisiae*. J. Biol. Chem. 267: 6646–6653

Höhfeld J, Veenhuis M & Kunau WH (1991) PAS3, a *Saccharomyces cerevisiae* gene encoding a peroxisomal integral membrane protein essential for peroxisome biogenesis. J. Cell Biol. 114: 1167–1178

Höhfeld J, Mertens D, Wiebel FF & Kunau WH (1992) Defining components required for peroxisome biogenesis in *Saccharomyces cerevisiae*. In: Neupert W & Lill R (Eds), New Comprehensive Biochemistry Series, Vol 22, December 1992. Membrane Biogenesis and Protein Targeting, Elsevier Science Publishing Co., New York.

Imanaka T, Small GM & Lazarow PB (1987) Translocation of acyl-CoA oxidase into peroxisomes requires ATP hydrolysis but not a membrane potential. J. Cell Biol. 105: 2915–2922

Kamiryo T, Parthasarathy S & Numa S (1976) Evidence that acyl coenzyme A synthetase activity is required for repression of yeast acetyl coenzyme A carboxylase by exogenous fatty acids. Proc. Natl. Acad. Sci. USA 73: 386–390

Keller GA, Warner TG, Steimer KS & Hallewell RA (1991) Cu,Zn superoxide dismutase is a peroxisomal enzyme in human fibroblasts and hepatoma cells. Proc. Natl. Acad. Sci USA 88: 7381–7385

Kindl H & Lazarow PB (1982) (Eds) Glyoxysomes and peroxisomes, Ann. NY Acad. Sci. 386, New York

Kirsch T, Löffler HG & Kindl H (1986) Plant acylCoA-oxidase: purification, characterization, and monomeric apoprotein. J. Biol. Chem. 261: 8570–8575

Klionsky DJ, Herman PK & Emr SD (1990) The fungal vacuole: composition, function, and biogenesis. Microbiol. Rev. 54: 266–292

Kunau WH, Kionka C, Ledebur A, Mateblowski M, Moreno de la Garza M, Schultz-Borchard U, Thieringer R & Veenhuis M (1987) β-oxidation systems in eukaryotic microorganisms. In: Fahimi HD & Sies H (Eds) Peroxisomes in Biology and Medicine (pp 128–140). Springer-Verlag, Berlin

Kunau WH, Bühne S, Moreno de la Garza M, Kionka C, Mateblowski M, Schultz-Borchard U & Thieringer R (1988) Comparative enzymology of β-oxidation. Biochem. Soc. Transactions 16: 418–420

Lageweg W, Tager JM & Wanders RJA (1991) Topography of very-long-chain-fatty-acid-activating activity in peroxisomes from rat liver. Biochem. J. 276: 53–56

Lazarow PB (1984) The peroxisomal membrane. In: Büler E (Ed) Membrane Structure and Function, Vol 5 (pp 1–31). John Wiley & Sons, Inc. New York

Lazarow PB & Fujiki Y (1985) Biogenesis of peroxisomes. Ann. Rev. Cell Biol. 1: 489–530

Lazarow PB & Moser HW (1989) Disorders of peroxisome biogenesis. In: Scriver CR, Beaudet AL, Sly WS & Valle D (Eds) The Metabolic Basis of Inherited Disease, 6th ed (pp 1479–1509). McGraw-Hill, Inc. New York

Lewin AS, Hines V & Small GM (1990) Citrate synthase encoded by the CIT2 gene of *Saccharomyces cerevisiae* is peroxisomal. Mol. Cell. Biol. 10: 1399–1405

Lüers G, Beier K, Hashimoto T, Fahimi HD & Völkl A (1990) Biogenesis of peroxisomes: sequential biosynthesis of the membrane and matrix proteins in the course of hepatic regeneration. Eur. J. Cell Biol. 52: 175–184

McCammon MT, Veenhuis M, Trapp SB & Goodman JM (1990a) Association of glyoxylate and beta-oxidation enzymes with peroxisomes of *Saccharomyces cerevisiae*. J. Bacteriol. 172: 5816–5827

McCammon MT, Dowds CA, Orth K, Moomaw CR, Slaughter CA & Goodman JM (1990b) Sorting of peroxisomal membrane protein PMP47 from *Candida boidinii* into peroxisomal membranes of *Saccharomyces cerevisiae*. J. Biol. Chem. 265: 20098–20105

Minard KI & McAllister-Henn L (1991) Isolation, nucleotide sequence analysis, and disruption of the MDH2 gene from *Saccharomyces cerevisiae:* evidence for three isozymes of yeast malate dehydrogenase. Mol. Cell. Biol. 11: 370–380

Mitraki A & King J (1989) Protein folding intermediates and inclusion body formation. Biotechnology 7: 690–697

Miyazawa S, Osumi T, Hashimoto T, Ohno K, Miura S & Fujiki Y (1989) Peroxisome targeting signal of rat liver acyl-coenzyme A oxidase resides at the carboxy terminus. Mol. Cell. Biol. 9: 83–91

Mori T, Tsukamoto T, Mori H, Tashiro Y & Fujiki Y (1991) Molecular cloning and deduced amino acid sequence of nonspecific lipid transfer protein (sterol carrier protein 2) of rat liver: a higher molecular mass (60 kD) protein contains the primary sequence of nonspecific lipid transfer protein as its C-terminal part. Proc. Natl. Acad. Sci. USA 88: 4338–4342

Neupert W, Hartl FU, Craig E & Pfanner N (1990). How do polypeptides cross the mitochondrial membranes? Cell 63: 447–450

Newman AP & Ferro-Novick S (1990) Defining components required for transport from the ER to the Golgi complex in yeast. BioEssays 12: 485–491

Numa S (1981) Two long-chain acylCoenzyme A synthetases: their different roles in fatty acid metabolism and its regulation. Trends in Biochem. Sci 6: 113–115

Osumi T & Fujiki Y (1990) Topogenesis of peroxisomal proteins. Bio Essays 12: 217–222

Parish RW (1975) The isolation and characterization of peroxisomes (microbodies) from baker's yeast, *Saccharomyces cerevisiae*. Arch. Microbiol. 105: 187–192

Pfanner N, Söllner T & Neupert W (1991) Mitochondrial import receptors for precursor proteins. Trends in Biochem. Sci. 16: 63–67

Pringle JR, Preston RA, Adams AEM, Drubin DG, Haarer BK & Jones EW (1989) Fluorescence microscopy methods for yeast. In: Tartakoff AM (Ed) Methods in Cell Biology, Vol 31 (pp 357–435). Academic Press, New York

Rose AH & Harrison JS (1987–1991) (Eds) The Yeasts, 2nd edition, Vol 1–4. Academic Press, London

Rosenkrantz M, Alam T, Kim KS, Clark BJ, Srere PA & Guarente LP (1986) Mitochondrial and nonmitochondrial citrate synthases in *Saccharomyces cerevisiae* are encoded by distinct homologous genes. Mol. Cell. Biol. 6: 4509–4515

Rothman JH, Yamashiro CT, Kana PM & Stevens TH (1989) Protein targeting to the yeast vacuole. Trends in Biochem. Sci. 14: 347–350

Rusnak N, Shackelford J, Gould SJ, Subramani S & Krisans SK (1990) Recognition of a peroxisomal form of HMG-CoA reductase protein by an antibody that detects a peroxisomal targeting signal. J. Cell Biol. 111: 195a

Rytka J, Sledziewski A, Lukaszkiewicz J & Bilinski T (1978) Haemoprotein formation in yeast III. The role of carbon catabolite repression in the regulation of catalase A and T formation. Mol. Gen. Genet. 160: 51–57

Sakuraba H, Fujiwara S & Noguchi T (1991) Purification and characterization of peroxisomal apo and holo alanine: glyoxylate aminotransferase from bird liver. Arch. Biochem. Biophys 286: 453–460

Santos MJ, Imanaka T, Shio H, Small GM & Lazarow PB (1988) Peroxisomal membrane ghosts in Zellweger Syndrome – aberrant organelle assembly. Science 239: 1536–1538

Sengstad C, Stirling C, Schekman R & Rine J (1990) Genetic and biochemical evaluation of eucaryotic membrane protein topology: multiple transmembrane domains of *Saccharomyces cerevisiae* 3-Hydroxy-3-Methylglutaryl Coenzyme A reductase. Mol. Cell. Biol. 10: 672–680

Silver PA (1991) How proteins enter the nucleus. Cell 64: 489–497

Simon M, Adam G, Rapatz W, Spevak W & Ruis H (1991) The *Saccharomyces cerevisiae* ADR1 gene is a positive regulator of transcription of genes encoding peroxisomal proteins. Mol. Cell. Biol. 11: 699–704

Simon M, Binder M, Adam G, Hartig A & Ruis H (1992) Control of peroxisome proliferation in *Saccharomyces cerevisiae* by ADR1, SNF1 (CAT1, CCR1) and SNF4 (CAT3). Yeast 8: 303–309

Skoneczny M, Chelstowska A & Rytka J (1988) Study of the coinduction by fatty acids of catalase A and acyl CoA oxidase in standard and mutant *Saccharomyces cerevisiae* strains. Eur. J. Biochem. 174: 297–302

Small GM, Imanaka T, Shio H & Lazarow PB (1987) Efficient association of *in vitro* translation products with purified, stable *Candida tropicalis* peroxisomes. Mol. Cell. Biol. 7: 1848–1855

Small GM, Szabo LJ & Lazarow PB (1988) Acyl-CoA oxidase contains two targeting sequences each of which can mediate protein import into peroxisomes. EMBO J. 7: 1167–1173

Spevak W, Fessl F, Rytka J, Traczyk A, Skoneczny M & Ruis H (1983) Isolation of the catalase T structural gene of *Saccharomyces cerevisiae* by functional complementation. Mol. Cell. Biol. 3: 1545–1551

Sulter GJ, Looyenga L, Veenhuis M & Harder W (1990) Occurence of peroxisomal membrane proteins in methylotrophic yeasts grown under different conditions. Yeast 6: 35–43

Susani M, Zimniak P, Fessl F & Ruis H (1976) Localization of catalase A in vacuoles of *Saccharomyces cerevisiae:* evidence for the vacuolar nature of isolated 'yeast peroxisomes'. Hoppe-Seyler's Z. Physiol. Chem. 357: 961–970

Swinkels BW, Gould SJ, Bodnar AG, Rachubinski RA & Subramani S (1990) Identification of a novel peroxisomal targeting signal in the amino-terminal prepiece of 3-ketoacyl thiolase. J. Cell Biol. 111, 386a

Szabo AS & Avers CJ (1969) Some aspects of regulation of peroxisomes and mitochondria in yeast. Ann. NY Acad. Sci. 168: 302–312

Szabo LJ, Small GM & Lazarow PB (1989) The nucleotide sequence of POX18, a gene encoding a small oleate-inducible peroxisomal protein from *Candida tropicalis.* Gene 75: 119–126

Takada Y & Noguchi T (1985) Characteristics of alanine: glyoxylate aminotransferase from *Saccharomyces cerevisiae,* a regulatory enzyme in the glyoxylate pathway of glycine and serine biosynthesis from tricarboxylic acid-cycle intermediates. Biochem. J. 231: 157–163

Tan H, Okazaki K, Kubota I, Kamiryo T & Utiyama H (1990) A novel peroxisomal nonspecific lipid-transfer protein from *Candida tropicalis.* Gene structure, purification and possible role in β-oxidation. Eur. J. Biochem. 190: 107–112

Tanaka A & Fukui S (1989) Metabolism of n-alkanes. In: Rose AH & Harrison JS (Eds) The Yeasts, 2nd edition, Vol 3 (pp 261–287). Academic Press, London

Thieringer R, Shio H, Han Y, Cohen G & Lazarow PB (1991) Peroxisomes in *Saccharomyces cerevisiae:* immunofluorescence analysis and import of catalase A into isolated peroxisomes. Mol. Cell. Biol. 11: 510–522

Tsukamoto T, Miura S & Fujiki Y (1991) Restoration by a 35 k membrane protein of peroxisome assembly in a peroxisome-deficient mammalian cell mutant. Nature 350: 77–81

van der Klei IJ, Harder W & Veenhuis M (1991) Selective inactivation of alcohol oxidase in two peroxisome deficient mutants of the yeast *Hansenula polymorpha.* Yeast (in press)

Veenhuis M & Goodman JM (1990) Peroxisomal assembly: membrane proliferation precedes the induction of the abundant matrix proteins in the methylotrophic yeast *Candida boidinii.* J. Cell Sci. 96: 583–590

Veenhuis M & Harder W (1991) Microbodies. In: Rose AH & Harrison JS (Eds) The Yeasts, 2nd edition, Vol 4 (pp 601–653). Academic Press, London

Veenhuis M, Mateblowski M, Kunau WH & Harder W (1987) Proliferation of microbodies in *Saccharomyces cerevisiae.* Yeast 3, 77–84

Wanner G & Theimer RR (1982) Two types of microbodies in *Neurospora crassa.* Ann. NY Acad. Sci. 386: 269–284

Yamada T, Nawa H, Kawamoto S, Tanaka A & Fukui S (1980) Subcellular localization of long-chain alcohol dehydrogenase and aldehyde dehydrogenase in n-alkane grown *Candida tropicalis.* Arch. Microbiol. 128: 145–151

Yoo HS & Cooper TG (1989) The DAL7 promoter consists of multiple elements that cooperatively mediate regulation of the gene's expression. Mol. Cell. Biol. 9: 3231–3243

Antonie van Leeuwenhoek **62**: 79–93, 1992.

Heterologous protein production in yeast

Gerd Gellissen, Karl Melber, Zbigniew A. Janowicz, Ulrike M.Dahlems, Ulrike Weydemann, Michael Piontek, Alexander W.M. Strasser & Cornelis P. Hollenberg[1]
Rhein Biotech GmbH, Eichsfelder Str. 11, 4000 Düsseldorf 13, Germany; [1]Institut für Mikrobiologie, Heinrich-Heine-Universität, Universitätsstr. 1, 4000 Düsseldorf 1, Germany

Key words: heterologous gene expression, non-*Saccharomyces* yeast, *Saccharomyces cerevisiae*, secretion, protein production

Abstract

The exploitation of recombinant DNA technology to engineer expression systems for heterologous proteins represented a major task within the field of biotechnology during the last decade. Yeasts attracted the attention of molecular biologists because of properties most favourable for their use as hosts in heterologous protein production. Yeasts follow the general eukaryotic posttranslational modification pattern of expressed polypeptides, exhibit the ability to secrete heterologous proteins and benefit from an established fermentation technology. Aside from the baker's yeast *Saccharomyces cerevisiae*, an increasing number of alternative non-*Saccharomyces* yeast species are used as expression systems in basic research and for an industrial application.

In the following review a selection from the different yeast systems is described and compared.

Introduction

The biotechnological potential of unicellular microorganisms and eukaryotic cells opened opportunities for application in protein production. An extensive knowledge of bacterial genetics and the rapidly growing progress in molecular biology provided the tools for an easy manipulation of the bacterium *Escherichia coli*. In 1977, a human somatostatin gene was expressed in this organism representing the first commercial example of a biologically active protein from a eukaryotic source (Itakura et al. 1977). In addition, fermentation to high cell densities provided the possibility to produce rare eukaryotic proteins in quantities sufficient for the establishment of industrial isolation and downstream processing procedures. The disadvantage of the *E. coli* system is the necessity to renature the heterologous polypeptides, since in most cases this organism was found to be unable to produce the proteins in a properly folded form. Moreover, the use of many complex mammalian proteins in medical therapy depends on a faithful addition of structures such as carbohydrate chains to the translated polypeptides to avoid severe therapy-associated complications. Since the prokaryotic *E. coli* host is unable to synthesize such eukaryotic structures, mammalian cells serve as hosts for complex recombinant pharmaceuticals like erythropoietin (EPO; Lai et al. 1986; Müller-Wiefel et al. 1988), tissue plasminogen activator (Robinson & Browne 1991) and factor VIII (Wood et al. 1984; Kaufman et al. 1988) which is currently under clinical trial for treatment of classic hemophilia. In addition, expression systems based on animal tissue cultures assure a proper conformation of the proteins.

Being non-pathogenic microorganisms with established applications in agriculture and food industry, several yeast species combine the ease of genetic manipulation and favourable fermentation properties with the advantage of being true eukaryotes with the ability to modify proteins in a suitable manner. Therefore, in many instances yeasts became preferred hosts for heterologous gene expression after techniques were developed to introduce exogenous DNA into the cells (Beggs 1978; Hinnen et al. 1978; Klebe et al. 1983; Dohmen et al. 1991). The best characterized species *Saccharomyces cerevisiae* was the first to be used for this aspect of biotechnology (for additional reading see some recent reviews by Kingsman et al. 1988 and Hinnen et al. 1989) leading to early successes in protein production, including human α-interferon (Hitzeman et al. 1981), hepatitis B surface antigens (Valenzuela et al. 1982; Harford et al. 1983), and enzymes like calf prochymosin (Mellor et al. 1983) and *Aspergillus* glucoamylase (Innis et al. 1985). This limitation to a single species is now extended by an increasing number of alternative non-*Saccharomyces* yeasts exhibiting particular favourable performances as production organisms for certain heterologous proteins (Reiser et al. 1990; Buckholz & Gleeson 1991).

Protein production in *S. cerevisiae*

The wealth of information about this traditional industrial organism and the availability of cloned genetic material led to the early development of *S. cerevisiae* as an eukaryotic host for heterologous protein production (for review see Sturley & Young 1988; Kingsman et al. 1988). Recombinant protein expression was first described in 1981 for human α-interferon (Hitzeman et al. 1981) followed by many other examples in the subsequent years. The majority of recombinant proteins expressed in yeasts until now, are *Saccharomyces*-derived compounds including hepatitis B surface antigens (see also the following sections; Valenzuela et al. 1982) and insulin (Stepien et al. 1983; Thim et al. 1986), whose production has been licensed by FDA for administration to humans and which are now world-wide available. A summary of the eukaryotic proteins expressed in this organism is given by Kingsman et al. 1988; see also Table 2. The performance of *S. cerevisiae* as a system for gene transfer and expression has extensively been reviewed during the recent past (see for example articles by different authors in books edited by Russell 1988 and by Barr et al. 1989). The aim of the following section is a brief summary of selected aspects to enable a comparison with other yeast species described in this article.

In the first transformations of yeast strains enzymic removal of the cell wall was employed. The sphaeroplasts took up DNA aided by calcium ions and polyethylene glycol (Beggs 1978; Hinnen et al. 1979). Subsequently, methods have been described in which protoplast formation is not required (Klebe et al. 1983; Dohmen et al. 1991). After uptake of DNA (and regeneration of the cell walls in the first method) recombinant strains were selected against non-recombinant host strains by genetic markers present on the recombinant plasmid. A variety of auxotrophic markers and dominant selectable markers which confer resistance to a specific drug are available now (for review see Reiser et al. 1990).

Five basic types of plasmid are used for transformation (for review see Sturley & Young 1988). In vectors of the YIp type (yeast integration plasmid), integration occurs by homologous recombination of complementation genes contained on the plasmid at the respective mutant loci of suitable auxotrophic acceptor strains. The first transformed yeast resulted from such a homologous recombination at the *leu2* mutant locus of a leucine-auxotrophic strain (Hinnen et al. 1978), other successful transformations have been achieved using either *HIS3* or *LEU2* for integration (Struhl et al. 1979). Complex plasmids encompassing several yeast genes can be targeted to a specific locus of the genome by linearizing them within the appropiate region selected for integration (Orr-Weaver et al. 1981; Bothstein & Shortle 1985). Integrative transformants are extremely stable and therefore often used as production strains containing or producing a heterologous protein. Since integrative transformation is a relatively rare event (1–10 transformants/μg DNA), autonomously replicating vectors have been designed for higher transformation frequencies. After transformation, these vectors of the YRp type (yeast replication plasmid) remain in

Table 1. Selection of promoters used for expression of heterologous proteins in *S. cerevisiae*.

Promoter	Reference
ADH1	Bennetzen & Hall 1982; Hitzeman et al. 1981
CUP1	Karin et al. 1984
ENO	Holland et al. 1981
GAL1/GAL10	Johnston & Davis 1984
GAP3	Holland & Holland 1980; Edens et al. 1984
HSP90	Finkelstein & Strausberg 1983
MFα1	Kurjan & Herskowitz 1982; Brake et al. 1984
PDC1	Schmitt et al. 1983; Kellermann et al. 1986
PGK	Dobson et al. 1982; Tuite et al. 1982
PHO5	Meyhack et al. 1982; Kramer et al. 1984
PYK	Burke et al. 1983
TRP1	Dobson et al. 1982

an extra-chromosomal state due to the presence of a sequence for autonomous replication (*ARS*). They are present in 3–20 copies in recombinant strains, but are found to be very unstable due to abnormal segregation properties at mitosis or meiosis (Struhl et al. 1979). Plasmid stability can be gained by inclusion of a further yeast centromeric sequence leading to so-called YCp (yeast centromeric plasmid) vectors (Clarke & Carbon 1980). The addition of telomeric structures in YAC (yeast artificial chromosome) lead to linear vectors which can harbour long DNA fragments of up to 200 kb in seize (Burke et al. 1987).

The most commonly used vectors in yeast genetic engineering are derived from the 2 μm plasmid, termed YEp (yeast episomal plasmid) vectors (Beggs 1978). They can transform yeast at a frequency of 5000–20000 recombinants per μg DNA (Struhl et al. 1979). Transformants contain an average of 40 copies per cell and exhibit a higher stability as compared to YRp-transformed strains. All plasmids contain the 2 μm *ARS* sequence plus *STB* – locus as minimal 2 μm – derived component enabling a stable replication independent of the yeast chromosome (Kojo et al. 1981).

For expression, a heterologous gene is fused to a promoter element obtained from a highly expressed *S. cerevisiae* gene, in many cases that of a glycolytic enzyme. In Table 1 a selection of available promoter elements is presented.

Correct termination of the heterologous transcript is necessary and provided by a termination sequence also derived from a yeast gene (*ADH1*, *CYC1* etc.) and present in a DNA fragment 3′ to the gene to be expressed.

The heterologous protein can be secreted if

Table 2. Examples of heterologous eukaryotic proteins expressed in *S. cerevisiae* (modified after Kingsman et al. 1988).

Protein	Promoter	References
Human IFN-α	*ADH1*	Hitzeman et al. 1981
	PGK	Tuite et al. 1982; Hitzeman et al. 1983
	TRP1	Dobson et al. 1982
	PHO5	Hinnen et al. 1983; Kramer et al. 1984
	MFα	Bitter et al. 1984
Human IFN-γ	*PGK*	Derynck et al. 1983
HBsAg	*ADH1*	Valenzuela et al. 1982
	PGK	Hitzeman et al. 1983
	ARG3	De Wilde et al. 1985
Calf chymosin	*PGK*	Mellor et al. 1983
	GAL1	Goff et al. 1984
Insulin	*GAL1*	Stepien et al. 1983
Human α1-antitrypsin	*PHO5*	Rosenberg et al. 1984
	ARG3	Cabezon et al. 1984
Human EGF	*MFα*	Brake et al. 1984
β-endorphin	*MFα*	Bitter et al. 1984
Wheat α-amylase	*PGK*	Rothstein et al. 1984
Plant thaumatin	GAP	Edens et al. 1984

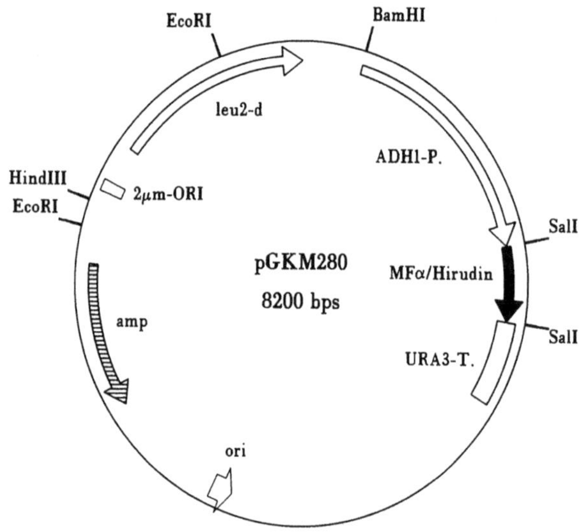

Fig. 1. Physical map of *S. cerevisiae* 2 μm-expression vector, pGKM280, designed for secretion of hirudin. A *SalI*-fragment encoding the hirudin gene fused to the *MFα* signal sequence is inserted into a multiple cloning site of the basic expression vector. This vector is composed of the following elements: ori and ampicillin-resistance (amp) for selection and propagation in *E. coli*, 2 μm-ORI for autonomous replication in *S. cerevisiae*, the *LEU2d* selection marker, the *ADH1*-P(romoter) and the terminator of the *URA3* gene (*URA3*-T) separated by a multiple cloning site for insertion of the gene to be expressed.

fused to a suitable signal sequence from secreted *S. cerevisiae* proteins. The three commonly used elements obtained from mating factor (*MFα*), acid phosphatase (*PHO5*) and invertase (*SUC2*) are briefly discussed in the section 'comparative aspects.'

A 2 μm-derived vector designed for the transformation of a *leu2*-deficient *S. cerevisiae* host strain with the ability to secrete hirudin is shown in Fig. 1.

Protein production in methylotrophic yeasts

Only a few yeast species can use methanol as a sole carbon and energy source. These methylotrophs belong to the genera *Candida*, *Torulopsis*, *Pichia* and *Hansenula* (Lee & Komagata 1980). Two representatives of this group, the ascomycetous yeasts *P. pastoris* and *H. polymorpha* have been studied extensively during the recent past (for review see Gleeson & Sudbery 1988) and have been developed as production organisms for heterologous proteins (Roggenkamp et al. 1986; Cregg et al. 1987; Gellissen et al. 1990; Buckholz & Gleeson 1991).

In both cases gene expression is based on fusion of heterologous gene sequences to strong methanol-inducible promoter structures originating from genes encoding key enzymes of methanol metabolism. Upon addition of methanol into the culture medium these enzymes are rapidly induced to constitute the bulk proteins within the cell. Their synthesis is repressed in cells grown on glucose.

The first step of the methanol pathway is catalyzed by alcohol oxidase (methanol oxidase) converting methanol to formaldehyde (Sahm 1977; Anthony 1982; Veenhuis et al. 1983). The genome of *P. pastoris* contains two highly homologous methanol oxidase genes designated *AOX1* and *AOX2* (Ellis et al. 1985; Koutz et al. 1989); from *H. polymorpha*, a single-copy methanol oxidase (*MOX*) gene has been isolated and characterized (Ledeboer et al. 1985). In both systems, expression of these genes is regulated primarily at the transcriptional level (Roggenkamp et al. 1984; Ellis et al. 1985). From *Hansenula*, two further inducible genes of the pathway encoding formate dehydrogenase (*FMD*) and dihydroxyacetonsynthase (*DAS*) have been cloned and analyzed (Janowicz et al. 1985; Hollenberg & Janowicz 1987). Promoter sequences which confer the repression-induction mechanism described above are used as components of expression vectors. In *Pichia*, the *AOX* promoter, in *Hansenula*, *MOX* and *FMD* promoters have been successfully used for protein production. Expression studies suggest differences between the regulatory mechanism of methanol pathway genes between the two methylotrophic organisms. In *Pichia*, methanol is the only carbon source to elicit high levels of methanol-inducible proteins (Tschopp et al. 1987). In a routine fermentation the cells are first grown to high cell densities on glucose or glycerol. Methanol is subsequently added to the medium to induce the production of a foreign protein. In *Hansenula*, high-level expression of these proteins is also observed in glycerol-grown cells providing an example of a repression/derepression mechanism (Z.A. Janowicz, unpublished results). Especially *FMD* promoter-driven production allows culture of recombinant strains on a mixture of methanol and glycerol or on glycerol alone avoiding a two-step fermentation with different carbon sources and resulting in a high speed of production.

Vectors for transformation contain nutritional markers to complement deficiencies in auxotrophic mutants. In *Pichia*, mutants defective in histidinol dehydrogenase are transformed with plasmids carrying either the homologous or a heterologous *HIS4* gene (Cregg et al. 1985); in *Hansenula*, transformation is achieved by complementing *URA3* (Roggenkamp et al. 1986) or *LEU2* auxotrophic mutants (Gleeson et al. 1986).

Expression vectors can be designed for an intracellular expression or for secretion of the heterologous protein. In both, *Pichia* and *Hansenula*, secretion signal sequences from different yeast and mammalian sources have been found to function efficiently (see Table 3 and the section 'comparative aspects').

In *Pichia*, recombinant strains are routinely obtained by integrating the foreign DNA into one of the two *AOX* genes or into the *HIS4* gene by homologous recombination (Sreekrishna et al. 1989; Buckholz & Gleeson 1991). This results in the integration of a single copy of the heterologous sequence. Recently, multiple integrated copies of a DNA fragment could be identified from transformants expressing a tetanus toxin fragment (Clare et al. 1991).

In *Hansenula*, one observes a high frequency of heterologous integration of the plasmid DNA. Transformations usually result in mitotically stable strains containing different copies of an intact integrated expression cassette; strains with up to hundred copies have been identified (Gellissen et al. 1991; Janowicz et al. 1991). Both, *Pichia* and *Hansenula*, exhibit the properties of an efficient production system for many proteins analyzed so far (see Table 3). They can be cultured to cell masses up to 130 g dry weight per liter (Vedvick et al. 1991; Gellissen et al. 1991).

H. polymorpha has recently been shown to be an ideal host for the expression of complex composite

Table 3. Expression of heterologous proteins in methylotrophic yeasts.

	Protein	Promoter	Reference
P. pastoris	HBsAg (S)*	*AOX1*	Cregg et al. 1987
	human EGF⁺		Siegel et al. 1989
	bovine lysozyme⁺		Digan et al. 1989
	streptokinase*		Hagenson et al. 1989
	human TNF*		Sreekrishna et al. 1989
	aprotinin⁺		Vedvick et al. 1991
	tetanus C*		Clare et al. 1991
	HSA⁺		S. Provow cited in Buckholz & Gleeson 1991
	human IGF1⁺		R. Brierley cited in Buckholz & Gleeson 1991
	Il-2⁺		G. Velicelebi cited in Buckholz & Gleeson 1991
	HIV gp120⁺		M. Romanos cited in Buckholz & Gleeson 1991
H. polymorpha	HBsAg (M)⁺	*MOX*	Shen et al. 1989
	HBsAg (S,S/L)*		Janowicz et al. 1991
	HSA⁺		Hodgkins et al. 1990
	α-galactosidase⁺		Fellinger et al. 1991
	A. niger glucose oxidase⁺		P. Sudberry cited in Buckholz & Gleeson, 1991
	human lipase⁺		P. Sudbery cited in Buckholz & Gleeson, 1991
	HBsAg (S)*	*FMD*	Janowicz et al. 1991
	glucoamylase⁺		Gellissen et al. 1991
	invertase⁺		Janowicz et al. 1988
	hirudin⁺		U. Weydemann, pers. comm.

* Intracellular expression; ⁺ Protein secreted into culture medium.

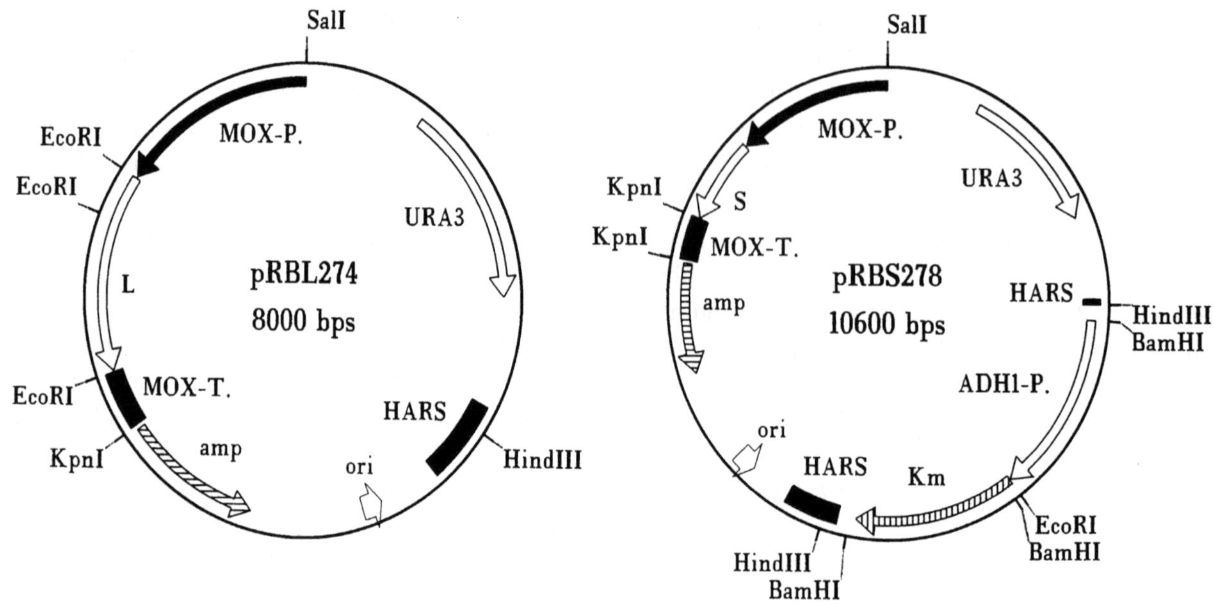

Fig. 2. Physical maps of *H. polymorpha* expression vectors, pRBL274 and pRBS278, designed for construction of strains producing HBsAg(S/L) particles. The following strategy was employed: First, *H. polymorpha* host strain was transformed with pRBL274, containing a L-antigen expression cassette (*MOX*-P(romoter)- L-gene - *MOX*-T(erminator)), HARS sequence for autonomous replication, *URA3* gene from *S. cerevisiae* as selection marker, ori and amp for selection and propagation in *E. coli*. In a second step, transformants with integrated L-cassettes showing high levels of L-antigen expression were transformed with pRBS278. pRBS278 contains a S-antigen expression cassette (*MOX*-P.- S-gene -*MOX*-T.), the HARS sequence, *URA3* gene and a Tn5 gene (Km) confering resistance against G418 under control of the *ADH1*-P(romoter), ori and amp sequences. After selection of supertransformants with G418 strains were obtained which express S- and L-antigens at different levels and ratios depending on the copy number of integrated expression cassettes of each respective antigen (Janowicz et al. 1991).

structures such as polyvalent vaccines. This yeast seems to be able to secrete immunogenic 22 nm particles composed of hepatitis B middle surface antigens, since recombinant proteins could be recovered from cell supernatants after treatment of the cell wall with β-glucanase (Shen et al. 1989). The engineering of strains enabled to co-express S and L surface antigens of the hepatitis B virus (HBV) intracellularly and to form mixed particles composed of the two antigens has been published recently (Janowicz et al. 1991) and is reviewed in the following to demonstrate the quality of *Hansenula* as expression system.

Recombinant *H. polymorpha* strains expressing composite mixed particles, derived from hepatitis B surface antigens

The envelope of hepatitis B virus consists of a structure composed of a major viral protein, S antigen, and two related minor proteins, the M and L antigens. S antigen has been successfully expressed in *S. cerevisiae* (Valenzuela et al. 1982) and in *P. pastoris* (Cregg et al. 1987). The recombinant protein is assembled *in vivo* into highly immunogenic 22nm particles similar to those observed in the serum of hepatitis B patients. Vaccines formulated with *S. cerevisiae*-derived hepatitis B surface antigens (HBsAg) with S-specific epitopes are used for formulation of a commercial vaccine since 1986 (Emini et al. 1986; Petre et al. 1987). Studies have shown that additional important protective epitopes are present in the L-antigen which are excluded from the present recombinant vaccines (Neurath et al. 1987). Therefore expression strains were constructed with a balanced, simultaneous production of L and S antigens in form of composite, mixed particles which mimic the composition of the natural viral coat.

Strains containing various copy numbers of both, the L- and S expression cassette, were constructed

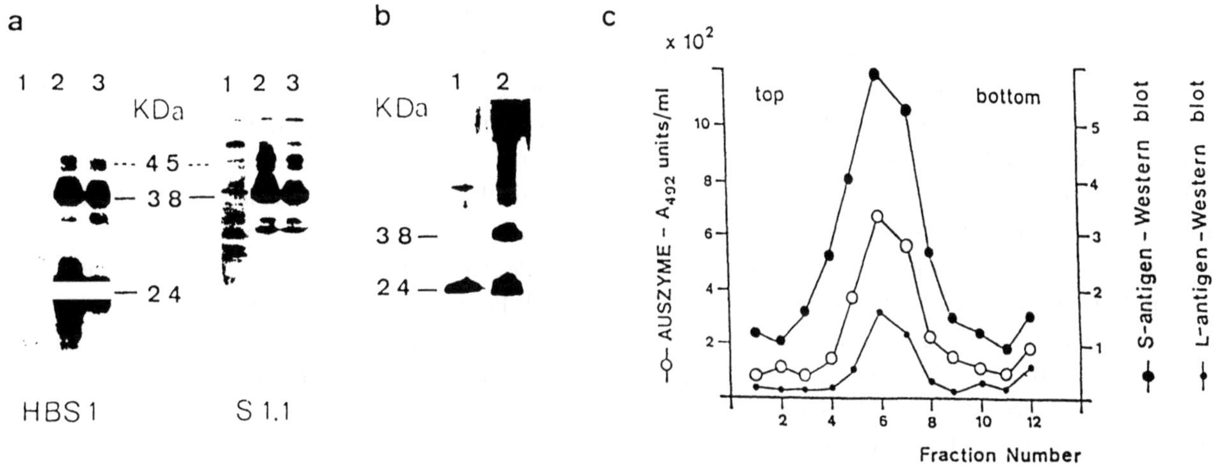

Fig. 3. Characterization of S and L hepatitis B surface antigens (HBsAg) produced in *H. polymorpha.* (a) Immunoblot analysis of protein crude extracts from strains expressing S-antigen only (lane 1) and S- and L-antigens simultaneously (lanes 2 and 3). Duplicate blots were reacted to either mAb specific for a S-epitope (HBS1) or mAb specific for a L-epitope (S1.1). S-antigen is detected as a 24 kDa protein while the 38 kDa protein represents L-antigen. A small portion of L-antigen (less than 5%) is produced as a glycosylated 45 kDa form. (b) Silver-stained PAGE/SDS-gel of purified HBsAg from a strain expressing S-antigen only (lane 1) and a strain expressing S- and L-antigens simultaneously (lane 2). (c) Formation of composite S/L-antigen particles. Crude extract from a strain expressing both S- and L-antigens was subjected to CsCl density gradient centrifugation. Gradient fractions were assayed by immunoblot analysis for presence of S- and L- antigens and by EIA specific for particulate HBsAg structures (AUSZYME, Abbott). Correlating peaks of antigenic material were observed demonstrating formation of highly immunogenic composite S/L-particles.

as shown in Fig. 2. Different recombinant strains were obtained with fixed L/S expression ratios ranging from 1:1 to 1:15. Centrifugation analysis of particles and immunological experiments demonstrated that the produced proteins were assembled into particles now containing both antigens (Fig. 3). During fermentation, the ratio of both antigens remained constant due to stable integration of both expression cassettes (Janowicz et al. 1991).

Protein production in *Kluyveromyces lactis* and other non-*Saccharomyces* yeast hosts

The adaptation of metabolism to efficiently utilize a special carbon source as has been described for the methylotrophs in the previous section is also observed in the budding yeast, *K. lactis*. Like other related species, *K. lactis* is able to grow on lactose as sole energy and carbon source. During growth on lactose-containing substrates like whey, enzymes of a lactose utilization pathway are strongly induced (for review see Phaff 1985). Lactose is taken up by a lactose permease (*LAC12*) and is then converted into glucose and galactose by an intracellular β-galactosidase which is encoded by the *K. lactis LAC4* gene (Dickson & Markin 1978). Again, having the strong lactose-inducible *LAC4* promoter in hand, a most promising component for heterologous gene expression became available.

The possibility to introduce foreign DNA into *K. lactis* strains has been described some ten years ago when vectors containing a bacterial kanamycin resistance gene as dominant selection marker were used for transformation (Das & Hollenberg 1982). In the presence of a *S. cerevisiae ARS* sequence on the plasmid a heterologous integration into the host's genome was observed. When a genuine sequence for autonomous replication (*KARS*) was present, vectors remained in a plasmid state. Because of possible strain instability under non-selective conditions, a chymosin production system was developed based on integrated heterologous DNA targeted to the lactase gene (van den Berg et al. 1990). Chymosin is an enzyme of commercial importance, since it represents the active constituent of cheese rennet. On the plasmid, a kanamycin-resistance gene expressed from a *S. cerevisiae*

ADH1 promoter (Reiss et al. 1984) served as dominant marker for G418 resistance. A prochymosin coding sequence was inserted into the expression cassette of the vector consisting of *LAC4* promoter and *LAC4* terminator; in other constructs, the enzyme sequence was fused to a variety of leader sequences including the *S. cerevisiae*-derived alpha-factor leader. High amounts of completely soluble and biologically active chymosin could be recovered from the culture supernatant in all the different recombinant strains analyzed. High yields were obtained in constructs with two genuine presequences and the alpha-factor fusion. A transformant containing several integrated copies of the later construct was selected for large scale production of chymosin and production at an industrial 41000 l scale has been reported (van den Berg et al. 1990). Thus, the described *K. lactis* strains exceeded by far other heterologous production systems for calf chymosin based on *E. coli* (Emtage et al. 1983), *S. cerevisiae* (Mellor et al. 1983; Moir et al. 1985), *Aspergillus nidulans* (Cullen et al. 1987) and *Trichoderma reesei* (Harkkii et al. 1989). It remains open whether the excellent expression properties hold for further heterologous proteins.

An alternative *K. lactis*-based expression system has been described recently. A circular plasmid with a functional organisation similar to the 2 μm DNA in *S. cerevisiae* has been isolated from a related species, *K. drosophilarum* which can replicate in *K. lactis* (Chen et al. 1986). Expression vectors derived from this 1.6 μm circular plasmid pKD1 have been developed using the *S. cerevisiae* *URA3* gene or a Tn903-derived kanamycin-resistance sequence as components for selection in suitable host strains (Bianchi et al. 1987).

As an additional third possibility, transformation with vectors has been described developed from linear double stranded DNA plasmids which confer on cells a 'killer' character (Gunge 1986). The favourable expression properties and the established tradition as safe organism in food industry make *K. lactis* a promising system for a commercial protein production.

H. polymorpha, *P. pastoris* and *K. lactis* are probably the most advanced alternatives to *S. cerevisiae*, but an increasing number of other hosts has been developed for transfer and expression of foreign genes. This group of non-*Saccharomyces* yeasts includes among others species the fission yeast *S. pombe* (Beach & Nurse 1981; Grimm & Kohli 1988), *Candida maltosa* (Kunze et al. 1987), *Yarrowia lypolytica* (Davidow et al. 1985) and *S. occidentalis* (Klein & Favreau 1988; Dohmen et al. 1989; Piontek et al. 1990). A comprehensive compilation of the different systems available has recently been given (Reiser et al. 1990).

Comparative aspects

In search for superior systems for the production of proteins, the baker's yeast *S. cerevisiae* was the initial choice to supplement prokaryotes. In particular this organism offers the possibility to secrete the heterologous compounds via a multi-component secretory apparatus (Schekman & Novick 1982; Smith et al. 1985; Moir 1989) providing the prerequisite for proper folding, for proteolytic processing and for posttranslational modifications like N- and O-glycosylation (see Innis 1989 for a recent review), phosphorylation and N-terminal acetylation (Hallewell et al. 1987). However, certain limitations and disadvantages have been encountered in industrial and commercial applications (Kingsman et al. 1987). These include instability of production strains (da Silva & Bailey 1991), hyperglycosylation of many secreted glycoproteins (Innis 1989) and a relatively poor secretion efficiency (Kingsman et al. 1987; Moir 1989) observed among other cases for calf prochymosin (Smith et al. 1985), human lysozyme (Jigami et al. 1986), interferon (Hitzeman et al. 1983) and human tissue plasminogen activator (Lemontt et al. 1985). The low productivity may be due to additional stress imposed on the cells by heterologous gene expression. This view was supported by observations, where a heterologous gene introduced into a multicopy vector exhibited a low secretion when fused to a strong constitutive promoter (Ernst 1986). It has been argued that high protein secretion might interfere with growth and that it might be necessary to decrease the expression of secretory proteins during the early phase of fermentation. However, this basic problem of strain stability in plasmid-based expression under non-selective conditions might be encountered in industrial scale-ups even when inducible promoters in 2 μm vectors are used (Da Silva & Bailey 1991). Despite of a recent publication reporting the high-yield production of an *Aspergillus* glucose oxidase in *S. cerevisiae* (de Baetselier et al. 1991), *Saccharomyces*-based expression was found be second to non-*Saccharomyces* alternatives in several instances when production of specific compounds were compared (van den Berg et al. 1990; Buckholz & Gleeson 1991). A general improvement of productivity seems to depend on the identification and characterization of *Saccharomyces* mutants with an increased secretion efficiency (Moir 1989). Alternatively, the described limitations have resulted in the search and identification of other yeast hosts providing improvements without greater efforts to modify and improve a basic system. A selection of these alternative yeasts have been presented in this review.

Development of the different yeast systems is based in many cases on complementation of nutritional deficiencies in mutants which have been identified in analogy to the respective strains of *S. cerevisiae*. In many cases genes originating from the baker's yeast were found to function in heterologous yeasts (for review see Reiser et al. 1990). For instance, *URA3* complements auxotrophies in *H. polymorpha* (Roggenkamp et al. 1986), *S. pombe, K. lactis*, *Rhodosporidium toruloides* and *Zygosaccharomyces rouxii* (cited in Reiser et al. 1990); *HIS4* is used as marker for complementation in *P. pastoris* (Cregg & Madden 1987), *TRP5* in *S. occidentalis* (Dohmen et al. 1989). Accordingly, the respective sequences serve as components in transformation vectors in addition to a selection of homologous markers (Reiser et al. 1990). After transformation the foreign DNA can persist due to the presence of autonomous replication sequences like in the 2 μm-derived plasmids in *S. cerevisiae*, for instance in *S. occidentalis* (Dohmen et al. 1989). Preferably, the DNA is stably integrated into the host's genome by homologous recombination as described for *P. pastoris* (Cregg & Madden 1987) or by heterologous integration as described for *H. polymorpha* (Janowicz et al. 1991; Gellissen et al. 1991) and *S. pombe* (Grimm & Kohli 1988).In *Pichia*, multiple copies of an expression cassette can be integrated (Vedvick et al. 1991); similarly, integration of several copies are described for a prochymosin-producing recombinant *K. lactis* strain (van den Berg et al. 1990). Most favourable is the simultaneous and stable integration of multiple vector copies in *Hansenula*, for which recombinants with up to 100 copies have been identified (Janowicz et al. 1991). In all three cases the transformed strains remain stable during fermentation. In particular *Hansenula* strains, stability has been observed after 400 generations of growth (Z.A. Janowicz, pers. comm.) The possibility to integrate multiple copies of an heterologous expression cassette without an obvious loss in vitality provides a means to identify strains with optimal productivity. In *Hansenula*, productivity was found to be correlated to a level of 20 copies in hirudin-secreting recombinant strains (U. Weydemann, pers. comm.). Similar observations are described for *Pichia* (Vedvick et al. 1991).

A variety of different strong promoters are used in expression vectors in *S. cerevisiae* as listed in Table 1 (see also Kingsman et al. 1988). Some of them function in heterologous yeasts; a bacterial cellulase was expressed in *S. occidentalis* and in *P. stipitis* when fused to *S. cerevisiae*-derived *PDC1* or *ADH1* promoter sequences (Piontek et al. 1990), an efficient production of chymosin was observed in *K. lactis*, when a *S. cerevisiae*-derived *GAP* promoter was used (van den Berg et al. 1990). In most cases, homologous promoter sequences from highly expressed genes are applied. The best yields obtained so far are reported for the *LAC4*-controlled expression of prochymosin in *K. lactis* (van den Berg et al. 1990), for fusions to the methanol-inducible *AOX* promoter in *Pichia* (Buckholz & Gleeson 1991) and its counterparts in the *Hansenula* system, the *MOX* and *FMD*-promoters (Gellissen et al. 1990; Gellissen et al. 1991; Janowicz et al. 1991). They can stringently be controlled by a suitable choice of medium components. If desired, recombinant strains can be grown to high cell densities before switching to inductive conditions for heterologous gene expression. This is advantageous in cases of interference of the expressed compounds with cell growth.

For secretion, it is necessary to fuse the polypeptide to a N-terminal leader sequence that directs the translated nascent polypeptide into the secretory apparatus. In *S. cerevisiae*, the signal sequences for the genuine export proteins mating pheromone (MFα1) (Brake et al. 1984), invertase (Taussig & Carlson 1983) and acid phosphatase (Meyhack et al. 1982) have been used for secretion of heterologous proteins (Kingsman et al. 1988; Hinnen et al. 1989). In case of MFα1, a two-step maturation takes place. First, a signal sequence is cleaved at the entry into the ER (Waters et al. 1988), the remaining propeptide is further cleaved by the proteolytic activity of the KEX2 protease, a dibasic endopeptidase located within the Golgi apparatus (Julius et al. 1984). In most constructs using this leader, the heterologous sequences are fused to the KEX2 recognition site. Some fusions with this prepro-segment were found to be faithfully processed in the other yeast expression systems analyzed so far indicating a similar mode of secretion and processing (for review see Buckholz & Gleeson 1991). In addition, signal sequences from sources others than *S. cerevisiae* can serve as suitable components for secretion. In *Hansenula*, the *GAM1* leader sequence from *S. occidentalis* was readily recognized and an authentic enzyme with a faithfully processed N-terminus was recovered from culture supernatants of recombinant strains (Gellissen et al. 1991). On the contrary, structural properties of the protein to be secreted might result in an inaccurate removal of a leader sequence. This unexpected result has recently been reported for a prepro-MFα aprotinin fusion in *Pichia*, where a lack of cleavage at the Lys-Arg processing site and inappropiate N-terminal extensions of the mature protein were observed (Vedvick et al. 1991).

Glycosylation

Many heterologous proteins are hyperglycosylated (>40 mannose residues) when secreted from a *S. cerevisiae* host (Innis 1989), but not in methylotrophic yeasts. Detailed studies demonstrated an average chain length of 8–14 mannose residues in oligosaccharide chains of glycoproteins from *P. pastoris* (Grinna & Tschopp 1989). Invertase is hyperglycosylated neither in *Pichia* (Tschopp et al. 1987) nor in *Hansenula* (Janowicz et al. 1988) in contrast to the situation in *S. cerevisiae*. In *Hansenula*, an authentic N- and O-glycosylation was observed in secreted *S. occidentalis* glucoamylase (Gellissen et al. 1991), no hyperglycosylation could be detected in a variety of other secreted polypeptides including hormones and enzymes (U. Weydemann, pers. comm.). However, overglycosylation has been reported for a recombinant α-galactosidase expressed in this organism (Sierkstra et al. 1991).

Lack of overglycosylation has also been shown for *Y. lypolytica* (A. Franke, cited by Buckholz & Gleeson 1991) and *S. occidentalis* (Piontek et al. 1990). There might exist yet undiscovered differences of glycosylation among the different yeasts, but the prospects are good that the growing number of non-*Saccharomyces* hosts might provide a spectrum of production systems which include a host of choice for optimal modification and maximal productivity for any given protein.

It would be desirable to develop universal vectors suitable for the transformation of several yeasts. In a comparative study transformants could simultaneously be screened and tested for optimal expression.

A comparison of the different systems demonstrate that several yeasts express and secrete heterologous proteins at least as efficient as *S cerevisiae*. After scale-up the methylotrophs can be fermented to very high cell densities accumulating 1 gram protein per liter or more in the culture supernatants. Examples for such maximal yields are recombinant *Pichia* strains producing invertase (Tschopp et al. 1987) and human serum albumin (Tschopp & Sreekrishna; cited by Wegner 1990) and *Hansenula* strains secreting hirudin (U. Weydemann, pers. comm.) or glucoamylase (Gellissen et al. 1991; see Table 3). In *Hansenula*, the accumulated heterologous glucoamylase comprised about 90% of all extracellular proteins, thus providing for an easy purification and downstream processing on an industrial scale.

In summary, non-*Saccharomyces* yeasts are useful alternatives for the expression of heterologous protein products of commercial and pharmaceutical interest.

References

Anthony C (1982) The Biochemistry of Methylotrophs. Academic Press, New York

Barr PJ, Brake AJ & Valenzuela P (1989) Yeast Genetic Engineering. Butterworths, Boston

Beach D & Nurse P (1981) High frequency transformation of the fission yeast *Schizosaccharomyces pombe*. Nature 290: 140–142

Beggs JD (1978) Transformation of yeast by a replicating hybrid plasmid. Nature 275: 104–109

Bennetzen JL & Hall BD (1982) The primary structure of the *Saccharomyces cerevisiae* gene for alcohol dehydrogenase I. J. Biol. Chem. 257: 3018–3025

Bianchi MM, Falcone C, Chen XJ, Wéslowski-Louvel M, Frontali L & Fukuhara H (1987) Transformation of the yeast *Kluyveromyces lactis* by new vectors derived from the 1,6 μm circular plasmid pKD1. Curr. Genet. 12: 185–192

Bitter GA, Chen KK, Banks AR & Lai PH (1984) Secretion of foreign proteins from *Saccharomyces cerevisiae* directed by α-factor fusions. Proc. Natl. Acad. Sci. USA 81: 5330–5334

Botstein D & Shortle D (1985) Strategies and applications of in vitro mutagenesis. Science 229: 1193–1201

Brake AJ, Merryweather JP, Coit DG, Heberlein UA, Masiarz FR, Mullenbach GT, Urdea MS, Valenzuela P & Barr PJ (1984) α-factor-directed synthesis and secretion of mature foreign proteins in *Saccharomyces cerevisiae*. Proc. Natl. Acad. Sci. USA 81: 4642–4646

Buckholz RG & Gleeson MAG (1991) Yeast systems for the commercial production of heterologous proteins. Bio/Technology 9: 1067–1072

Burke ET, Carle GS & Olson MV (1987) Cloning of large fragments of exogenous DNA into yeast by means of artifical chromosome vector. Science 236: 806–812

Burke RL, Tekamp-Olson P & Najarian R (1983) The isolation, characterization, and sequence of the pyruvate kinase gene of *Saccharomyces cerevisiae*. J. Biol. Chem. 258: 2193–2201

Cabezon T, DeWilde M, Herion P, Loriau R & Bollen A (1984) Expression of human α1-antitrypsin cDNA in the yeast *Saccharomyces cerevisiae*. Proc. Natl. Acad. Sci. USA 81: 6594–6598

Chen XJ, Saliola M, Falcone C, Bianchi MM & Fukuhara H (1986) Sequence organization of the circular plasmid pKD1 from the yeast *Kluyveromyces drosophilarum*. Nucleic Acids Res. 14: 4471–4481

Clare JJ, Rayment FB, Ballantine SP, Sreekrishna K & Romanos MA (1991) High-level expression of tetanus toxin fragment C in *Pichia pastoris* strains containing multiple tandem integrations of the gene. Bio/Technology 9: 455–460

Clarke L & Carbon J (1980) Isolation of a yeast centromere and construction of functional small circular chromosomes. Nature 257: 504–509

Cregg JM & Madden PA (1987) Development of yeast transformation systems and construction of methanol-utilizing-defective mutants of *Pichia pastoris* by gene disruption. In: Stewart GG, Russell I, Klein RD & Hiebsch RR (Eds) Biological Research on Industrial Yeast, Vol. II (pp 1–18). CRC Press, Boca Raton, Florida

Cregg JM, Barringer KJ, Hessler AY & Madden KR (1985) *Pichia pastoris* as a host system for transformations. Mol. Cell. Biol. 5: 3376–3385

Cregg JM, Tschopp JF, Stillman C, Siegel R, Akong M, Craig WS, Buckholz RG, Madden KR, Kellaris PA, Davis GR, Smiley BL, Cruze J, Torregrossa R, Velicelebi G & Thill GP (1987) High-level expression and efficient assembly of hepatitis B surface antigen in the methylotrophic yeast, *Pichia pastoris*. Bio/Technology 5: 479–485

Cullen D, Gray GL, Wilson LJ, Hayenga KJ, Lamsa MH, Rey MW, Norton S & Berka RM (1987) Controlled expression and secretion of bovine chymosin in *Aspergillus nidulans*. Bio/Technology 5: 369–376

Da Silva NA & Bailey JE (1991) Influence of dilution rate and induction of cloned gene expression in continuous fermentations of recombinant yeast. Biotechnol. Bioeng. 37: 309–317

Da Silva NA & Bailey JE (1991) Influence of plasmid origin and promoter strength in fermentations of recombinant yeast. Biotechnol. Bioeng. 37: 318–324

Das S & Hollenberg CP (1982) A high-frequency transformation system for the yeast *Kluyveromyces lactis*. Curr. Genet. 6: 123–128

Davidow LS, Apostolakos D, O'Donell MM, Proctor AR, Ogrydziak DM, Wing RA, Stasko I & DeZeeuw JR (1985) Integrative transformation of the yeast *Yarrowia lypolytica*. Curr. Genet. 10: 39–48

De Baetselier A, Vasavada A, Dohet P, Ha-Thi V, De Beukelaer M, Erpicum T, De Clerck L, Hanotier J & Rosenberg S (1991) Fermentation of a yeast producing *A. niger* glucose oxidase: scale-up, purification and characterization of the recombinant enzyme. Bio/Technology 9: 559–561

Derynck R, Singh A & Goeddel DV (1983) Expression of the human interferon-α cDNA in yeast. Nucleic Acids Res. 11: 1819–1837

DeWilde M, Cabezon T, Harford N, Rutgers T, Simoen E & Van Wijnendaele F (1985) Production in yeast of hepatitis B surface antigen by r-DNA technology. 18.Congress. Joint IABS/WHO symposium on standardization and control of biologicals produced by recombinant DNA technology, Geneva , Switzerland, 1983. Develop. Biol. Standard 59: 99–107

Dickson RC & Markin JS (1978) Molecular cloning and expression in *E. coli* of a yeast gene coding for β-galactosidase. Cell 15: 123–130

Digan ME, Lair SV, Brierley RA, Siegel RS, Williams ME, Ellis SB, Kellaris PA, Provow SA, Craig WS, Velicelebi G, Harpold MM & Thill GP (1989) Continuous production of a novel lysozyme via secretion from the yeast, *Pichia pastoris*. Bio/Technology 7: 160–164

Dobson MJ, Tuite MF, Roberts NA, Kingsman AJ, Kingsman SM, Perkins RE, Conroy SC, Dunbar B & Fothergill LA (1982) Conservation of high efficiency promoter sequences in *Saccharomyces cerevisiae*. Nucleic Acids Res. 10: 2625–2637

Dohmen RJ, Strasser AWM, Zitomer RS & Hollenberg CP (1989) Regulated overproduction of α-amylase by transformation of the amylolytic yeast *Schwanniomyces occidentalis*. Curr. Genet. 15: 319–325

Dohmen RJ, Strasser AWM, Höner CB & Hollenberg CP (1991) An efficient transformation procedure enabling long-term storage of competent cells of various yeast genera. Yeast 7: 691–692

Edens L, Bom I, Ledeboer AM, Maat J, Toonen MY, Visser C & Verrips CT (1984) Synthesis and processing of the plant protein thaumatin in yeast. Cell 37: 629–633

Ellis SB, Brust PF, Koutz PJ, Waters AF, Harpold MM & Gingeras TR (1985) Isolation of alcohol oxidase and two other methanol regulatable genes from the yeast, *Pichia pastoris*. Mol. Cell. Biol. 5: 1111–1121

Emini EA, Ellis RW, Miller WJ, McAleer WJ, Scolnick EM & Gerety RJ (1986) Production and analysis of recombinant hepatitis B vaccine. Journal of Infection 13(suppl A): 3–9

Emtage JS, Angal S, Doel MT, Harris TJR, Jenkins B, Lilley G & Lowe PA (1983) Synthesis of calf prochymosin (prorennin) in *Escherichia coli*. Proc. Natl. Acad. Sci. USA 80: 3671–3675

Ernst JF (1986) Improved secretion of heterologous proteins by *Saccharomyces cerevisiae*: effects of promoter substitution in alpha-factor fusions. DNA 5: 483–491

Fellinger AJ, Verbakel JMA, Veale RA, Sudbery PE, Bom IJ, Overbeke N & Verrips CT (1991) Expression of the α-galactosidase from *Cyamopsis tetragnoloba* (guar) by *Hansenula polymorpha*. Yeast 7: 463–474

Finkelstein DB & Strausberg S (1983) Heat shock-regulated production of *Escherichia coli* β-galactosidase in *Saccharomyces cerevisiae*. Mol. Cell. Biol. 3: 1625–1633

Gellissen G, Strasser AWM, Melber K, Merckelbach A, Weydemann U, Keup P, Dahlems U, Piontek M, Hollenberg CP & Janowicz ZA (1990) Die methylotrophe Hefe *Hansenula polymorpha* als Expressionssystem für heterologe Proteine. BioEngineering 5: 20–26

Gellissen G, Janowicz ZA, Merckelbach A, Piontek M, Keup P, Weydemann U, Hollenberg CP & Strasser AWM (1991) Heterologous gene expression in *Hansenula polymorpha*: efficient secretion of glucoamylase. Bio/Technology 9: 291–295

Gleeson MA & Sudbery PE (1988) The methylotrophic yeasts. Yeast 4: 1–15

Gleeson MA, Ortori GS & Sudbery PE (1986) Transformation of the methylotrophic yeast *Hansenula polymorpha*. J. Gen. Microbiol. 132: 3459–3465

Goff CG, Moir DT, Kohno T, Gravius TC, Smith RA, Yamasaki E & Taunton-Rigby A (1984) The expression of calf prochymosin in *Saccharomyces cerevisiae*. Gene 27: 35–46

Grimm C & Kohli J (1988) Observations on integrative transformation in *Schizosaccharomyces pombe*. Mol. Gen. Genet. 215: 87–93

Grinna LS & Tschopp JF (1989) Size distribution and general structural features of N-linked oligosaccharides from the methylotrophic yeast, *Pichia pastoris*. Yeast 5: 107–115

Gunge N (1986) Linear DNA killer plasmids from the yeast *Kluyveromyces*. Yeast 2: 153–162

Hagenson MJ, Holden KA, Parker KA, Wood PJ, Cruze JA, Fuke M, Hopkins TR & Stroman DW (1989) Expression of streptokinase in *Pichia pastoris* yeast. Enzyme Microb. Technol. 11: 650–656

Hallewell RA, Mills R, Tekamp-Olson P, Blacher R, Rosenberg S, Otting F, Masiarz FR & Scandella CJ (1987) Amino terminal acetylation of authentic human Cu,Zn superoxide dismutase produced in yeast. Bio/Technology 5: 363–366

Harford N, Cabezon T, Crabeel M, Simoen E, Rutgers T & de Wilde M (1983) Expression of hepatitis B surface antigen in yeast. Second WHO/IABS Symposium on Viral Hepatitis: standardization on immunoprophylaxis of infections by hepatitis B viruses, Athens, Greece 1982. Develop. Biol. Standard 54: 125–130

Harkkii A, Uusitabo J, Bailey M, Pentillä M & Knowles JKC (1989) A novel fungal expression system: secretion of active calf chymosin from the filamentous fungus *Trichoderma reesei*. Bio/Technology 7: 596–603

Hinnen A, Hicks JB & Fink GR (1978) Transformation of yeast. Proc. Natl. Acad. Sci. USA 75: 1929–1933

Hinnen A, Meyhack B & Tsapis R (1983) High expression and secretion of foreign proteins in yeast. In: Korhola M & Vaisanen E (Eds) Gene Expression in Yeast, Foundation for Biotechnical and Industrial Fermentation Research, Vol. 1 (pp 157–166). Kauppakirjapaino, Helsinki

Hinnen A, Meyhack B & Heim J (1989) Heterologous gene expression in yeast. In: Barr PJ, Brake AJ & Valenzuela P (Eds) Yeast Genetic Engineering (pp 193–214). Butterworths, Boston

Hitzeman RA, Hagie FF, Levine HL, Goeddel DV, Ammerer G & Hall BD (1981) Expression of a human gene for interferon in yeast. Nature 293: 717–722

Hitzeman RA, Leung DW, Perry LJ, Kohr WJ, Levine HL & Goeddel DV (1983) Secretion of human interferons by yeast. Science 219: 620–625

Hodgkins MA, Sudbery PE, Perry-Williams S & Goodey A (1990) Secretion of human serum albumin from *Hansenula polymorpha*. Yeast 6S: 435

Holland JP & Holland MJ (1980) Sructural comparison of two nontandemly repeated yeast glyceraldehyde-3-phosphate dehydrogenase genes. J. Biol. Chem. 255: 2596–2605

Holland MJ, Holland JP, Thill GP & Jackson KA (1981) The primary structure of two yeast enolase genes. Homology between the 5′noncoding flanking regions of yeast enolase and glyceraldehyde-3-phosphate dehydrogenase gene. J. Biol. Chem. 256: 1385–1395

Hollenberg CP & Janowicz ZA (1987) DNA molecules coding for FMDH control region. Europan patent Application, EPA No. 0299108

Innis MA (1989) Glycosylation of heterologous proteins in *Saccharomyces cerevisiae*. In: Barr PJ, Brake AJ & Valenzuela P (Eds) Yeast Genetic Engineering (pp 233–246). Butterworths, Boston

Innis MA, Holland MJ, McCabe PC, Cole GE, Wittman VP, Tal R, Watt KWK, Gelfand DH, Holland JP & Maede JH (1985) Expression, glycosylation, and secretion of an *Aspergillus* glucoamylase by *Saccharomyces cerevisiae*. Science 228: 21–26

Itakura K, Hirose T, Crea R, Riggs AD, Heynecker H, Bolivar F & Boyer HW (1977) Expression in *Escherichia coli* of a chemically synthesized gene for the hormone somatostatin. Science 198: 1056–1063

Janowicz ZA, Eckart MR, Drewke C, Roggenkamp RO & Hollenberg CP (1985) Cloning and characterization of the DAS gene encoding the major methanol assimilatory enzyme from the methylotrophic yeast *Hansenula polymorpha*. Nucleic Acids Res. 13: 2043–3062

Janowicz ZA, Merckelbach A, Eckart M, Weydemann U, Roggenkamp R & Hollenberg CP (1988) Expression system based on the methylotrophic yeast *Hansenula polymorpha*. Yeast 4S: 155

Janowicz ZA, Melber K, Merckelbach A, Jacobs E, Harford N, Comberbach M & Hollenberg CP (1991) Simultaneous expression of the S and L surface antigens of hepatitis B, and formation of mixed particles in the methylotrophic yeast, *Hansenula polymorpha*. Yeast 7: 431–443

Jigami Y, Muraki M, Harada N & Tanaka H (1986) Expression of synthetic human-lysozyme gene in *Saccharomyces cerevisiae*: use of a synthetic chicken-lysozyme signal sequence for secretion and processing. Gene 43: 273–279

Johnston M & Davis RW (1984) Sequences that regulate the divergent *GAL1-GAL10* promoter in *Saccharomyces cerevisiae*. Mol. Cell. Biol. 4: 1440–1448

Julius D, Brake A, Blair L, Kunisawa R & Thorner J (1984) Isolation of the putative structural gene for the lysine-arginine-cleaving endopeptidase required for processing of yeast prepro-α-factor. Cell 37: 1075–1083

Karin M, Najarian R, Haslinger A, Valenzuela P, Welch J & Fogel S (1984) Primary structure and transcription of an amplified genetic locus: the *CUP1* locus of yeast. Proc. Natl. Acad. Sci. USA 81: 337–341

Kaufman RJ, Wasley LC & Dorner AJ (1988) Synthesis, processing and secretion of recombinant human factor VIII expressed in mammalian cells. J. Biol. Chem. 263: 6352–6362

Kellermann E & Hollenberg CP (1988) The glucose- and ethanol-dependent regulation of *PDC1* from *Saccharomyces cerevisiae* are controlled by two distinct promoter regions. Curr. Genet. 14: 337–344

Kingsman SM, Kingsman AJ & Mellor J (1987) The production of mammalian proteins in *Saccharomyces cerevisiae*. Tibtech. 5: 53–57

Kingsman SM, Kingsman AJ, Dobson MJ, Mellor J & Roberts NA (1988) Heterologous gene expression in *Saccharomyces cerevisiae*. In: Russell GE (Ed) Yeast Biotechnology (pp 113–152). Intercept Ltd, Wimborne, Dorset

Klebe RJ, Harriss JV, Sharp ZD & Douglas MG (1983) A general method for polyethylene-glycol-induced transformation of bacteria and yeast. Gene 25: 333–341

Klein RD & Favreau MA (1988) Transformation of *Schwanniomyces occidentalis* with *ADE2* gene cloned from *S. occidentalis*. J. Bacteriol. 170: 5572–5578

Kojo H, Greenberg BD & Sugino A (1981) Yeast 2μm plasmid DNA replication in vitro: origin and direction. Proc. Natl. Acad. Sci. USA 78: 7261–7265

Koutz P, Davis GR, Stillman C, Barringer K, Cregg J & Thill G (1989) Structural comparison of the *Pichia pastoris* alcohol oxidase genes. Yeast 5: 167–177

Kramer RA, DeChiara TM, Schaber MD & Hilliker S (1984) Regulated expression of a human interferon gene in yeast: control by phosphate concentration or temperature. Proc. Natl. Acad. Sci. USA 81: 367–370

Kunze G, Bode R & Birnbaum D (1987) Autonomous and integrative transformation of the yeast *Candida maltosa*. In: Neijssel OM, Van der Meer RR & Luyben KCAM (Eds) Proc 4th Congr of Biotechnol, Vol. I (pp 398–401). Elsevier, Amsterdam

Kurjan J & Herskowitz I (1982) Structure of a yeast pheromone gene (MFα): a putative α-factor precursor contains four tandem copies of mature α-factor. Cell 30: 933–943

Lai PH, Everett R, Wang FF, Arakawa T & Goldwasser E (1986) Structural characterisation of human erythropoietin. J. Biol. Chem. 291: 3116–3121

Ledeboer AM, Edens L, Maat J, Visser C, Bos JW, Verrips CT, Janowicz ZA, Eckart M, Roggenkamp R & Hollenberg CP (1985) Molecular cloning and characterization of a gene coding for methanol oxidase in *Hansenula polymorpha*. Nucleic Acids Res. 13: 3063–3082

Lee JD & Komagata K (1980) Taxonomic study of methanol-assimilating yeasts. J. Appl. Microbiol. 26: 133–158

Lemontt JF, Wei C-M & Dackowski WR (1985) Expression of active human uterine tissue plasminogen activator in yeast. DNA 4: 419–428

Mellor J, Dobson MJ, Roberts NA, Tuite MF, Emtage JS, White S, Lowe PA, Patel T, Kingsman AJ & Kingsman SM (1983) Efficient synthesis of enzymatically active calf chymosin in *Saccharomyces cerevisiae*. Gene 24: 1–14

Meyhack B, Bajwa W, Rudolph H & Hinnen A (1982) Two yeast acid phosphatase structural genes are the result of a tandem duplication and show different degrees of homology in their promoter and coding sequences. EMBO J. 1: 675–680

Moir DT (1989) Yeast mutants with increased secretion efficiency. In: Barr PJ, Brake AJ & Valenzuela P (Eds) Yeast Genetic Engineering (pp 215–231). Butterworths, Boston

Moir DT, Mao J, Duncan MJ, Smith RA & Kohne T (1985) Production of calf chymosin by the yeast *Saccharomyces cerevisiae*. In: Developments in Industrial Microbiology, Vol. 26 (pp 75–85). Society for industrial microbiology, Arlington, VA

Müller-Wiefel DE, Feist K, Twittenhof W & Brune T (1988) One year experience with the treatment of renal anaemia by recombinant human erythropoietin. Monatsschrift Kinderheilkunde 136: 514

Neurath AR, Jameson BA & Huima T (1987) Hepatitis B virus proteins eliciting protective immunity. Microbiol. Sci. 4: 45–51

Orr-Weaver TL, Szostak JW & Rothstein RJ (1981) Yeast transformation: a model system for the study of recombination. Proc. Natl. Acad. Sci. USA 78: 6354–6358

Petre J, van Wijnendaele F, de Neys B, Conrath K, van Opstal O, Hauser P, Rutgers T, Cabezon T, Capiau C, Harford N, de Wilde M, Stephenne J, Carr S, Hemling H & Swadish J (1987) Development of a hepatitis B vaccine from transformed yeast cells. Postgrad. Med. J. 63(suppl 2): 73–81

Phaff HJ (1985) Biology of yeasts other than *Saccharomyces*. In: Demain AL, Solomon NA (Eds) Biology of Industrial Microorganisms (pp 537–562). Benjamin/Cumming, Menlo Park

Piontek M, Hollenberg CP & Strasser AWM (1990) *Schwanniomyces occidentalis*: a promising system for expression of foreign proteins. Yeast 6S: 422

Reiser J, Glumoff V, Kälin M & Ochsner U (1990) Transfer and expression of heterologous genes in yeasts other than *Saccharomyces cerevisiae*. In: Reiser J (Ed) Advances in Biochemical Engineering/Biotechnology, Vol. 43 Applied Molecular Genetics (pp 75–102). Springer, Berlin

Reiss B, Sprengel R & Schaller H (1984) Protein fusions with the kanamycin resistance gene from transposon Tn5. EMBO J. 3: 3317–3322

Robinson JH & Browne MJ (1991) Redesigning t-PA for improved thrombolytic therapy. Tibtech. 9: 86–90

Roggenkamp RO, Janowicz ZA, Stanikowski B & Hollenberg CP (1984) Biosynthesis and regulation of peroxysomal methanol oxidase from the methylotrophic yeast *Hansenula polymorpha*. Mol. Gen. Genet. 194: 489–493

Roggenkamp RO, Hansen H, Eckart M, Janowicz ZA & Hollenberg CP (1986) Transformation of the methylotrophic yeast *Hansenula polymorpha* by autonomous replication and integration vectors. Mol. Gen. Genet. 202: 302–308

Rosenberg S, Barr PJ, Najarian RC & Hallewell RA (1984) Synthesis in yeast of a functional oxidation mutant of human α1-antitrypsin. Nature 312: 77–80

Rothstein SJ, Lazarus CM, Smith WE, Baulcombe DC & Gatenby AA (1984) Secretion of a wheat α-amylase expressed in yeast. Nature 308: 662–665

Russell GE (1988) Yeast Biotechnology. Intercept, Wimborne, Dorset

Sahm H (1977) Metabolism of methanol by yeast. In: Ghose TK, Frechter A & Blankenbrough H (Eds) Advances in Microbiological Engineering, Vol. 6 (pp 77–103). Springer, Berlin

Schekman R & Novick P (1982) The secretory process and yeast cell-surface assembly. In: Strathern JN, Jones EW & Broach JR (Eds) The Molecular Biology of the Yeast *Saccharomyces cerevisiae* (pp 361–393). Cold Spring Harbour Laboratory, Cold Spring Harbour, New York

Schmitt HD, Ciriacy M & Zimmermann FK (1983) The synthesis of yeast pyruvate decarboxylase is regulated by large variations in the messenger RNA level. Mol. Gen. Genet 192: 247–252

Shen SH, Bastien L, Nguyen T, Fung M & Slilaty SN (1989) Synthesis and secretion of hepatitis B middle surface antigen by the methylotrophic yeast *Hansenula polymorpha*. Gene 84: 304–309

Siegel RS, Buckholz R & Thill GP (1989) Production of epidermal growth factor in *Pichia pastoris* yeast cells. European patent application W090/10697

Sierkstra LN, Verbakel JMA & Verrips CT (1991) Optimisation of a host/vector system for heterologous gene expression by *Hansenula polymorpha*. Curr. Genet. 19: 81–87

Smith RA, Duncan MJ & Moir DT (1985) Heterologous protein secretion from yeast. Science 229: 1219–1224

Sreekrishna K, Nelles L, Potenz R, Cruze J, Mazzaferro P, Fish W, Fuke M, Holden K, Phelps D, Wood P & Parker K (1989) High-level expression, purification, and characterisation of recombinant human tumor necrosis factor synthesized in the methylotrophic yeast, *Pichia pastoris*. Biochemistry 28: 4117–4125

Stepien PP, Brousseau R, Wu R, Narang S & Thomas DY (1983) Synthesis of a human insulin gene VI. Expression of the synthetic proinsulin gene in yeast. Gene 24: 289–297

Struhl K, Stinchkomb DT, Scherer S & Davis RW (1979) High frequency transformation of yeast: autonomous replication of hybrid molecules. Proc. Natl. Acad. Sci. USA 76: 1035–1039

Sturley SL & Young TW (1988) Genetic manipulation of commercial yeast strains. In: Russell GE (Ed) Yeast Biotechnology (pp 1–38). Intercept, Wimborne, Dorset

Taussig R & Carlson M (1983) Nucleotide sequence of the yeast *SUC2* gene for invertase. Nucleic Acids Res. 11: 1943–1954

Thim L, Hansen MT, Norris K, Hoegh I, Boel E, Forstrom J, Ammerer G& Fiil NP (1986) Secretion and processing of insulin precursor in yeast. Proc. Natl. Acad. Sci. USA 83: 6766–6770

Tschopp JF, Brust BF, Cregg JM, Stillman CA & Gingeras TR (1987) Expression of the lacZ gene from two methanol-regulated promoters in *Pichia pastoris*. Nucleic Acids Res. 9: 3859–3876

Tschopp JF, Sverlow G, Kosson R, Craig W & Grinna L (1987) High-level secretion of glycosylated invertase in the methylotrophic yeast, *Pichia pastoris*. Bio/Technology 5: 1305–1308

Tuite MF, Dobson MJ, Roberts NA, King RM, Burke DC, Kingsman SM & Kingsman AJ (1982) Regulated high efficiency expression of human interferon-alpha in *Saccharomyces cerevisiae*. EMBO J. 1: 603–608

Valenzuela P, Medina A, Rutter WJ, Ammerer G & Hall BD (1982) Synthesis and assembly of hepatitis B virus surface antigen particles in yeast. Nature 298: 347–350

van den Berg JA, van der Laken KJ, van Ooyen AJJ, Renniers TCHM, Rietveld K, Schaap A, Brake AJ, Bishop RJ, Schultz K, Moyer D, Richman M & Shuster JR (1990) *Kluyveromyces* as a host for heterologous gene expression: expression and secretion of prochymosin. Bio/Technology 8: 135–139

Vedvick T, Buckholz RG, Engel M, Urcan M, Kinney J, Provow S, Siegel RS & Thill GP (1991) High-level secretion of biologically active aprotinin from the yeast *Pichia pastoris*. J. Industrial Microbiol. 7: 197–202

Veenhuis M, van Dijken JP & Harder W (1983) The significance of peroxysomes in the metabolism of one-carbon compounds in yeasts. Adv. Microb. Physiol. 24: 1–82

Waters MG, Evans EA & Blobel G (1988) Prepro α-factor has a cleavable signal sequence. J. Biol. Chem. 263: 6209–6214

Wegner GH (1990) Emerging applications of the methylotrophic yeasts. FEMS Microbiology Reviews 87: 279–284

Wood WI, Capon DJ, Simonsen CC, Eaton DL, Gitschier J, Keyt B, Seeburg PH, Smith DH, Hollingshead P, Wion KL, Delwart E, Tuddenham GD, Vehar GA & Lacon RM (1984) Expression of active human factor VIII from recombinant DNA clones. Nature 312: 330–337

Antonie van Leeuwenhoek **62**: 95–108, 1992.

The pheromone signal pathway in *Saccharomyces cerevisiae*

James B. Konopka & Stanley Fields
Department of Microbiology, State University of New York at Stony Brook, Stony Brook, New York 11794-5222, USA

Key words: cell cycle control, G protein, hormone signaling, mating, transcription, yeast

Abstract

Haploid cells of the yeast *Saccharomyces cerevisiae* normally undergo a budding life cycle, but after binding the appropriate mating pheromone they undergo a different developmental pathway that leads to conjugation. This intercellular communication between the two mating types activates a signal transduction pathway that stimulates the diverse physiological changes required for conjugation, such as induction of cell surface agglutinins, cell division arrest in G1, morphogenesis to form a conjugation tube, and cell fusion. The components of this pathway include a G protein-coupled receptor, several protein kinases, and a pheromone-responsive transcription factor. The molecular mechanisms that transduce the pheromone signal are remarkably similar to the mechanisms of hormone signaling used in multicellular organisms. Thus, the analysis of the pheromone signal pathway in yeast directly contributes to the study of cell growth and development in other eukaryotic organisms.

Introduction

During the life cycle of the budding yeast *Saccharomyces cerevisiae*, cells can exist as one of three distinct cell types. The haploid cells occur as mating type **a** or α. Haploid cells of opposite mating type can undergo conjugation to form the third cell type (the diploid **a**/α cell). Diploid cells grow vegetatively under favorable conditions but when starved for nutrients they will undergo sporulation to produce four haploid spores. This pattern contrasts with the life cycle of the fission yeast *Schizosaccharomyces pombe*, during which conjugation occurs under starvation conditions and the zygotes proceed directly to sporulation (Egel et al. 1990).

In *S. cerevisiae*, conjugation requires intercellular communication between cells of opposite mating type. Cells signal each other with mating pheromones to stimulate the signal transduction pathway that induces the diverse physiological changes required for mating. Cytological analysis (Baba et al. 1989; Byers 1981) has revealed some of the landmark events in the conjugation process (illustrated in Fig. 1). One of the earliest responses is the induction of cell surface agglutinins which facilitate mating by holding the non-motile yeast cells in place. The cells then synchronize their cell cycles by arresting in the G1 phase. Localized morphogenesis forms a conjugation tube which connects the two cells and enables the cell to fuse nuclei. In this review we will focus on the molecular mechanisms of the pheromone signal transduction pathway which enables cells to mate efficiently. Recent studies on this pathway have gained widespread interest because yeast cells employ functional analogs of the signal transduction components used by multicellular organisms, including receptors, G proteins, protein kinases and transcription factors. Due to space limitations, we are only able to cite selective references and encourage readers to find additional references within the citations listed.

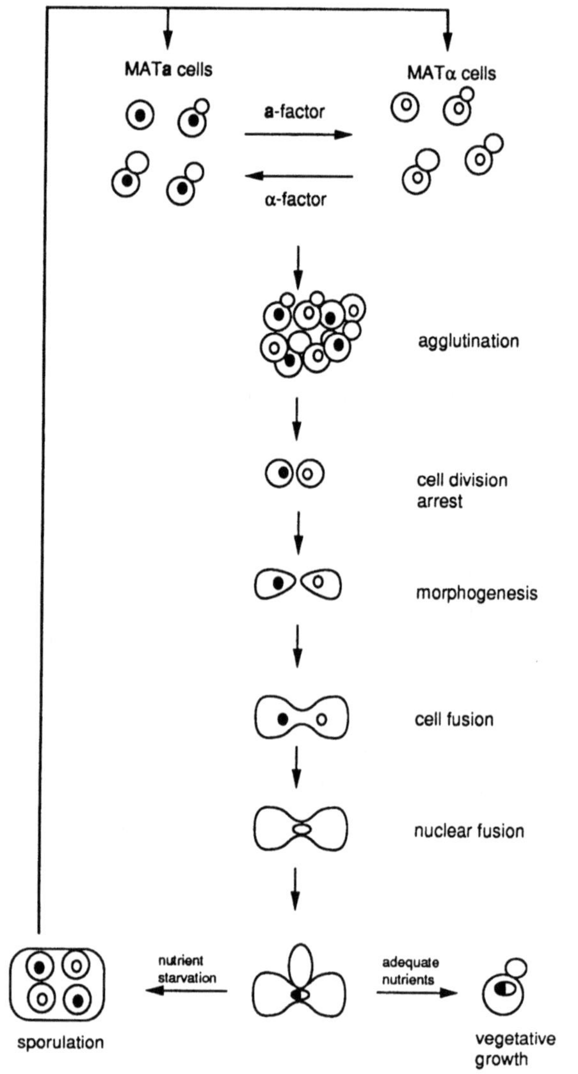

Fig. 1. Life cycle of *S. cerevisiae* with emphasis on the morphological events that occur during conjugation.

Cell-type identity

The haploid mating type is identified by the pheromone and receptor combination that a particular cell produces. **a** cells produce **a**-factor and have cell surface receptors for α-factor, while α cells produce α-factor and have cell surface receptors for **a**-factor. This mating type identity is controlled at the transcriptional level; **a** cells uniquely express **a**-specific genes such as those encoding **a**-factor, and α cells uniquely express α-specific genes such as those encoding α-factor. This transcriptional regulation is primarily determined by the products of the *MAT* locus. In α cells, the *MAT*α locus encodes two transcriptional regulators: α1, an activator of α-specific gene transcription and α2, a repressor of **a**-specific gene transcription. In **a** cells, the *MAT***a** locus consists of two genes which are not needed for the **a**-cell identity, since **a**-specific genes will be expressed in the absence of α2 and α-specific genes will not be expressed in the absence of α1. The *MAT***a** locus is needed, however, to confer the diploid cell identity. In **a**/α diploids, the products of the *MAT***a**1 and *MAT*α2 genes (**a**1 and α2, respectively) act together to repress the transcription of genes needed to confer **a**- or α-specific identity and to allow the expression of genes needed for the diploid cells to undergo sporulation in response to nutrient starvation. The regulation of cell type identity by the *MAT* locus is a complex topic that is beyond the scope of this review and which has been reviewed by others (Nasmyth & Shore 1987; Herskowitz 1989; Sprague 1990).

Pheromones and receptors are the determinants of intercellular communication

In spite of the fact that millions of yeast cells may be crowded together in a mating reaction, mating nearly always occurs between two cells of opposite mating type. How do they maintain the mating specificity under these conditions? Studies have shown that there are no physical barriers to prevent mis-mating, but instead the main determinants of mating specificity are the pheromones and receptors (Bender and Sprague 1989). One way this specificity was demonstrated was to show that the mating behavior of each cell type could be switched if the cell was engineered to produce the opposite combination of pheromone and receptor. Thus, no other form of intercellular communication has been established. Other components such as agglutinins may facilitate mating, but are not critical factors in determining whether a particular cell will mate as an **a** or α.

The essential role of pheromone signaling in mating partner selection has also been demonstrated in competitive mating assays in which cells of one mating type are given a choice of several potential mating partners. These studies have shown that cells select as their partner the opposite mating type cell producing the highest level of pheromone (Jackson & Hartwell 1990b). For example, wild-type cells can efficiently distinguish between pheromone-producing and non-producing cells as potential mating partners. This discrimination occurs even though both the pheromone-producing and non-producing cells are fully capable of responding to pheromone. Thus, pheromone-mediated intercellular communication enables cells to sense the presence and to identify the location of an appropriate partner cell; this process has been termed courtship (Jackson & Hartwell 1990a). How is the spatial location of a partner cell communicated through pheromone signaling? The post-receptor G protein signaling pathway (see below) is implicated because mutants that are constitutively activated or are hypersensitive to pheromone are defective in discriminating between pheromone-producing and non-producing cells as mating partners (Jackson & Hartwell 1990b). In addition, the peromone receptors appear to play a separate role in mating partner selection because the presence of these receptors partially restores the ability to discriminate between potential mating partners of a strain in which pheromone signaling is constitutively activated (Jackson et al. 1991).

α-Factor is a model secretory protein; a-factor appears to be secreted by a novel mechanism

Analysis of pheromone structure and production has provided some fascinating details about hormone processing. The pheromones are similar in that they are produced as larger precursor polypeptides which are subsequently modified by proteolysis to their mature form before being emitted to the extracellular environment. Both pheromones are also proteolytically cleaved by the receiving cell, but since this cleavage inactivates the pheromone it is probably important only for adaptation and recovery (MacKay et al. 1988; Steden et al. 1989). Although the pheromones are extensively modified, it appears that only one functional form of each pheromone is produced by each cell type. In addition, pheromone production increases during mating due to induced transcription of the pheromone genes and this quantitative change is probably important for attracting a partner cell.

α-factor is encoded by two genes, *MFα1* and *MFα2*, either of which is sufficient for mating (Kurjan & Herskowitz 1982). Their primary translation products contain a leader sequence to target the pheromone to the endoplasmic reticulum for entry into the secretory pathway. In the secretory pathway the *MFα1* translation product is proteolytically processed to yield four units of mature α-factor and the *MFα2* product two units of α-factor. The KEX2 protease cleaves at lys-arg dipeptides to liberate the individual units of pro-α-factor (Julius et al. 1984; Achstetter and Wolf 1985). The STE13 protease then removes x-ala dipeptides from the N-terminus (Julius et al. 1983) and the KEX1 protease removes the lys-arg dipeptide from the C-terminus (Dmochowska et al. 1987) to yield the mature α-factor which is released into the extracellular environment. The post-translational processing of α-factor has been of broad interest because many human hormones are processed in a similar manner (Marx 1991).

Recent studies on the structure and production of **a**-factor have uncovered some unexpected results. Comparison of the predicted translation products of the two **a**-factor genes, *MF***a***1* and *MF***a***2* (Brake et al. 1985), with the structure of purified mature **a**-factor (Anderegg et al. 1988) indicates that each translation product contains one unit of **a**-factor which must be liberated by proteolytic action. **a**-factor is also modified by a *RAM1*-mediated modification of the C-terminal cysteine with a thioether linkage to farnesyl (Powers et al. 1986). The same *RAM1*-mediated function is also necessary for the farnesylation of the yeast *RAS* proteins. **a**-factor is further modified by carboxymethylation of the C-terminal cysteine mediated by STE14 (Hrycina & Clarke 1990). Both of these modifications are needed for full biological activity (Anderegg et al. 1988). An interesting question is how does the **a**-factor get out of the cell? The predicted primary translation product does not contain an identifiable leader sequence to target it to the endoplasmic reticulum for secretion (Brake et al. 1985). Determination of the sequence of *STE6* (McGrath & Varshavsky 1989; Kuchler et al. 1989), a gene needed for **a**-factor production, has suggested one possibility. The STE6 protein is predicted to have twelve membrane spanning domains and to have some similarity to the mammalian multiple-drug-resistance protein (P-glycoprotein) which acts by pumping the drugs out of cells. The model proposed for STE6 action is that it transports **a**-factor out of the cell by a mechanism that is independent of the standard secretory pathway. The chemical modification of **a**-factor and unusual method of transport will need to be further investigated to determine if they have specific consequences for intercellular communication.

Pheromone receptor structure reveals similarity to the large family of seven transmembrane segment receptors

The cell surface receptor for α-factor is encoded by *STE2* (Burkholder & Hartwell 1985; Nakayama et al. 1985) and for **a**-factor by *STE3* (Nakayama et al. 1985; Hagen et al. 1986). Although both receptors are apparently able to activate the same intracellular signaling pathway (Bender & Sprague 1986; Nakayama et al. 1987), their predicted proteins do not share significant sequence similarity. However, the pheromone receptors are structurally similar in that they appear to contain seven transmembrane domains (Cartwright & Tipper 1991). In recent studies, a broad range of receptors have been identified with a similar serpentine structure including rhodopsin, β-adrenergic receptor and olfactory receptors (Dohlman et al. 1991). This family of receptors binds a wide range of ligands but all share the common functional property of transducing their signal through activation of an intracellular G protein. In fact, *S. cerevisiae* cells that express the heterologous *S. kluyveri* α-factor receptor (Marsh & Herskowitz 1988) or the β-adrenergic receptor (King et al. 1990) have both been shown to activate mating responses in response to their agonists.

The pheromone receptors play a complex role in coordinating the proper transmission of the pheromone signal. Ligand binding studies indicate that there are approximately 10,000 α-factor receptors per **a** cell with an affinity for α-factor of $Kd = 6 \times 10^{-9}$ M (Jenness et al. 1983). Association of the receptor with the G protein, the next component in the signaling pathway (see below), enhances the affinity for ligand (Blumer & Thorner 1990) as has been observed for other receptors. The C-terminus of the α-factor receptor mediates an adaptive response to pheromone that is probably similar to the receptor desensitization reactions mediated by the C-termini of rhodopsin and the β-adrenergic receptors (Reneke et al. 1988; Konopka et al. 1988). Ligand binding also induces the down regulation of α-factor receptors by endocytosis (Jenness and Spatrick 1986; Chvatchko et al. 1986). New receptor synthesis replaces the internalized receptors which are degraded in the vacuole. The stimulation of *STE2* transcription observed in response to α-factor exposure probably acts to maintain the level of receptors (Hartig et al. 1986). The function of receptor endocytosis is not known but it may be needed for regulating the quantitative level of signaling or for regulating the spatial distribution of signal in the cell since α-factor receptors become clustered to the site of morphogenesis and cell fusion during mating (Jackson et al. 1991).

How do the pheromone receptors appropriately activate the wide range of mating functions? All the necessary physiological responses for mating are apparently induced by the same G protein pathway since mutations that constitutively activate the pheromone signaling pathway downstream of the receptor allow cells to mate in the absence of pheromone receptors (Blinder et al. 1989). However, restoring the pheromone receptors improved the ability of the constitutively activated cells to discriminate between partner cells indicating that pheromone receptors may activate an alternate signal transduction pathway such as the ion transport activities observed for the opsins and the β-adrenergic receptor (Jackson et al. 1991). It remains to be determined if pheromone receptors have such an activity; an attempt to detect ion channel activity for α-factor receptors expressed in frog oocytes was not successful (Yu et al. 1989).

G protein signaling is highly regulated

Ligand bound pheromone receptors transduce the signal across the plasma membrane through activation of a heterotrimeric G protein. The G protein is composed of an α-subunit encoded by *GPA1* (also known as *SCG1*) (Dietzel & Kurjan 1987a; Miyajima et al. 1987), a β subunit encoded by *STE4*, and a γ subunit encoded by *STE18* (Whiteway et al. 1989). The pheromone signaling G protein is thought to be structurally and functionally similar to the G proteins found in many other diverse signaling systems. The model for G protein action is that ligand-bound receptors stimulate the α subunit to exchange bound GDP for GTP which causes its dissociation from the βγ subunits. The α subunit slowly hydrolyzes the GTP to GDP and then reassociates with the βγ subunits to return to the unstimulated state. In the well-studied cases of the β-adrenergic receptor or rhodopsin, GTP-bound G_α goes on to interact with the effector of the next step of the signal transduction pathway (Kaziro et al. 1991). Thus it was a surprising result that deletion of the gene encoding the yeast G_α subunit results in constitutive activation of the pheromone signaling pathway (Dietzel & Kurjan 1987a; Miyajima et al. 1987). Further studies have demonstrated that overproduction of the β subunit (Whiteway et al. 1990; Nomoto et al. 1990; Cole et al. 1990) and specific mutations in its gene ($STE4^{HPL}$; Blinder et al. 1989) will also cause constitutive signaling. These genetic studies indicate that the βγ subunits actively transmit the pheromone signal to the as yet unidentified next step in the signaling pathway. The βγ subunits of some other G proteins have also been reported to activate effectors such as phospholipaseA_2 (Kaziro et al. 1991).

Genetic analysis of the function of the α subunit supports the idea that the pheromone signaling G protein functions in a similar manner to other G proteins, although technical difficulties have prevented a direct *in vitro* demonstration. Heterologous G_α subunits from other organisms expressed in yeast can suppress the constitutive signaling of a *GPA1* deletion, indicating that they can interact with βγ to attenuate signaling (Dietzel & Kurjan 1987a; Kang et al. 1990). However, these heterologous G_α proteins cannot couple to the pheromone receptors and consequently the cells are sterile. *In vitro* mutagenesis of *GPA1* has implicated the C-terminus as being necessary for receptor interaction (Hirsch et al. 1991).

GTPase activity has been analyzed by site-directed mutation of conserved residues that are thought to be important for hydrolysis of GTP to GDP. These mutations have the effect of making cells hypersensitive to pheromone, as would be expected for a mutation that would prolong the activated state of G_α (Miyajima et al. 1989; Stone & Reed 1990; Kurjan et al. 1991). One of these mutants, $GPA1^{val50}$, corresponds to a mutation that blocks the GTPase activity of the human ras protein and unmasks the cancer-causing potential of this proto-oncogene product. Cells carrying the $GPA1^{val50}$ mutation are initially hypersensitive but do not maintain this state and eventually adapt to higher pheromone concentrations than wild-type cells (Miyajima et al. 1989; Kurjan et al. 1991). Some investigators have interpreted these results to indicate that the G_α subunit is not merely a negative regulator of the βγ subunits, but that GTP-bound G_α activates adaptation (Miyajima et al. 1989; Stone & Reed 1990). Matsumoto and colleagues have demonstrated that this adaptation is independent of the previously identified *SST2* adaptation pathway and is dependent on the function of the *SGV1* gene which encodes a putative protein kinase (Irie et al. 1991). In contrast, Kurjan and colleagues interpret the hyperadaptation of $GPA1^{val50}$ strains to be due to a dominant defect in pheromone signaling because it is dependent on gene doseage and because it is not observed with other mutations that are also expected to be defective in GTP hydrolysis. The different interpretations for the $GPA1^{val50}$ mutation highlight the need to develop *in vitro* functional assays for the pheromone stimulated G protein.

G protein function is regulated by several mechanisms in addition to GTP-binding. The stoichiometry of the G protein subunits is so tightly regulated that overproduction of G_α makes cells more resistant to pheromone (Dietzel & Kurjan 1987a) and even one extra copy of the G_β gene (*STE4*) causes constitutive pheromone signaling (Whiteway et al. 1990; Nomoto et al. 1990; Cole et al. 1990). This sensitive balance in G protein subunit stoichiometry may be used to regulate pheromone signaling since transcription of *GPA1* is stimulated by pheromone (Jahng et al. 1988). Phosphorylation also regulates the signaling activity of the βγ subunits because the β subunit is rapidly phosphorylated in response to pheromone and a deletion mutation that prevents this phosphorylation results in increased sensitivity to pheromone (Cole & Reed 1991). The possibility has been raised that G_β phosphorylation may be mediated by the SGV1 protein kinase (Irie et al. 1991) since the phosphorylation of G_β depends on the presence of G_α (Cole & Reed 1991). The *SST2* gene, which functions to permit adaptation to pheromone, must act at least in part by antagonizing the G protein signal because it is necessary for recovery from the effects of constitutively signaling alleles of G_β ($STE4^{HPL}$; Blinder & Jenness 1990). Additional regulation of G protein function may be mediated by the *CDC36, CDC39, CDC72, CDC73* and *SRM1* genes because mutation of these genes results in constitutive G protein signaling (de Barros Lopes et al. 1990; Clark & Sprague 1989; Neiman et al. 1990). It is not known why G protein function is regulated by so many different mechanisms; this regulation may be necessary to facilitate intercellular communication between mating cells.

The signal pathway

Activation of the βγ subunits of the G protein triggers an unknown effector; the involvement of a small molecule as in other signaling systems is still an open question. Since G protein subunits are thought to be membrane associated due to lipid modification (Whiteway et al. 1989; Blumer & Thorner 1990), it seems one possibility is that the next step in the signaling pathway will also be in the membrane. Components of this signal pathway have been identified by several different genetic strategies, and include STE5, STE7, STE11, STE12, FUS3 and KSS1 (MacKay & Manney 1974; Hartwell 1980; Courchesne et al. 1989; Elion et al. 1990). Mutations in genes encoding these components lead to the inability to respond to pheromone and thus a sterile phenotype. FUS3 and KSS1 have overlapping functions, and the double mutation is necessary to observe the defect in signal transduction (Elion et al. 1991). Figure 2 shows a schematic of the response pathway for α-factor, and the functions of these pathways components are described below. The pathway for **a**-factor response, apart from the pheromone and receptor, is the same.

An intact response pathway is required for pheromone to elicit any changes, as mutation of either *STE4, STE5, STE7, STE11* or *STE12* restores viability to a *gpa1* strain (Nakayama et al. 1988). One of the most sensitive indicators of pheromone response is transcriptional induction, and in particular induction of the *FUS1* gene (McCaffrey et al. 1987; Trueheart et al. 1987). Thus while pheromone binding appears to lead to changes that act both through and independent of new transcription, many studies have analyzed alterations in pathway components by their effects on gene induction. These alterations include loss of function and gain of function alleles and overproduction via high copy plasmids or a galactose-inducible promoter.

The function of STE5 is currently unknown, although genetic studies indicate that it acts after the G protein (Nakayama et al. 1988; Blinder et al. 1989). The *STE5* gene sequence predicts a protein of 917 amino acids, which is rapidly phosphorylated upon α-factor addition (M. Hasson, R. Freedman & J. Thorner, personal communication). In addition, these workers have shown that overproduction of STE5 leads to transcriptional activiation of a pheromone-responsive gene, and the requirement for the other pathway components in this activation is consistent with the order of function described above.

STE7, STE11, FUS3 and *KSS1* all encode protein kinases, based on the presence of conserved sequence motifs (Teague et al. 1986; Courchesne et al. 1989; Rhodes et al. 1990; Elion et al. 1990). A critical issue is what are the particular targets of specific kinases; while this is not yet known, many

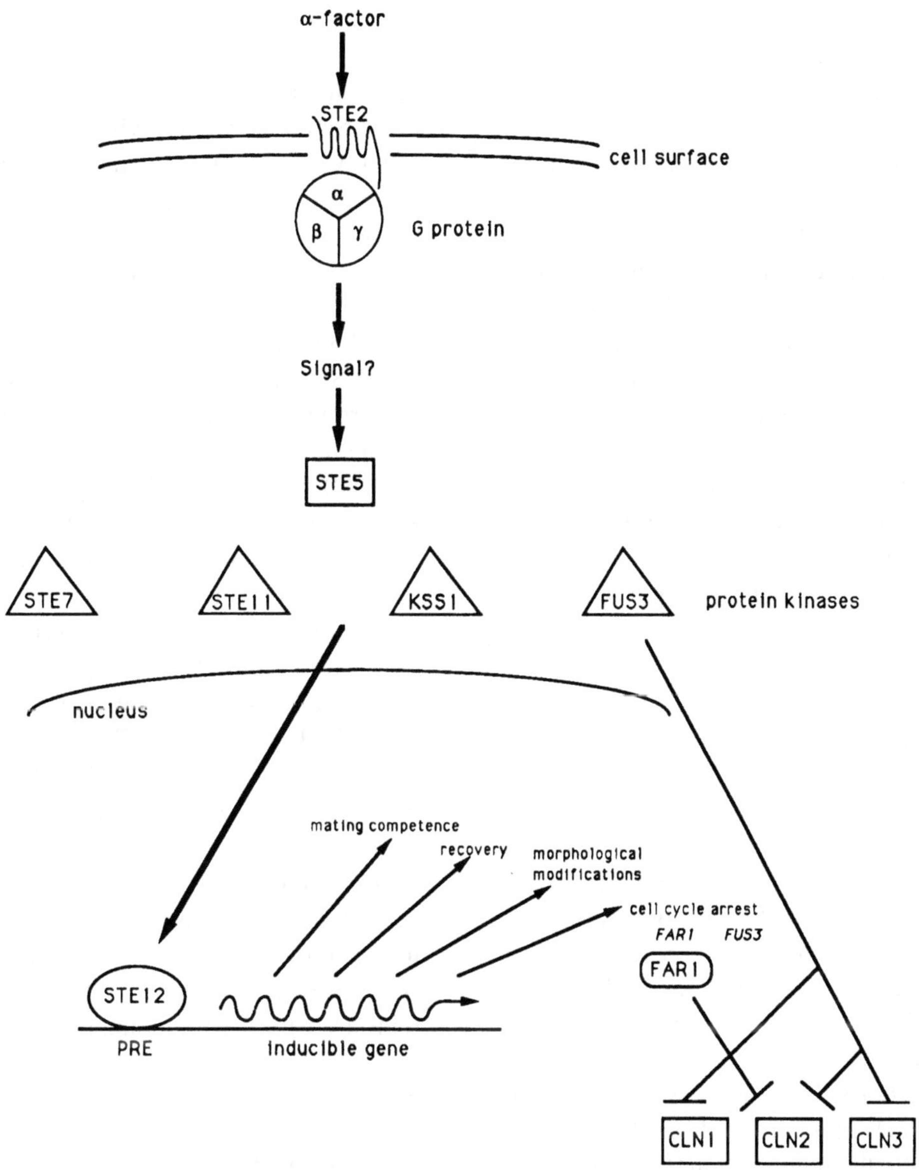

Fig. 2. Model of the pheromone response pathway in **a** cells; the pathway is essentially the same in α cells except for the pheromone and receptor. The arrows indicate activation by pheromone; the lines terminating with bars indicate repression by pheromone.

of the components of the pathway have been shown to be phosphorylated. For STE11, direct evidence for its role as a protein kinase has been obtained by showing that a mutation in the conserved lysine residue of protein kinases that is essential for ATP-binding abolishes function. In addition, an unknown protein of 78 kD appears to be a target for phosphorylation (Rhodes et al. 1990).

FUS3 was identified as a gene required for the cell fusion process (Elion et al. 1990), and *KSS1* as a gene that when overexpressed will suppress the adaptation defect of an *sst2* mutant (Courchesne et al. 1989). While a mutation in either gene alone allows α-factor induction of transcription, the double mutant completely fails to induce (Elion et al. 1991). For both FUS3 and KSS1, the proteins resemble the ERK family of protein kinases involved in other signal transduction processes (Boulton et al. 1991). Recently, these two yeast protein kinases have been reported to be phosphorylated upon both tyrosine and threonine upon pheromone addition (A. Gartner & G. Ammerer, personal communication). This result thus directly implicates tyrosine phosphorylation in the yeast response pathway, in parallel with its role in growth control in many animal systems.

One outstanding question regarding the multiple protein kinases in this pathway concerns their activation by pheromone: how does the signal from G protein dissociation get conveyed to these kinases? Other questions revolve around the hierarchy of response: does one kinase activate the next, are there multiple branches involving different kinases, or does more than one kinase target a particular substrate? While genetic evidence addressing some of these points has recently been obtained, biochemical experiments will be necessary to demonstrate changes in kinase activity upon pheromone addition and phosphorylation of particular substrates by a given kinase.

Cell cycle arrest

Progression through Start requires the protein kinase activity of the CDC28 protein and the G1 cyclins, CLN1, CLN2 and CLN3 (reviewed in Reed 1991). Cells are viable if any one of the three cyclins is present, but loss of all three leads to large, shmoo-like, G1-arrested cells (Richardson et al. 1989). Therefore pheromone must somehow eliminate the function of all three cyclins. α-Factor treatment leads to the rapid loss of *CLN1* and *CLN2* mRNA (Wittenberg et al. 1990), but *CLN3* mRNA is actually slightly induced (Nash et al. 1988).

The phenotype that has been used to identify mutants specifically affected in the arrest process is that of being defective for cell cycle arrest upon α-factor treatment, but proficient for transcriptional induction. Three mutants are known to show this pattern: *far1*, *fus3* and *CLN3-1*. The *FAR1* gene product acts to eliminate CLN2 function: a *far1* mutant fails to arrest in the presence of α-factor, whereas a *far1 cln2* double mutant does arrest (Chang & Herskowitz 1990). Although the *FAR1* gene is induced by α-factor, thus providing a link between transcriptional induction and cell cycle arrest, this induction alone is insufficient for mediating FAR1 inhibition of CLN2. Thus the FAR1 protein may require posttranslational modifications that are the result of pheromone treatment. Similarly, while FAR1 is required for the loss of *CLN2* mRNA upon α-factor treatment, its role in cell cycle arrest appears to be through post-transcriptional changes of the CLN2 protein (F. Chang and I. Herskowitz, personal communication).

FUS3 is required both for cell cycle arrest and, with KSS1, transcriptional induction. The defect in G1 arrest in response to pheromone of a *fus3* mutant can be suppressed by a *cln3* null mutation (Elion et al. 1990). Therefore, FUS3 appears to function to inactivate CLN3, most likely by a post-transcriptional effect. Since the FUS3 sequence indicates a likely protein kinase, inactivation of CLN3 may occur by FUS3-mediated phosphorylation. In addition, FUS3 is required for repression of CLN1 and CLN2 mRNAs in response to pheromone (Elion et al. 1991). However, the CLN1 and CLN2 proteins also appear to be inactivated posttranscriptionally by a FUS3-independent mechanism (Elion et al. 1991).

Hyperactive alleles of *CLN3* were identified that led to α-factor resistance and to small cell size, due to shortening of the G1 phase of the cell cycle (Cross 1988; Nash et al. 1988). These CLN3 proteins are truncated, missing the C-terminal region which may signal the degradation of the wild type protein. Therefore the long-lived cyclin may be a powerful activator of Start, and hence incapable of being inactivated by α-factor.

It is predicted that an as yet unidentified product is responsible for the inactivation of CLN1, which might be uncovered in a *cln2 cln3* mutant that is resistant to pheromone (Chang & Herskowitz 1990). Of interest will be how FAR1, FUS3 and this additional component coordinate their activities to inactivate the three G1 cyclins and bring about cell cycle arrest. It is also possible that in addition to the cyclins, the CDC28 protein is also a target for inactivation by pheromone. With respect to initiating a new cell cycle, the effects of all these inactivating components must be reversed upon either mating or adaptation to pheromone.

Transcriptional induction

Many cell-type-specific transcripts produced by **a**-, α- and haploid-specific genes are induced upon pheromone addition. Besides being necessary for this induction, the pheromone response pathway plays a major role in setting the uninduced level of transcription of these genes (Hartig et al. 1986; McCaffrey et al. 1987; Fields et al. 1988). Thus it is the combination of this pathway acting with the *MAT*-encoded proteins that determines the amount of cell-type-specific expression (reviewed in Dolan & Fields 1991). While the mating type locus is the primary determinant of cell identity, there is no evidence that transcription of the *MAT* genes themselves or the activity of the encoded transcriptional regulators is responsive to pheromone treatment.

The α1 and α2 proteins function by binding in cooperation with a ubiquitous transcriptional regulator, MCM1 (Bender & Sprague 1987; Keleher et al. 1988; Passmore et al. 1988; Ammerer 1990). MCM1 is required for activation of α-specific and **a**-specific transcription in α and **a** cells, respectively, and for repression of **a**-specific transcription in α cells. Like the *MAT*-encoded proteins, MCM1 activity has not been demonstrated to respond to pheromone. The MCM1 sequence has extensive homology to the mammalian serum response factor (SRF) as well as to other DNA-binding proteins. In parallel with MCM1, SRF also appears to function in cooperation with transcription factors that bind adjacent to it (Ryan et al. 1989; Shaw et al. 1989).

The STE12 protein is the DNA-binding component that has been shown to be necessary for pheromone induction of transcription. STE12 binds to the pheromone response element (PRE) which mediates α-factor induction of **a**-specific genes and α- and **a**-factor induction of haploid-specific genes (Dolan et al. 1989; Errede and Ammerer 1989). STE12 is also necessary for **a**-factor induction of α-specific genes, although its role in this process is less defined. In addition to evidence that it binds to the PRE, a direct role for STE12 in pheromone-induced transcription was shown by generating a fusion of this protein with the DNA-binding domain of the yeast protein GAL4 (Song et al. 1991). This fusion protein can activate transcription of a gene under GAL4 regulation, but only if cells are treated with pheromone. This experiment indicates that STE12 is sufficient to mediate induction and does not absolutely require any other proteins bound to DNA. In addition, STE12 becomes rapidly hyperphosphorylated upon pheromone treatment, and this modification correlates with the ability of STE12 to function in induced transcription (Song et al. 1991).

The binding of STE12 to a PRE is highly dependent on adjacent sequences. Cooperative binding has been demonstrated for STE12 with MCM1 and a Ty1-binding factor (Errede and Ammerer, 1989) as well as with itself (Yuan and Fields, 1991). Thus the cell can differentially modulate both the uninduced and induced expression of the appropriate cell-type-specific genes. It remains to be established how the phosphorylation of STE12 is regulated by pheromone, and how the different modified forms of this protein function in the transcriptional process.

The time course of transcriptional induction is very rapid and this induction does not require new protein synthesis (see for example, McCaffrey et al. 1987). The induction also appears to be dose-responsive, with higher levels of pheromone leading to greater increases in transcription (as assayed by increased agglutination in Moore 1983). The expression of a large number of genes is pheromone-inducible, and their encoded products function to facilitate cell-cell communication, to generate changes in the cell surface, to mediate cell cycle arrest, and to promote recovery and adaptation. Thus as in many other eukaryotic signaling systems, a major result of ligand binding is a change in gene expression which can affect numerous cellular processes.

Signal transduction coordinately induces physiological responses required for mating

Pheromone signaling must coordinate the intercellular communication between cells with the induction of mating responses within cells. The mating cells presumably sense a pheromone gradient to spatially locate their partner, but the molecular mechanisms by which they organize the future site of cell fusion within the cell are not known. Cytological analysis of mating cells indicates that this is a highly ordered process during which cell growth becomes polarized to the future site of cell fusion (Baba et al. 1989; Byers 1981). The polarization of cell growth is probably mediated by changes in cytoskeletal elements such as actin and SPA2 (Drubin 1991). The polarization of cell growth that occurs during budding in the vegetative life cycle has many similarities to pheromone-induced morphogenesis (Drubin 1991). In the case of budding, specific genes involved in bud site selection have been identified, but they do not appear to be necessary for mating (Chant and Herskowitz 1991; Chant et al. 1991). Determining the mechanism by which signal transduction coordinates cell polarization will be of great interest since this is a basic question that pertains not only to the development of yeast but to complex multicellular organisms as well.

Once the decision to polarize cell growth is made, new cell growth and secretion are targeted in that direction. Several proteins known to be targeted to the future site of cell fusion facilitate mating. Cell surface agglutinins are produced which facilitate the mating process by holding the non-motile yeast cells together (Lipke et al. 1989). New cell wall growth fuses the cell walls of the partner cells, making it safe for the cells to degrade a portion of cell wall. This degradation allows the plasma membranes to fuse and to create a pore that connect the two cells to form a zygote. Two genes, *FUS1* and *FUS2*, are known to be needed for this cell fusion (Trueheart et al. 1987). The FUS1 protein is an O-glycosylated integral membrane protein that is localized to the projection of pheromone-induced morphogenesis (Trueheart & Fink 1989). In addition to coordinating morphogenesis, polarization of secretion may also reinforce intercellular communication. One can imagine that pheromone secretion emanating from a site opposed to the partner cell would provide that cell with a strong signal. In the receiving cell, localization of pheromone receptors to the site of morphogenesis, as has been observed for the α-factor receptors (Jackson et al. 1991), may also act to reinforce the spatial sensing.

The final step in mating is nuclear fusion. Early in the mating process the nuclei are observed to be in position near the site of morphogenesis, with the spindle pole facing toward the future site of cell fusion (Baba et al. 1989; Byers 1981). Nuclear fusion is a pheromone-stimulated event and not simply a natural consequence of two nuclei present in the same cell (Rose et al. 1986). Studies on nuclear fusion indicate the process is mediated by microtubules and that the KAR3 protein may be responsible for moving the nuclei together (Meluh & Rose 1990).

Adaptation

Haploid cells that are not successful in finding a mating partner are able to adapt to pheromone and resume growth (Moore 1984). Adaptation is known to be carried out by mechanisms described in the previous sections that operate on the pheromone receptors and G protein subunits. The strongest known contributor to adaptation, *SST2*, is not required for signal transduction (Dietzel & Kurjan 1987b). The mechanism of SST2 action is not known but it appears to act at least in part by antagonizing the G protein signal (Blinder & Jenness 1989). The mechanisms needed for adaptation are also needed for regulating pheromone signaling because cells that are defective in adaptation show a defect in mating partner selection proportional to their degree of hypersensitivity (Jackson et al. 1990b). In addition to having adaptation mechanisms, cells stimulate recovery by secreting a protease that degrades the pheromone produced by the opposite cell type. The protease for α-factor produced by **a** cells is known as BAR1 (MacKay et al. 1988), and the protease for **a**-factor produced by α cells is probably encoded by *SSL1* (Steden et al. 1989). It seems likely that as additional signaling components are analysed, other regulatory mechanisms will be identified that contribute to adaptation.

Cells that have successfully mated recover and resume cell division because thay take on a new identity as an **a**/α diploid which is insensitive to pheromone. The acquisition of the diploid identity is rapid since *MAT*-determined regulation can begin as soon as cell fusion occurs. Since some signal transduction components will be rapidly depleted, such as by endocytosis of pheromone receptors, the pre-existing signal transduction components should not continue to send signal for very long. This form of recovery is probably sufficient because there appears to be no long-lived pheromone signal intermediate, given that haploid cells recover quickly after pheromone is removed (Konopka et al. 1988).

Concluding remarks

S. cerevisiae conjugation is a highly ordered process which requires the spatial and temporal coordination of a wide range of physiological responses. The pheromone signal transduction pathway mediates the intercellular communication leading to the intracellular induction of mating responses. Many of the genes that mediate pheromone signaling have been identified because of the amenability of this yeast to genetic and molecular biological approaches. Remarkably, the yeast cells employ signal transduction components analogous to those used by more complex eukaryotes (Boulton et al. 1990; Dohlman et al. 1991; Kaziro et al. 1991; Marx 1991). As a result, yeast conjugation has become a uniquely appropriate system for the study of complex cell biological processes such as hormone signaling, differentiation, cell cycle control, cell polarization, and induction of gene expression. Future studies in these areas will undoubtedly benefit from the scientific interaction of yeast researchers with those studying similar problems in a wide range of experimental organisms.

References

Achstetter T & Wolf DH (1985) Hormone processing and membranebound proteinases in yeast. EMBO J. 4: 173–177

Ammerer G (1990) Identification, purification, and cloning of a polypeptide (PRTF/GRM) that binds to mating-specific promoter elements in yeast. Genes Dev. 4: 299–312

Anderegg RJ, Betz R, Carr SA, Crabb JW & Duntze W (1988) Structure of *Saccharomyces cerevisiae* mating hormone **a**-factor. J. Biol. Chem. 263: 18236–18240

Baba M, Baba N, Ohsumi Y, Kanaya K & Ohsumi M (1989) Three dimensional analysis of morphogenesis induced by mating pheromone α-factor in *Saccharomyces cerevisiae*. J. Cell Science 94: 207–216

de Barros Lopes M, Ho J-Y & Reed SI (1990) Mutations in cell division cycle genes *CDC36* and *CDC39* activate the *Saccharomyces cerevisiae* mating pheromone response pathway. Mol. Cell. Biol. 10: 2966–2972

Bender A & Sprague GF (1986) Yeast peptide pheromones, **a**-factor and α-factor, activate a common response mechanism in their target cells. Cell 47: 929–937

—(1987) MATα1 protein, a yeast transcription activator, binds synergistically with a second protein to a set of cell-type-specific genes. Cell 50: 681–691

—(1989) Pheromones and pheromone receptors are the primary determinants of mating specificity in the yeast *Saccharomyces cerevisiae*. Genetics 121: 463–476

Blinder D & Jenness D (1989) Regulation of post-receptor signaling in the pheromone response pathway of *Saccharomyces cerevisiae*. Mol. Cell. Biol. 9: 3720–3726

Blinder D, Bouvier S & Jenness DD (1989) Constitutive mutants in the yeast pheromone response: ordered function of the gene products. Cell 56: 479–486

Blumer KJ & Thorner J (1990) β and γ subunits of a yeast guanine nucleotide-binding protein are not essential for membrane association of the α subunit but are required for receptor coupling. Proc. Natl. Acad. Sci. U.S.A. 87: 4363–4367

Boulton TG, Nye SH, Robbins DJ, Ip NY, Radziejewska E, Morgenbesser SD, DePinho RA, Panayotatos N, Cobb MH & Yancopoulos GD (1991) ERKs: a family of protein-serine/threonine kinases that are activated and tyrosine phosphorylated in response to insulin and NGF. Cell 65: 663–675

Brake AJ, Brenner C, Najarian R, Laybourn P & Merryweather J (1985) Structure of genes encoding precursors of yeast peptide mating pheromone **a**-factor. In: Gething MJ (Ed) Current Communications in Molecular Biology (pp. 103–108). Cold Spring Harbor Laboratory, New York

Burkholder AC & Hartwell LH (1985) The yeast α-factor receptor: structural properties deduced from the sequence of the *STE2* gene. Nucleic Acids Res. 13: 8463–8475

Byers B (1981) Cytology of the yeast life cycle. In: Strathern JN, Jones EW & Broach JR (Eds) The Molecular Biology of the Yeast Saccharomyces. I. Life Cycle and Inheritance (pp. 59–96). Cold Spring Harbor Laboratory, New York

Cartwright CP & Tipper DJ (1991) *In vivo* analysis of Ste2, a yeast plasma membrane protein, by using β-lactamase gene fusions. Mol. Cell. Biol. 11: 2620–2628

Chang F & Herskowitz I (1990) Identification of a gene necessary for cell cycle arrest by a negative growth factor of yeast: FAR1 is an inhibitor of a G1 cyclin, CLN2. Cell 63: 999–1011

Chant J & Herskowitz I (1991) Genetic control of bud site selection in yeast by a set of gene products that constitute a morphogenetic pathway. Cell 65: 1203–1212

Chant J, Corrado K, Pringle JR & Herskowitz I (1991) Yeast

BUD5, encoding a putative GDP-GTP exchange factor, is necessary for bud site selection and interacts with bud formation gene *BEM1*. Cell 65: 1213–1224

Chvatchko Y, Howald I & Reizman H (1986) Two yeast mutants defective in endocytosis are defective in pheromone response. Cell 46: 355–364

Clark KL & Sprague Jr GF (1989) Yeast pheromone response pathway: characterization of a suppressor that restores mating to receptorless mutants. Mol. Cell. Biol. 9: 2682–2694

Cole GM & Reed SI (1991) Pheromone induced phosphorylation of a G protein β subunit in *S. cerevisiae* is associated with an adaptive response to mating pheromone. Cell 64: 703–716

Cole GM, Stone DE & Reed SI (1990) Stoichiometry of G protein subunits affects the *Saccharomyces cerevisiae* mating pheromone signal transduction pathway. Mol. Cell. Biol. 10: 510–517

Courchesne WE, Kunisawa R & Thorner J (1989) A putative protein kinase overcomes pheromone-induced arrest of cell cycling in *S. cerevisiae*. Cell 58: 1107–1119

Cross F (1988) *DAF1*, a mutant gene affecting size control, pheromone arrest, and cell cycle kinetics of *Saccharomyces cerevisiae*. Mol. Cell. Biol. 8: 4675–4684

Dietzel C & Kurjan J (1987a) The yeast *SCG1* gene: a G_α-like protein implicated in the **a**- and α-factor response pathway. Cell 50: 1001–1010

—(1987b) Pheromonal regulation and sequence of the *Saccharomyces cerevisiae SST2* gene: a model for desensitization to pheromone. Mol. Cell. Biol. 7: 4169–4177

Dmochowska A, Dignard D, Henning D, Thomas DY & Bussey H (1987) Yeast *KEX1* gene encodes a putative protease with a carboxypeptidase B-like function involved in killer toxin and α-factor precursor processing. Cell 50: 573–584

Dohlman HG, Thorner J, Caron MG & Lefkowitz RJ (1991) Model systems for the study of 7-transmembrane-segment receptors. Ann. Rev. Biochem. 60: 653–688.

Dolan JW & Fields S (1991) Cell-type-specific transcription in yeast. Biochim. et Biophys. Acta 1088: 155–169

Dolan JW, Kirkman C & Fields S (1989) The yeast STE12 protein binds to the DNA sequence mediating pheromone induction. Proc. Natl. Acad. Sci. U.S.A. 86: 5703–5707

Drubin DG (1991) Development of cell polarity in budding yeast. Cell 65: 1093–1096

Egel R, Nielson O & Weilguny D (1990) Sexual differentiation in fission yeast. Trends in Genet. 6: 369–373

Elion EA, Grisafi PL & Fink GR (1990) *FUS3* encodes a cdc2+/CDC28-related kinase required for the transition from mitosis into conjugation. Cell 60: 649–664

Elion EA, Brill JA & Fink GR (1991) FUS3 represses CLN1 and CLN2 and in concert with KSS1 promotes signal transduction. Proc. Natl. Acad. Sci. U.S.A. 88: 9392–9396

Errede B & Ammerer G (1989) STE12, a protein involved in cell-type-specific transcription and signal transduction in yeast, is part of protein-DNA complexes. Genes Dev. 3: 1349–1361

Fields S, Chaleff DT & Sprague GF Jr (1988) Yeast *STE7*, *STE11* and *STE12* genes are required for expression of cell-type-specific genes. Mol. Cell. Biol. 8: 551–556

Hagen DC, McCaffrey G & Sprague GF (1986) Evidence the yeast *STE3* gene encodes a receptor for the peptide pheromone **a** factor: gene sequence and implications for the structure of the presumed receptor. Proc. Natl. Acad. Sci. U.S.A. 83: 1418–1422

Hartig A, Holly J, Saari G & MacKay VL (1986) Multiple regulation of *STE2*, a mating-type-specific gene of *Saccharomyces cerevisiae*. Mol. Cell. Biol. 6: 2106–2114

Hartwell L (1980) Mutants of *Saccharomyces cerevisiae* unresponsive to cell division control by polypeptide mating hormone. J. Cell Biol. 85: 811–822

Herskowitz I (1989) A regulatory hierarchy for cell specialization in yeast. Nature 342: 749–757

Hirsch JP, Dietzel C & Kurjan J (1991) The carboxyl-terminus of Scg1, the Gα subunit involved in yeast mating, is implicated in interactions with the pheromone receptors. Genes Dev. 5: 467–474

Hyrcina CA & Clarke S (1990) Farnesyl cysteine C-terminal methyltransferase activity is dependent upon the *STE14* gene product in *Saccharomyces cerevisiae*. Mol. Cell. Biol. 10: 5071–5076

Irie K, Nomoto S, Miyajima I & Matsumoto K (1991) *SGV1* encodes a *CDC28/cdc2*-related kinase required for a Gα subunit-mediated adaptive response to pheromone in *S. cerevisiae*. Cell 65: 785–795

Jackson CL & Hartwell LH (1990a) Courtship in *Saccharomyces cerevisiae*: an early cell-cell interaction during mating. Mol. Cell. Biol. 5: 2202–2213

Jackson CL & Hartwell LH (1990b) Courtship in *Saccharomyces cerevisiae*: both cell types choose mating partners by responding to the strongest pheromone signal. Cell 63: 1039–1051

Jackson CL, Konopka JB & Hartwell LH (1991) *S. cerevisiae* α-pheromone receptors activate a novel signal transduction pathway for mating partner discrimination. Cell 67: 389–402

Jahng K-Y, Ferguson J & Reed SI (1988) Mutations in a gene encoding the α subunit of a *Saccharomyces cerevisiae* G protein indicate a role in mating pheromone signaling. Mol. Cell. Biol. 8: 2484–2493

Jenness DD & Spatrick P (1986) Down regulation of the α-factor pheromone receptor in *Saccharomyces cerevisiae*. Cell 46: 345–353

Jenness DD, Burkholder AC & Hartwell LH (1983) Binding of α-factor pheromone to yeast **a** cells: chemical and genetic evidence for an α-factor receptor. Cell 35: 521–529

Julius D, Blair L, Brake A, Sprague G & Thorner J (1983) Yeast α-factor is processed from a larger precursor polypeptide: the essential role of a membrane-bound dipeptidyl aminopeptidase. Cell 32: 839–852

Julius D, Brake A, Blair L, Kunisawa R & Thorner J (1984) Isolation of the putative structural gene for the lysine-arginine-cleaving endopeptidase required for the processing of yeast pre-pro-alpha-factor. Cell 37: 1075–1089

Kang Y-S, Kane J, Kurjan J, Stadel JM & Tipper DJ (1990)

Effects of expression of mammalian Gα and hybrid mammalian-yeast Gα proteins on the yeast pheromone response signal transduction pathway. Mol. Cell. Biol. 10: 2582–2590

Kaziro Y, Itoh H, Kozasa T, Nakafuku M & Satoh T (1991) Structure and function of signal-transducing GTP-binding proteins. Ann Rev. Biochem. 60: 349–400

Keleher CA, Goutte C & Johnson AD (1988) The yeast cell-type-specific repressor α2 acts cooperatively with a non-cell-type-specific protein. Cell 53: 927–936

King K, Dohlman HG, Thorner J, Caron MG & Lefkowitz RJ. Control of yeast mating signal transduction by a mammalian β_2-adrenergic receptor and the G_S α-subunit. Science 250: 121–123

Konopka JB, Jenness DD & Hartwell LH (1988) The C-terminus of the *Saccharomyces cerevisiae* α-pheromone receptor mediates an adaptive response to pheromone. Cell 54: 609–620

Kuchler K, Sterne RE & Thorner J (1989) *Saccharomyces cerevisiae STE6* gene product: a novel pathway for protein export in eukaryotic cells. EMBO J 8: 3973–3984

Kurjan J & Herskowitz I (1982) Structure of a yeast pheromone gene (*MF*α): a putative α-factor precursor contains four tandem copies of mature α-factor. Cell 30: 933–943

Kurjan J, Hirsch JP & Dietzel C (1991) Mutations in the guanine nucleotide-binding domains of a yeast Gα protein confer a constitutive or uninducible state to the pheromone response pathway. Genes Dev. 5: 475–483

Lipke PN, Wojciechowicz D & Kurjan J (1989) *AG*α*1* is the structural gene for the *Saccharomyces cerevisiae* α-agglutinin, a cell-surface glycoprotein involved in cell-cell interactions during mating. Mol. Cell. Biol. 9: 3155–3165

MacKay VL & Manney TR (1974) Mutations affecting sexual conjugation and related processes in *Saccharomyces cerevisiae*. II. Genetic analysis of nonmating mutants. Genetics 76: 272–282

MacKay VL, Welch SK, Insley MY, Manney TR, Holly J, Saari GC & Parker ML (1988) The *Saccharomyces cerevisiae BAR1* gene encodes an exported protein with homology to pepsin. Proc. Natl. Acad. Sci. USA 85: 55–59

Marsh L & Herskowitz I (1988) STE2 protein of *Saccharomyces kluyveri* is a member of the rhodopsin/β-adrenergic receptor family and is responsible for recognition of the peptide ligand α-factor. Proc. Natl. Acad. Sci. USA. 85: 3855–3859

Marx J (1991) How peptide hormones get ready for work. Science 252: 779–780

McCaffrey G, Clay FJ, Kelsay K & Sprague GF Jr (1987) Identification and regulation of a gene required for cell fusion during mating of the yeast *Saccharomyces cerevisiae*. Mol. Cell. Biol. 7: 2680–2690

McGrath JP & Varshavsky A (1989) The yeast *STE6* gene encodes a homologue of the mammalian multidrug resistance P-glycoprotein. Nature 340: 400–404

Meluh P & Rose MD (1990) *KAR3*, a kinesin-related gene required for yeast nuclear fusion. Cell 60: 1029–1041

Miyajima I, Nakafuku M, Nakayama N, Brenner C, Miyajima A, Kaibuchi K, Arai K, Kaziro Y & Matsumoto K (1987) *GPA1*, a haploid-specific essential gene, encodes a yeast homolog of mammalian G protein which may be involved in mating factor signal transduction. Cell 50: 1011–1019

Miyajima I, Arai K & Matsumoto K (1989) $GPA1^{val50}$ mutation in the mating-factor signaling pathway in *Saccharomyces cerevisiae*. Mol. Cell. Biol. 9: 2289–2297

Moore SA (1983) Comparison of dose-response curves for α factor-induced cell division arrest, agglutination, and projection formation of yeast cells. J. Biol. Chem. 258: 13849–13856

—(1984) Yeast cells recover from α-factor-induced division arrest in the absence of α-factor destruction. J. Biol. Chem. 259: 1004–1010

Nakayama N, Miyajima A & Arai K (1985) Nucleotide sequences of *STE2* and *STE3*, cell type specific sterile genes from *Saccharomyces cerevisiae*. EMBO J. 4: 2643–2648

Nakayama N, Miyajima A & Arai K (1987) Common signal transduction system shared by *STE2* and *STE3* in haploid cells of *Saccharomyces cerevisiae*: autocrine cell-cycle arrest results from forced expression of *STE2*. EMBO J. 6: 249–254

Nakayama N, Kaziro Y, Arai K-I & Matsumoto K (1988) Role of *STE* genes in the mating factor signaling pathway mediated by *GPA1* in *Saccharomyces cerevisiae*. Mol. Cell. Biol. 8: 3777–3783

Nash R, Tokiwa G, Anand S, Erickson K & Futcher AB (1988) The $WHI1^+$ gene of *Saccharomyces cerevisiae* tethers cell division to cell size and is a cyclin homolog. EMBO J. 7: 4335–4346

Nasmyth K & Shore D (1987) Transcriptional regulation in the yeast life cycle. Science 237: 1162–1170

Neiman AM, Chang F, Komachi K & Herskowitz I (1990) CDC36 and CDC39 are negative elements in the signal transduction pathway of yeast. Cell. Reg. 1: 391–401

Nomoto S, Nakayama N, Arai K & Matsumoto K (1990) Regulation of the yeast pheromone response pathway by G protein subunits. EMBO J. 9: 691–696

Passmore S, Maine GT, Elble R, Christ C & Tye B (1988) *Saccharomyces cerevisiae* protein involved in plasmid maintenance is necessary for mating of *MAT*α cells. J. Mol. Biol. 204: 593–606

Powers S, Michaelis S, Broek D, Santa Anna-A S, Field J, Herskowitz I & Wigler M (1986) *RAM*, a gene of yeast required for a functional modification of RAS proteins and for production of mating pheromone **a**-factor. Cell 47: 413–422

Reed SI (1991) G1-specific cyclins: in search of an S-phase-promoting factor. Trends in Genetics 7: 95–99

Reneke JE, Blumer KJ, Courchesne WE & Thorner J (1988) The carboxyl-terminal segment of the yeast α-factor receptor is a regulatory domain. Cell 55: 221–234

Rhodes N, Connell L & Errede B (1990) STE11 is a protein kinase required for cell-type-specific transcription and signal transduction in yeast. Genes Dev. 4: 1862–1874

Richardson HE, Wittenberg C, Cross F & Reed SI (1989) An essential G1 function for cyclin-like proteins in yeast. Cell 59: 1127–1133

Rose MD, Price BR & Fink GR (1986) *Saccharomyces cere-*

visiae nuclear fusion requires prior activation by alpha-factor. Mol. Cell. Biol. 6: 3490–3497

Ryan WA Jr, Franza BR Jr & Gilman MZ (1989) Two distinct cellular phosphoproteins bind to the *c-fos* serum response element. EMBO J. 6: 1785–1792

Shaw PE, Schroter H & Nordheim A (1989) The ability of a ternary complex to form over the serum response element correlates with serum inducibility of the human *c-fos* promoter. Cell 56: 563–572

Song O, Dolan JW, Yuan YO & Fields S (1991) Pheromone-dependent phosphorylation of the yeast STE12 protein correlates with transcriptional activation. Genes Dev. 5: 741–750

Sprague GF Jr (1990) Combinatorial associations of regulatory proteins and the control of cell type in yeast. Adv. Genet. 27: 33–67

Steden M, Betz R & Duntze W (1989) Isolation and characterization of *Saccharomyces cerevisiae* mutants supersensitive to G1 arrest by the mating hormone **a**-factor. Mol. Gen. Genet. 219: 439–444

Stone DE & Reed SI (1990) G protein mutations that alter the pheromone response in *Saccharomyces cerevisiae*. Mol. Cell. Biol. 10: 4439–4446

Teague MA, Chaleff DT & Errede B (1986) Nucleotide sequence of the yeast regulatory gene *STE7* predicts a protein homologous to protein kinases. Proc. Natl. Acad. Sci. USA 83: 7371–7375

Trueheart & Fink GR (1989) The yeast cell fusion protein FUS1 is O-glycosylated and spans the plasma membrane. Proc. Natl. Acad. Sci. USA. 86: 9916–9920

Trueheart J, Boeke JD & Fink GR (1987) Two genes required for cell fusion during yeast conjugation: evidence for a pheromone-induced surface protein. Mol. Cell. Biol. 7: 2316–2328

Whiteway M, Hougan L, Dignard D, Thomas DY, Bell L, Saari GC, Grant FJ, O'Hara P & MacKay VL (1989) The *STE4* and *STE18* genes of yeast encode potential β and γ subunits of the mating factor receptor-coupled G protein. Cell 56: 467–477

Whiteway M, Hougan L & Thomas DY (1990) Overexpression of the *STE4* gene leads to mating response in haploid *Saccharomyces cerevisiae*. Mol. Cell. Biol. 10: 217–222

Wittenberg C, Sugimoto K & Reed SI (1990) G1-specific cyclins of *S. cerevisiae*: cell cycle periodicity, regulation by mating pheromones and association with the $p34^{CDC28}$ protein kinase. Cell 62: 225–237

Yu L, Blumer KJ, Davidson N, Lester HA & Thorner J (1989) Functional expression of the yeast α-factor receptor in *Xenopus* oocytes. J. Biol. Chem. 264: 2087–20850

Yuan YO & Fields S (1991) Properties of the DNA-binding domain of the *Saccharomyces cerevisiae* STE12 protein. Mol. Cell. Biol. 11: 5910–5918

Antonie van Leeuwenhoek **62**: 109–130, 1992.

The RAS-adenylate cyclase pathway and cell cycle control in *Saccharomyces cerevisiae*

Johan M. Thevelein
Laboratorium voor Moleculaire Celbiologie, Katholieke Universiteit te Leuven, Kardinaal Mercierlaan 92, B-3001 Leuven-Heverlee, Flanders, Belgium

Key words: yeast, cAMP, growth control, RAS-oncogene, general glucose sensor, signal transduction, nutrients

Abstract

The cell cycle of *Saccharomyces cerevisiae* contains a decision point in G_1 called 'start', which is composed of two specific sites. Nutrient-starved cells arrest at the first site while pheromone-treated cells arrest at the second site. Functioning of the RAS-adenylate cyclase pathway is required for progression over the nutrient-starvation site while overactivation of the pathway renders the cells unable to arrest at this site. However, progression of cycling cells over the nutrient-starvation site does not appear to be triggered by the RAS-adenylate cyclase pathway in response to a specific stimulus, such as an exogenous nutrient. The essential function of the pathway appears to be limited to provision of a basal level of cAMP. cAMP-dependent protein kinase rather than cAMP might be the universal integrator of nutrient availability in yeast. On the other hand stimulation of the pathway in glucose-derepressed yeast cells by rapidly-fermented sugars, such as glucose, is well documented and might play a role in the control of the transition from gluconeogenic growth to fermentative growth. The initial trigger of this signalling pathway is proposed to reside in a 'glucose sensing complex' which has both a function in controlling the influx of glucose into the cell and in activating in addition to the RAS-adenylate cyclase pathway all other glucose-induced regulatory pathways in yeast. Two crucial problems remaining to be solved with respect to cell cycle control are the nature of the connection between the RAS-adenylate cyclase pathway and nitrogen-source induced progression over the nutrient-starvation site of 'start' and second the nature of the downstream processes linking the RAS-adenylate cyclase pathway to Cyclin/CDC28 controlled progression over the pheromone site of 'start'.

Abbreviations: cAMP-PK – cAMP-dependent protein kinase

Introduction

Interest in the yeast *RAS* genes originates from the oncogenic properties of specific alleles of the mammalian *ras* genes. These *ras* oncogenes are the main type of oncogene found in naturally occurring and artificially induced mammalian tumors (Barbacid 1987). The *ras* gene products belong to the group of G proteins. These proteins are active in the GTP-bound state and inactive in the GDP-bound state. Exchange of GDP for GTP, i.e. activation of the G protein, is triggered by specific exchange factors while conversion of bound GTP to GDP, i.e. inactivation of the G protein, is triggered by the intrinsic GTPase activity of the G protein (Gilman 1987). *Ras* oncogene products are hyper-active, i.e. most of the time they are in the GTP-bound form. This can be caused mainly by two mechanisms : a strong decrease in intrinsic GTPase activity or a strong decrease in the affinity of guanine nucleotide binding. Since the level of GTP in the cytosol is much higher than the level of GDP, increased spontaneous exchange of guanine nucleotides leads to a higher proportion of GTP-bound RAS protein compared to GDP-bound RAS protein. Because they are G proteins and because of their membrane localization the RAS proteins are generally considered to be signal transmitters (Barbacid 1987). Although the importance of the RAS proteins for induction of proliferation in quiescent mammalian cells and for their oncogenic transformation by several other types of oncogenes is well documented (Barbacid 1987), their precise physiological function, e.g. the signal transduction pathway in which they would act as signal transmitter, remains unclear.

The RAS-adenylate cyclase pathway in yeast

Components of the pathway

The two yeast *RAS* genes, *RAS1* and *RAS2*, were isolated by hybridization with a mammalian ras gene as probe. In their N-termini they show about 80% homology with the mammalian ras genes while the C-termini are more divergent and also appreciably longer (DeFeo-Jones et al. 1983; Dhar et al. 1984; Powers et al. 1984). Yeast *RAS* genes and mammalian *ras* genes are functionally interchangeable (Kataoka et al. 1985b). Deletion of one of the *RAS* genes in yeast does not affect viability while deletion of both *RAS* genes is lethal (Kataoka et al. 1984; Tatchell et al. 1984). Deletion of *RAS2* and expression of a yeast equivalent, *RAS2*^{val19}, of a mammalian ras oncogene, H-*ras*^{val12}, produces phenotypic changes which are very similar to those produced by genetic manipulation of cAMP metabolism. Deletion of RAS2 causes increased levels of storage carbohydrates (glycogen and trehalose) and in diploid cells sporulation in rich media (Matsumoto et al. 1983; Fraenkel 1985; Tatchell et al. 1985; François et al. 1991). These phenotypic characteristics are also observed in yeast mutants with lowered cAMP levels or lower activity of cAMP-dependent protein kinase (cAMP-PK) (Matsumoto et al. 1983, 1985; Nakajima et al. 1987; Shin et al. 1987a). Deletion of RAS2 also causes inability to grow on non-fermentable carbon sources but this is only true in minimal media and it is mechanistically not related to the effect on cAMP metabolism (Fraenkel 1985; Breviario et al. 1986). On the other hand, expression of the yeast oncogene equivalent, *RAS2*^{val19} and similar *RAS1* oncogene equivalents, causes very low levels of glycogen and trehalose, heat shock and nitrogen starvation sensitivity and in diploid cells, very poor sporulation capacity (Kataoka et al. 1984; Toda et al. 1985; Marshall et al. 1987). These phenotypic characteristics are also observed in yeast mutants with enhanced cAMP levels or elevated activity of cAMP-PK (Matsumoto et al. 1983, 1985; Toda et al. 1985; Tatchell et al. 1985; Toda et al. 1987a; Hottiger et al. 1989; Cannon et al. 1990). The similarity between the phenotypes of *RAS* mutants and of mutants in cAMP metabolism led to the discovery that the RAS proteins of yeast function as the G_s proteins of yeast adenylate cyclase (Broek et al. 1985; Toda et al. 1985). It is important however to emphasize that the RAS proteins are required not only for specific stimulation of adenylate cyclase but also for maintenance of enough basal activity to overcome hydrolysis of cAMP by phosphodiesterase activity (Nikawa et al. 1987b). Hence, in the absence of *ras* function wild type cells are unable to maintain a cAMP level high enough to support growth.

Yeast adenylate cyclase, which is encoded by the *CYR1* (= CDC35) gene, is composed of a C-terminal domain which acts as catalytic site, a leucine-rich repeat in the middle and a N-terminal domain which appears to act as an inhibitory regulator of the catalytic domain (Boutelet et al. 1985; Kataoka et al. 1985a). The central leucine rich repeat domain was initially thought to act as plasma membrane anchor but results of recent work are more consistent with adenylate cyclase being anchored to the membrane by means of another protein, possibly the IRA1 protein (Mitts et al. 1990; Mitts et al. 1991). This difference might have physiological relevance because the binding of the cyclase to the membrane is reversible and appears to change with the growth conditions (Mitts et al. 1990). Adenylate cyclase in yeast contains an associated protein which is encoded by the *SRV2* or *CAP* gene (Fedor-Chaiken et al. 1990; Field et al. 1990). The N-terminal domain of this bifunctional protein appears to be responsible for interaction with RAS while the C-terminal domain appears to be involved in the control of cytoskeleton remodeling through interaction with profilin (Gerst et al. 1991; Vojtek et al. 1991).

Yeast cAMP-PK is composed of two catalytic subunits, encoded by the genes *TPK1*, *TPK2* and *TPK3*, and two regulatory subunits, encoded by the gene *BCY1*. Deletion of the three *TPK* genes is lethal while deletion of the *BCY1* gene causes to a very severe extent the typical phenotype associated with high activity of protein kinase A (Toda et al. 1987a; Toda et al. 1987b; Cannon et al. 1990).

Other components of the RAS-adenylate cyclase pathway include the *CDC25* gene product, the *IRA1* and *IRA2* gene products, the *RPI1* gene product and the phosphodiesterases encoded by *PDE1* and *PDE2*. Genetic and recent biochemical evidence indicates that CDC25 acts as a stimulator of GDP/GTP exchange on the RAS proteins. An important point again is that CDC25 is required not only for specific stimulation of RAS activity but also for its basal activity. Hence, in the absence of CDC25 activity, there is not enough RAS activity and therefore also not enough adenylate cyclase activity and not enough cAMP for viability. Deletion of *CDC25* is therefore lethal. (Camonis et al. 1986; Martegani et al. 1986; Broek et al. 1987; Robinson et al. 1987) The *IRA1* and *IRA2* gene products are stimulators of the intrinsic GTPase activity of the RAS proteins. They act as inhibitors of the pathway. Their deletion causes increased RAS activity, increased cAMP levels and it is also able to suppress CDC25 deficiency (Tanaka et al. 1989, 1990b). The *IRA* gene products are unable to down regulate the *RAS2*^{val19} oncogene product (Tanaka et al. 1990a). Recently, an additional component of the RAS-adenylate cyclase pathway, called RPI1, has been discovered which appears to act as a positive effector of both *IRA* gene products (Kim and Powers 1991). The *PDE1* and *PDE2* genes encode respectively the low- and high-affinity phosphodiesterase (Sass et al. 1986; Nikawa et al. 1987b). The K_m of the high-affinity phosphodiesterase (0.17 μM at pH 8) (Suoranta & Londesborough 1984) lies in the concentration range of the estimated intracellular cAMP level ($\pm$ 0.1–1 μM). The K_m of the low-affinity phosphodiesterase (120 μM at pH 8) (Londesborough & Lukkari 1980) on the other hand is not only much higher than the estimated intracellular cAMP concentration but also much higher than the peak values obtained immediately after addition of a rapidly-fermented sugar to cells growing on a non-fermentable carbon source (e.g. van der Plaat 1974). One possible explanation for this discrepancy is that the K_m of the so-called low-affinity phosphodiesterase is actually much lower *in vivo* than what is measured *in vitro* (see further). That the phosphodiesterases play a significant role in cAMP breakdown *in vivo* is shown by the fact that their deletion causes the typical phenotype associated with high activity of cAMP-PK (Nikawa et al. 1987b). Deletion of both *PDE* genes also suppresses deletion of both *RAS* genes and the resulting strain shows the typical phenotype associated with high activity of cAMP-PK, indicating that in the absence of RAS and phosphodiesterase activity adenylate cyclase is able to synthesize more than enough cAMP to keep the cells viable (Nikawa et al. 1987b). Deletion of the *PDE* genes causes only a very modest ($\pm$ two-fold) increase in the basal level of cAMP (Nikawa et al. 1987b). This is also true for other strains with reportedly 'enhanced' cAMP levels : strains carrying the *RAS2*^{val19} oncogene or strains with a deletion of the N-terminal inhibitory domain of adenylate cyclase (Toda et al. 1985; Nikawa et al. 1987a; Mbonyi et al. 1988; M. Beullens & K. Mbonyi, unpublished results). It is important however that in all these cases this modest increase in the cAMP level leads to a dramatic increase in the level of cAMP-PK activity, as inferred from the typical phenotype associated with high activity of this enzyme. Although it might not be impossible that a two-fold increase in the basal cAMP level (as measured in extracts as a mean value for the whole cell) causes such a strong increase in the activity of cAMP-PK, this observation might also point to the existence of an additional unknown signal amplification mechanism operating on cAMP-PK. Deletion of the *PDE* genes has already led to the identification of one additional control mechanism operating on the pathway: feedback-inhibition of cAMP synthesis by cAMP-PK (Nikawa et al. 1987a).

Feedback-inhibition of the pathway

The RAS-adenylate cyclase pathway is inhibited by feedback-inhibition of cAMP synthesis through activation of cAMP-PK (Nikawa et al. 1987a). Yeast strains with two of the three *TPK* genes deleted and the remaining *TPK* gene partially inactivated by a point mutation (*tpk*^{w}: 'weakened *TPK* allele') show extremely high cAMP levels, up to 100–1000 fold higher than the normal cAMP level in a wild type cell. These mutants display the same phenotype as those with lowered cAMP levels (Cameron et al. 1988). Although it has been claimed that deletion of the *BCY1* gene in such strains does not affect the phenotype (this is also what normally would be expected because of the extremely high cAMP level in the cells) this has since been contradicted (Mbonyi et al. 1990). Concerning the actual mechanism of feedback-inhibition the first question appears to be the identity of the phosphorylated substrate. CDC25 (Munder & Küntzel 1989), RAS (Resnick & Racker 1988), IRA (Tanaka et al. 1989, 1990b), and adenylate cyclase (De Vendittis et al. 1986) have all been proposed as potential candidates for inhibition by phosphorylation and it remains unclear which one is the true physiologically relevant substrate. In addition however there are other interesting problems. It has been reported that the *RAS2*^{val19} oncogene product is not sensitive to feedback-inhibition since deletion of both *PDE* genes in a *RAS2*^{val19} strain also causes 1000-fold higher cAMP levels (Nikawa et al. 1987a). This raises the obvious question why the phosphodiesterases are able to keep the intracellular cAMP level close to the normal wild type level in a *RAS2*^{val19} strain while they are unable to do so in a strain with reduced activity of cAMP-PK. A possible explanation is that phosphodiesterase activity *in vivo* is stimulated by cAMP-PK. An interesting candidate in this respect is the *PDE1* encoded 'low-affinity' phosphodiesterase whose function is at present unclear because of the very high K_m of its *in vitro* activity. If the K_m of this enzyme would be lowered *in vivo* directly or indirectly by cAMP-dependent protein phosphorylation, this would not only give an explanation for the very high *in vitro* K_m value but also for the hyperaccumulation of cAMP in yeast strains which have reduced cAMP-PK activity and where both phosphodiesterases are active. Indirect evidence for post-translational regulation of this high-K_m phosphodiesterase was already provided by Suoranta (1985) who demonstrated that the increase in activity of this enzyme during starvation was not inhibited by cycloheximide. *In vitro* evidence for regulation by cAMP-PK however could not be obtained.

An important consequence of the strong feedback-inhibition mechanism, which apparently has not been realized before, is the enormous discrepancy between the capacity of a yeast cell to synthesize cAMP and the actual intracellular concentration of cAMP. Time course experiments on glucose-induced hyperaccumulation of cAMP in the *tpk*^{w1} yeast strains with very low activity of protein kinase A reveal that yeast cells have the capacity to synthesize the total intracellular cAMP content of wild type cells at least 200 times (and probably much more) in one minute. This dramatic overcapacity of adenylate cyclase makes the relevance of any *in vitro* activity measurement of the enzyme very questionable when it is being used as an indirect indicator for the cAMP concentration in the cell. Several striking examples illustrating this point are available in the literature. E.g. Fasano et al. (1988) reported that *ras2-ts1* strains, as opposed to *ras2-ts31* strains, are not temperature-sensitive for growth on glucose, i.e. their cAMP level remains high enough for growth at high temperature. On the other hand Mg^{2+}-GTP dependent adenylate cyclase activity in membranes prepared from such glucose-grown cells was below detection for both strains. This clearly indicates that the difference between adenylate cyclase activity sufficient for growth or insufficient for growth is situated at a level of adenylate cyclase activity which is below detection in *in vitro* experiments. Hence, a decrease, even to a level below detection, of adenylate cyclase activity measured *in vitro* cannot be taken as a reliable indicator for cAMP deficiency *in vivo*. Petitjean et al. (1990) have reported that Mg^{2+}/GTP-dependent adenylate cyclase activity in membrane preparations from the temperature-sensitive *cdc25–10* mutant was tenfold lower in cells incubated at the restrictive temperature compared to cells incubated at the permissive temperature. However, in membrane preparations of the *cdc25–1*, *cdc25–5* and *cdc35–10* strains, adenylate cyclase activity had already dropped to the same extent and to the same low level for cells incubated at the permissive temperature. Mn^{2+}-dependent adenylate cyclase activity displayed the same behavior. These data confirm that the level of adenylate cyclase activity in purified membranes is not a reliable indicator for cAMP deficiency *in vivo*, since growing and arrested cells repeatedly display the same level of activity *in vitro*.

Another consequence of the feedback-inhibition mechanism is that cells with a deletion of the three *TPK* genes should have extremely high cAMP levels, at least 1000-fold higher than the wild type level. The fact that such cells still grow, albeit extremely slowly, whereas cells without *RAS* genes do not grow at all, has been taken as an indication for the existence of an alternative RAS pathway (Toda et al. 1987b). However, the extremely high cAMP level might also cause to a certain extent activation of e.g. a cGMP-dependent protein kinase which otherwise would never be activated by normal cAMP levels.

Nutrient-control of the RAS-*adenylate cyclase pathway*

Demonstration of nutrient-control of the RAS-adenylate cyclase pathway did not originate from the studies on the involvement of the pathway in control of cell proliferation but rather on its involvement in control of cell metabolism. Several lines of research on the mechanism responsible for the rapid changes in enzyme activity triggered by addition of glucose (or related rapidly-fermented sugars) to yeast cells grown on non-fermentable carbon sources (e.g. ethanol, glycerol) had converged in the discovery that cAMP-dependent protein phosphorylation was involved (reviews: Holzer 1984, Thevelein 1984b, 1988). Subsequent experiments showed that both the *RAS* and *CDC25* gene products were involved in triggering the rapid transient increase observed in the cAMP level after addition of glucose. The product of the *RAS2*^{val19} oncogene was unable to mediate glucose-induced cAMP signaling (Mbonyi et al. 1988; Munder & Küntzel 1989; Van Aelst et al. 1990, 1991b). These studies were complicated by the fact that both RAS and CDC25 are also required for basal cAMP synthesis. The use of the *RAS* oncogene alleles, *RAS2*^{val19} and *RAS2*^{ile152}, has helped to circumvent this problem. Recent identification of a RAS2 mutant which only supports basal cAMP synthesis but not induction of a cAMP increase confirms the functioning of RAS as a signal transducer and not merely as a component required for adenylate cyclase activity. This mutant also does not show elevated protein kinase activity which precludes any possible interference from the feedback-inhibition mechanism (E. Oris & P. Durnez, unpublished results). The inability of the *RAS2*^{val19} oncogene product to act as a signal transducer in glucose-induced cAMP signaling has also been confirmed recently (Kim & Powers 1991). Experiments with the double mutants *cdc25 ira1* and *cdc25 ira2* have confirmed the requirement of CDC25 for signal transduction (Tanaka et al. 1989; L. Van Aelst, E. Boy-Marcotte, M. Jacquet & J. Thevelein, unpublished results). The pathway leading from glucose to CDC25 largely remained unclear until the recent breakthrough with the discovery that the *GGS1* (= *FDP1*, = *BYP1*, = *CIF1*) gene product is involved in this part of the pathway (see further). It was already known that glucose phosphorylation is required and that it may be carried out by any one of the three sugar kinases: glucokinase, hexokinase 1 or hexokinase 2 (or any one of the two hexokinases in the case of fructose). Further metabolism of the sugar does not seem to be required (Beullens et al. 1988).

It has recently been shown that the *fdp1* mutant is deficient in glucose-induced activation of cAMP synthesis (Van Aelst et al. 1991a). The *fdp1* mutant is unable to grow on rapidly-fermented sugars like glucose, fructose, sucrose and mannose. It grows on non-fermentable carbon sources like glycerol and ethanol and also on slowly assimilated sugars like galactose and maltose. In this case it displays high (probably cAMP-dependent) protein kinase activity which is manifested e.g. in overphosphorylation of glycogen synthase (Van de Poll & Schamhart 1977) and very low trehalose levels (Panek et al. 1979; Charlab et al. 1985; Van Aelst et al. 1991a). The *fdp1* mutant is deficient in glucose (and fructose)-induced inactivation of fructose-1,6-bisphosphatase (Van de Poll et al. 1974; Banuelos & Fraenkel 1982, Van Aelst et al. 1991a), inactivation of phosphoenolpyruvate carboxykinase and malate dehydrogenase (Gancedo & Schwerzmann 1976), activation of trehalase (Van Aelst et al. 1991a), plasma membrane H^+-ATPase (J. Becher dos Passos & R. Brandao, unpublished results), cAMP synthesis (Van Aelst et al. 1991a), fructose-2,6-bisphosphate synthesis (Panek et al. 1988), potassium uptake (J. Ramos, unpublished results) and glucose-induced stimulation of PDC transcription (L. Sierkstra, unpublished results). Addition of a rapidly-fermented sugar to *fdp1* cells grown on non-fermentable carbon sources causes a rapid drop in the ATP and free phosphate level, intracellular acidification and hyperaccumulation of sugar phosphates, especially fructose-1,6-bisphosphate (Van de Poll & Schamhart 1977; Banuelos & Fraenkel 1982; Van Aelst et al. 1991a; L. Van Aelst, unpublished results). Because of these observations, it was suggested that the *fdp1* mutant is deficient in a hypothetical feedback-inhibition mechanism of glycolysis on sugar transport (van de Poll & Schamhart 1977).

It has always remained somewhat unclear whether the regulatory deficiencies in the *fdp1* mutant were not a secondary effect of the rapid deregulation of metabolism, e.g. the rapid drop in the ATP level. It is already known that, depending on the genetic background of the strain, the *fdp1* mutation can give rise to a spurious glucose-induced cAMP increase and activation of trehalase due to intracellular acidification (M. Beullens, unpublished results). This can probably explain certain contradictory results in the literature on the properties of the *fdp1* mutant (e.g. Panek et al. 1988). However, the recent isolation of two suppressor genes, *FPS1* and *FPS2*, which restore the growth defect on rapidly-fermented sugars but none of the regulatory defects, strongly supports the idea that all the regulatory deficiencies are direct consequences of the *fdp1* mutation (Van Aelst et al. 1991a; B. Bulaya, L. Van Aelst and S. Hohmann, unpublished results). Moreover, the availability of the *FPS1* and *FPS2* suppressors has allowed investigation of whether the *fdp1* mutant is deficient in long-term glucose-induced regulatory responses for which obviously growth on glucose is a prerequisite. Investigation of glucose repression of invertase, maltase and mitochondrial respiration activity and of the *SUC2* and *GAL10* genes at the level of transcription clearly showed that the *fdp1* mutation causes complete derepression of the main glucose repression pathway (L. Van Aelst and L. Sierkstra, unpublished results). This means that all glucose-induced regulatory responses investigated up to now are deficient in the *fdp1* mutant. In this respect, it is important to emphasize again the difference between glucose-induced stimulation and basal activity. E.g. the *fdp1* mutant is deficient in the induction of the cAMP signal by glucose but not in basal adenylate cyclase activity (Van Aelst et al. 1991a). Because the name FDP1 (which stands for 'deficient in *f*ructose *dip*hosphatase inactivation') is likely to cause confusion we suggest a change of name to *GGS1* (which stands for '*G*eneral *G*lucose *S*ensor'). The names *fdp1* and *byp1* will be kept for the specific point mutants.

The *GGS1* (*FDP1*) gene has recently been cloned in our laboratory. The sequence of the 1.5 kb long open reading frame does not show homology to other genes in the databases and the hydrophobicity plot does not show clear indications of membrane localization (L. Van Aelst, B. Bulaya and S. Hohmann, unpublished results). In addition, the *byp1* mutant (Breitenbach-Schmitt et al. 1984) was shown to be allelic to the *fdp1* mutant. *Byp1* cells grow on glucose media but with a very long lag phase. After growth on non-fermentable carbon sources, they are deficient in glucose-induced activation of cAMP synthesis and fructose-2,6-bisphosphate synthesis, inactivation of fructose-1,6-bisphosphatase, activation of plasma membrane H^+-ATPase and *PDC* transcription but not in trehalase activation and probably not in glucose repression. They also have a normal trehalose level which points to a normal level of cAMP-PK activity. Hence, the absence of the glucose-induced cAMP signal cannot be due to enhanced feedback-inhibition by cAMP-PK (Hohmann et al. 1991; J. Becher dos Passos & L. Van Aelst, unpublished results). Recently, a glutamine-accepting tRNA gene, which is known to suppress nonsense mutations, has been identified as a suppressor of *byp1* (S. Hohmann, unpublished results). This suggests that *byp1* is a truncated version of *GGS1* (*FDP1*) and that different domains of *GGS1* (*FDP1*) are responsible for induction of different glucose-dependent regulatory responses.

Based on these data we propose that the *GGS1* (*FDP1*) gene product forms part of a glucose sensing complex together with a low-affinity glucose carrier and a sugar kinase and that this complex has both a function in restraining the influx of sugars transported by the carrier and in triggering different glucose-induced signalling pathways (Fig. 2). The suggestion for a regulatory constraint on glucose influx into yeast cells is not only based on the phenotype of the *ggs1* (*fdp1* and *byp1*) mutants. Yeast cells need some kind of control system on glucose influx since its sugar kinase enzymes are not sensitive to feedback-inhibition by sugar phosphate as is the case with the mammalian enzyme (Sols 1976). Because of the very high amount of sugar kinase activity in yeast, uncontrolled glucose influx would rapidly deplete the ATP level. This is in fact exactly what is observed in *ggs1* mutants. Further support for the idea that glycolytic initiation is too rapid in *ggs1* mutants comes from the observation that the *fdp1* mutant is able to grow on a glucose or fructose medium containing an excess of D-xylose (Alonso et al. 1984). D-xylose reduces glycolytic influx both by competitive inhibition of glucose and fructose transport and by specific inactivation of the glucose and fructose phosphorylating enzymes (DelaFuente 1970; Fernandez et al. 1985; Schuddemat et al. 1986). Control of sugar influx by the glucose sensing complex also introduces the possibility for different types of physiological regulation, e.g. regulation by the well-known, but still poorly understood Pasteur effect (Sols 1976). The association of the sugar kinase and the glucose carrier in the complex implies that the sugar transported is immediately phosphorylated upon entry into the cell, i.e. that the sugar is not released in the cytosol as free sugar to be phosphorylated only subsequently by a kinase. Pool labeling experiments with 2-deoxyglucose have already provided indirect evidence for the existence of such a mechanism (Van Steveninck 1968; Jaspers & Van Steveninck 1975; Meredith & Romano 1977; Franzusoff & Cirillo 1982; Beullens & Thevelein 1990). However this does not necessarily imply that the complex is stable. The different components might only associate or start to interact with each other when glucose metabolism is being initiated, i.e. when the sugar kinases become active. It is important to stress that the present model also does not imply that *all* transported sugar passes through the glucose sensing complex, nor that all sugar kinase molecules are present in such complexes. The latter is virtually impossible because of the very large amount of sugar kinase protein in yeast cells (Fraenkel 1982). Yeast cells seem to contain a relatively large (6–8) and even strain-dependent number of glucose carriers or at least putative glucose carriers, based on sequence homology (Kruckeberg & Bisson 1990). Not all carriers are necessarily involved in glucose-induced signalling. In addition, a sophisticated mechanism for restraining/controlling sugar influx might only be needed/useful for the carriers with high activity. In addition, the model is not meant to imply that the individual complexes all have exactly the same function. One possibility for differentiation is that complexes with hexokinase PII as sugar kinase are involved in activating the glucose repression pathway while complexes with hexokinase PI are much less potent in this respect and complexes with glucokinase not active at all (Rose et al. 1991). For other glucose-induced regulatory phenomena, such as activation of the RAS-adenylate cyclase pathway (Beullens et al. 1988) and inactivation of fructose-1,6-bisphosphatase and phosphoenolpyruvate carboxykinase (Gancedo & Gancedo 1979; Entian et al. 1983) the exact nature of the sugar kinase does not appear to be important.

The high protein kinase activity in the *fdp1* mutant is not able to suppress *cdc25*^{ts} or *cdc35*^{ts} mutations or a *ras1 ras2* double mutation which shows that, if it is indeed cAMP-PK activity, the component responsible for the overactivation must be located upstream of CDC25 (M. Vanhalewyn and P. Durnez, unpublished results). One possibility is that the *GGS1* (*FDP1*) gene product is an inhibitor of the CDC25-RAS-adenylate cyclase pathway and that therefore its inactivation causes a somewhat increased cAMP level and concomitantly enhanced cAMP-PK activity. The observation that *cdc25 fdp1* double mutants are very unstable as opposed to e.g. *cdc35 fdp1* strains could point to interaction between the GGS1 (FDP1) and CDC25 proteins. In many respects the model of a 'General Glucose Sensing complex' forms a good basis for future experimentation. For instance, it is already known that overexpression of the CDC25 protein, which is normally present in very low levels in wild type cells, abolishes glucose repression (Van Aelst et al. 1991b). If one assumes interaction between GGS1 (FDP1) and CDC25, tieing up of GGS1 (FDP1) by CDC25 might render it unavailable for stimulation of the glucose repression pathway. In addition, inhibition of glucose repression and overstimulation of the RAS-adenylate cyclase pathway by overexpression of the Hsp90 heat shock protein (Piper 1990) might also be due to interference with the general glucose sensor. Hsp90 is known to associate with regulatory proteins in mammalian cells (Piper 1990). How many signalling pathways are triggered by the general glucose sensing complex is unclear, but at least several have to be involved. It is known that glucose-induced inactivation of fructose-1,6-bisphosphatase consists of two processes: a rapid partial inactivation due to cAMP-dependent phosphorylation of the protein and a much slower inactivation due to proteolysis (Müller & Holzer 1981; Mazon et al. 1982; Holzer 1984). The latter process occurs independently of the former (Rose et al. 1988). In the *fdp1* mutant both processes are absent. In addition, it is known that the main glucose repression pathway, glucose-induced activation of plasma membrane H^+-ATPase and potassium uptake, and also nitrogen-source induced activation of trehalase (which requires the presence of glucose in the medium), are not triggered by the RAS-adenylate cyclase pathway (Matsumoto et al. 1982; Becher dos Passos et al. 1991; K. Hirimburegama & P. Durnez, unpublished results). Hence, at least 3–4, and probably more, separate pathways are involved.

The proposed model of a 'general glucose sensing complex' resembles the PTS system (phosphoenolpyruvate-dependent carbohydrate : phosphotransferase system) or 'group translocation' system present in obligate and facultative anaerobic bacteria (reviews: Postma & Lengeler 1985; Saier et al. 1990) in two ways. First the sugar is phosphorylated during transport and second the system exerts widespread effects on the regulation of metabolism. The most important difference however would be that in yeast the association of the sugar kinase with the carrier serves to restrain the inflow of sugar at high extracellular sugar concentrations while in the PTS system it serves to improve at low extracellular sugar concentrations and under anaerobic conditions the efficiency of transport and in this way the energy yield of sugar catabolism (it saves 1 ATP molecule per glucose molecule taken up).

A subsequent component of the glucose-induced signalling pathway might be a glucose-repressible protein since there is a strong effect of the glucose repression state of the cells on induction of the cAMP signal. Glucose-repressed cells or mutants deficient in derepression do not show a glucose-induced cAMP signal while on the other hand mutants without glucose repression also show the signal when the cells are grown on glucose (Beullens et al. 1988; Mbonyi et al. 1990; Argüelles et al. 1990; Van Aelst et al. 1991b). The putative glucose-repressible protein might also be a component of the general glucose sensing complex or might even be the GGS1 (FDP1) protein itself.

The study of the glucose-induced signalling pathway leading to activation of RAS-adenylate cyclase is not only complicated by the fact that glucose also functions as carbon and energy source, but also by the fact that intracellular acidification is a very potent activator of cAMP synthesis *in vivo*. This activation is also dependent on both RAS and CDC25, but not on sugar kinase activity or the

GGS1 (*FDP1*) gene product (Fig. 1). In this way the acidification effect is also useful in that it allows to locate components of the glucose-induced signalling pathway upstream or downstream of the point where acidification interferes with the pathway. In fact, glucose also induces a small, transient drop in the intracellular pH which was believed to act as the trigger of the cAMP signal before the evidence for a specific signalling pathway was obtained. Intracellular acidification might constitute a true physiological trigger of the RAS-adenylate cyclase pathway. This is at least consistent with the stimulating effect of the pathway on the mobilization of reserve carbohydrates and the generation of energy. Increased energy production is an obvious requirement under stress conditions and intracellular acidification is well known to occur in stressed yeast cells (Weitzel et al. 1987; Leao & Van Uden 1984). In addition, intracellular acidification has been suggested to act as triggering mechanism for the acquisition of thermotolerance (Weitzel et al. 1987; Coote et al. 1991). It is extremely important that for studies on glucose-induced cAMP signaling the experiments are performed under conditions which preclude interference from either the presence or the generation of low intracellular pH. Such interference can easily occur for instance in glucose-repressed cells which are suspended in a medium without glucose (Argüelles et al. 1990). In several cases in the literature, it is not clear whether true glucose-induced cAMP signaling has been studied because the glucose-repression state of the cells was not clear (e.g. Arkinstall et al. 1991). An extensive discussion of the activation of the RAS-adenylate cyclase pathway by fermentable sugar and intracellular acidification has recently been published (Thevelein 1991).

Studies on the glucose-induced cAMP signal have repeatedly shown that it is not required for growth on glucose (Thevelein 1991). In addition, since glucose-induced activation of the RAS-adenylate cyclase pathway appears to be shut off when the cells are glucose-repressed, its function appears to be limited to the control of metabolism during the transition from gluconeogenic growth to fermentative growth (Thevelein 1991). Even in this case its role is more limited than previously thought. It has been suggested that glucose-induced activation of trehalase is triggered by the glucose-induced cAMP signal since a temperature-sensitive mutant in cAMP synthesis (*cdc35–10*) failed to show glucose-induced activation of trehalase at the restrictive temperature (François et al. 1984). However, at the restrictive temperature such a mutant is not only deficient in the cAMP signal but also in the basal level of cAMP. In mutants which are only deficient in the cAMP signal and not in the basal cAMP level, there is still glucose-induced activation of trehalase (Hohmann et al. 1991; L. Van Aelst, M. Vanhalewyn & J.C. Argüelles, unpublished results). In this case, the nitrogen signalling pathway might be responsible for the activation (see further) since the cells used in these experiments have ample internal nitrogen stores.

A second system of glucose control through the RAS-adenylate cyclase pathway has been described. Studies with different mutants in the pathway have provided evidence that glucose repression of the genes *ADH2*, *CTT1*, *HSP12*, *SSA3* and *UBI4* is mediated by the RAS-adenylate cyclase pathway (Cherry et al. 1989; Bissinger et al. 1989; Belazzi et al. 1991; Praekelt & Meacock 1990; Werner-Washburne et al. 1989; Tanaka et al. 1988). However, careful examination of the data, e.g. the effect of nitrogen starvation on the transcription of these genes and taking into consideration that glucose-activation of the RAS-adenylate cyclase pathway appears to be glucose-repressible, have led us to the suggestion that this apparent glucose repression effect is not mediated by the RAS-adenylate cyclase pathway but rather by a nitrogen-source induced signalling pathway which is dependent on glucose and also leads to activation of protein kinase A (Thevelein 1991). For the latter mechanism, we have suggested two possibilities: either synergistic activation of cAMP-PK by both cAMP and a signal from the nitrogen-source pathway or specific activation of free catalytic subunits of the protein kinase by the signal derived from the nitrogen-source-induced pathway. Recent results on the effect of nitrogen on *CTT1* transcription in yeast strains with a deletion of the *BCY1* gene encoding the regulatory subunit of cAMP-PK (Belazzi et al.

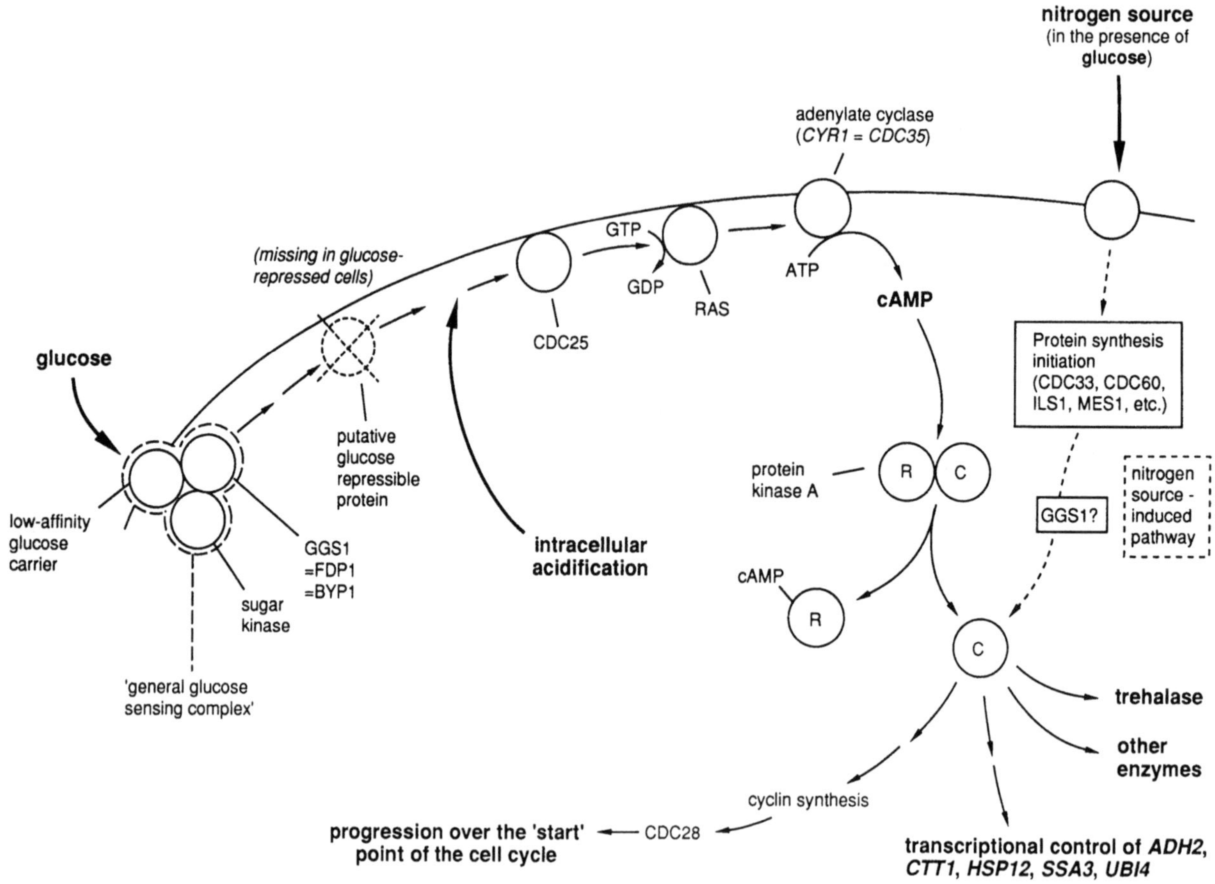

Fig. 1. Tentative model of the glucose-and acidification-induced signal-transmission pathway, leading to activation of the RAS-adenylate cyclase system in yeast and its relationship to the nitrogen-source induced, glucose-dependent signalling pathway. CDC25 and RAS activity are required both for glucose- and acidification-induced cAMP signalling. The 'general glucose sensing complex' and the putative glucose-repressible protein, which might also be a component of the same complex, are only required for activation by glucose. Intracellular acidification activates the pathway downstream of the putative glucose-repressible protein. Because of the presence of the glucose-repressible protein, glucose-induced activation of the RAS-adenylate cyclase system is no longer functional once the cells have become glucose-repressed. Hence, only during the transition period from derepression to repression is cAMP-dependent protein kinase being activated by the glucose-dependent RAS-adenylate cyclase pathway. Nitrogen signalling uses a separate pathway which appears to involve the initiation of protein synthesis and which is suggested to cause activation of free and only free catalytic subunits of cAMP-dependent protein kinase. This nitrogen-induced pathway is dependent on the presence of glucose in the medium which might indicate, as appears to be the case for all other glucose-dependent regulatory pathways, that it requires activity of the general glucose sensing complex. In glucose-repressed cells the nitrogen signalling pathway is suggested to be responsible for nutrient control of progression over the start point of the cell cycle, transcription of *ADH2*, *CTT1*, *HSP12*, *SSA3* and *UBI4* and post-translational regulation of enzyme activity. (Only components involved in activation of the pathway are shown. Not shown are the *IRA1* and *IRA2* gene products which stimulate RAS-GTPase activity, the *RPI1* gene product which acts as a stimulator of the IRA proteins, the *PDE1*- and *PDE2*- encoded low- and high-affinity phosphodiesterases, feedback inhibition of protein kinase A on cAMP synthesis, the SRV2 or 'CAP' subunit of adenylate cyclase and the SCH9 and YAK1 protein kinases which appear to be involved downstream of cAMP-dependent protein kinase.)

1991) and on nitrogen-source induced activation of trehalase in nitrogen-starved cells of such strains (P. Durnez, unpublished results), point to the second mechanism as the most likely one (Fig. 1). Recent results by François et al. (1991) point to the possibility that the genes encoding trehalose-6-phosphate synthase and trehalose-6-phosphate phosphatase are under similar transcriptional control as the *ADH2*, *CTT1*, *HSP12*, *SSA3* and *UBI4* genes.

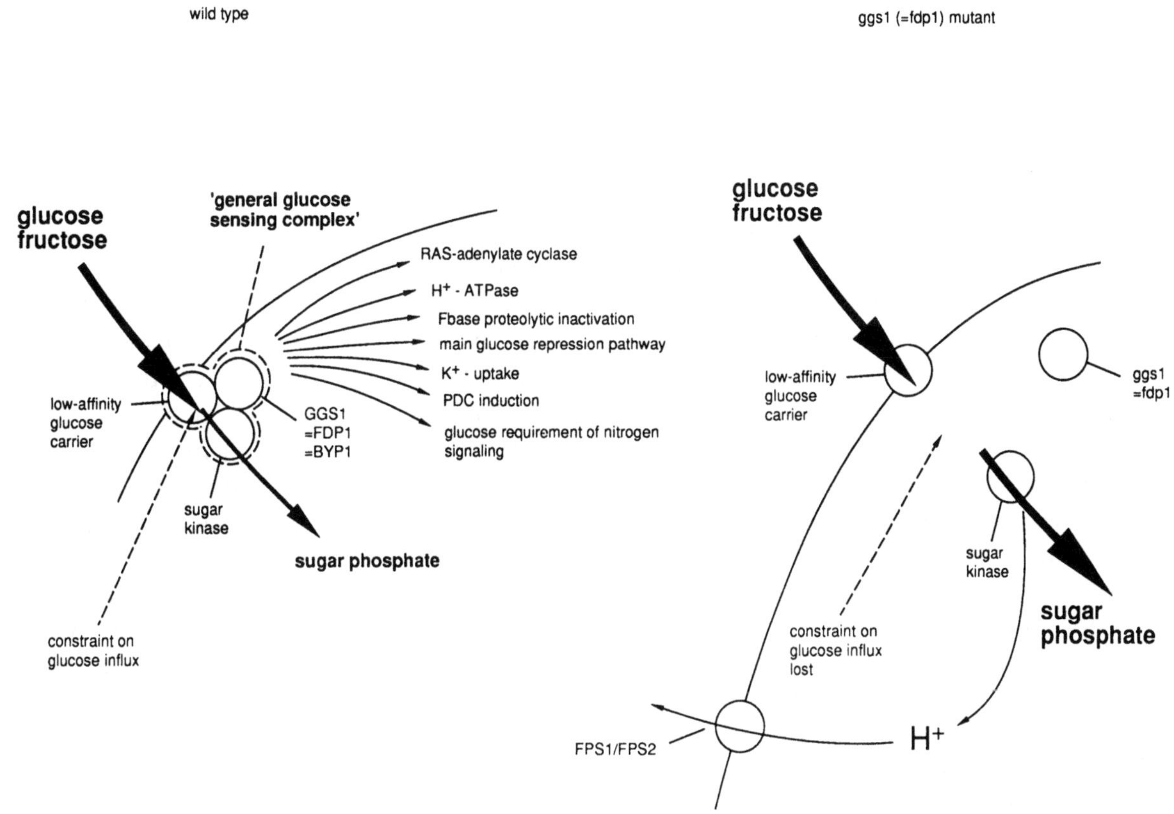

Fig. 2. Tentative model for the functioning of the 'general glucose sensing complex' in yeast cells. It is proposed that in wild type cells, glucose influx is mediated by a complex consisting of (at least) a low-affinity glucose carrier, a sugar kinase and the *GGS1* gene product (or a protein controlled by GGS1). The complex is suggested to have two main functions. The first one consists of restraining and regulating glucose influx. The second one consists of activating a whole series and maybe all glucose-induced signalling pathways. In the *ggs1* mutants the complex is not functional anymore. The constraint on glucose influx has been lost which results upon addition of glucose in rapid hyperaccumulation of sugar phosphate with concomitant depletion of the ATP level. Activation of the signalling pathways by glucose is lost. Depending on whether the *GGS1* gene product acts as an activator or an inhibitor of the first component of the signalling pathway, the pathway remains at a low basal level of activity or acquires a constitutively enhanced level. The latter situation is suggested to be responsible for the enchanced level of cAMP-dependent protein kinase activity in the *fdp1* mutant. Restoration of growth by the *FPS1* and *FPS2* suppressors is thought to be due to suppression of glucose-induced intracellular acidification caused by the overactive initiation of glycolysis.

A third system of glucose control through the RAS-adenylate cyclase pathway has been suggested by Kaibuchi et al. (1986). Their data pointed to involvement of this pathway in the stimulation of phosphatidylinositol turnover by glucose. However, this has not been confirmed by other groups and a recent paper by Frascotti et al. (1990) showed that functioning of the RAS-adenylate cyclase pathway is not required for this phenomenon. On the other hand, the latter authors confirmed the result by Kaibuchi et al. (1986) that strains containing the *RAS2*^{val19} allele show a higher level of glucose-induced PI turnover than wild type strains. This observation fits with the data of Kato et al. (1989) that the kinases responsible for PIP_2 formation are stimulated by cAMP-PK.

No evidence for control of the RAS-adenylate cyclase pathway by other nutrients besides glucose has been published. This is especially important in view of the fact that specific starvation for several other nutrients, like nitrogen, phosphate and sulfate causes cells to arrest at exactly the same point in the cell cycle as cells which are deficient in the RAS-adenylate cyclase pathway (see further).

The RAS-adenylate cyclase pathway and the yeast cell cycle

The 'start' point in the G1 phase and control of the yeast cell cycle

The main control point in the eukaryotic cell cycle is the so-called restriction point in the G_1 phase (Pardee 1974). Usually, non-proliferating eukaryotic cells have arrested at the decision point in G_1 of the cell cycle and have subsequently entered a resting state called G_0, in which the cells display increased resistance. In yeast the decision point has been called 'start' (Hartwell et al. 1974). At the start point, the yeast cell checks a number of parameters and integrates the results into the decision to start with the next cell cycle or to follow a specific differentiation pathway depending on the actual information received, i.e. entry into the resting state G_0 in case of insufficient nutrient supply, mating with a cell of the opposite mating type in case a mating pheromone has been detected, or for diploid cells to initiate the sporulation pathway in case of nitrogen-deficiency and abundant non-fermentable carbon sources (Pringle & Hartwell 1981).

Among the cell cycle mutants present in the Hartwell collection of temperature sensitive mutants, a few were identified as start mutants, i.e. at the restrictive temperature they were unable to proceed beyond the start point and arrested as unbudded cells (Hartwell et al. 1973; Hartwell 1974). These mutants could be divided into two types, those that arrested at the same point as nutrient-starved cells, i.e. *cdc25*, *cdc35* and *cdc33* and those that arrested at the same point as pheromone-treated cells, i.e. *cdc28*, *cdc36*, *cdc37*, *cdc39* (Reed 1980; Pringle & Hartwell 1981). Because the studies with adenylate-cyclase mutants had shown that cAMP deficiency causes specific cell cycle arrest at the start point, these mutants were investigated for possible defects in cAMP metabolism. In this way it was found that *CDC35* was allelic with *CYR1*, the gene for adenylate cyclase (Boutelet et al. 1985) and CDC25 was discovered as a new component of the RAS-adenylate cyclase pathway (Camonis et al. 1986; Martegani et al. 1986; Broek et al. 1987; Robinson et al. 1987). Subsequently *ras1Δ ras2*^{ts} strains were also shown to arrest specifically at the same site in the start point of the cell cycle (Fasano et al. 1988). Except for one report (Liao & Thorner 1980), which could not be confirmed later on (Casperson et al. 1983; Perlman et al. 1989), and the recent report on α-mating factor inhibition of glucose-induced cAMP signalling (Arkinstall et al. 1991), no interference has been found between cAMP metabolism and the pheromone site of start. The discovery that *cdc25* and *cdc35* are mutants in cAMP metabolism did not only confirm that the RAS-adenylate cyclase pathway is required for progression over the start point of the cell cycle but it also created a new approach to identify other components in this pathway. Cloning of the *CDC33* gene however showed that it encoded the gene for protein synthesis initiation factor 4E (Brenner et al. 1988). No effect on cAMP metabolism of the *cdc33*^{ts} mutation could be found (Verdier et al. 1989). The discovery that the *CDC33* gene product is required for protein synthesis initiation fitted with earlier observations that temperature-sensitive mutants in methionyl- (Unger & Hartwell 1976) and isoleucyl-tRNA synthetase (Hartwell & McLaughlin 1968) and in the *PRT1* gene product, which is also involved in protein synthesis initiation (Hanic-Joyce et al. 1987; Linder & Prat 1990) also arrest at the start point when shifted to the restrictive temperature. This correlation has been substantiated further with the recent discovery that the *CDC60* gene encodes cytoplasmic leucyl-tRNA synthetase (S. Hohmann, R. Singer & J.M. Thevelein, unpublished results). The *cdc60*^{ts} mutant also arrests at the nutrient starvation site of start (Bedard et al. 1981). Very few studies have been conducted on a possible connection between cAMP metabolism and protein synthesis initiation. One important result however is that overactivation of the RAS-adenylate cyclase pathway suppresses both the prolonged G_1 phase at the permissive temperature and the specific start arrest at the restrictive temperature of the *cdc33* mutant (Brenner et al. 1988).

It has been claimed that the RAS-adenylate cyclase pathway would not be directly required for cell cycle progression but only indirectly, through its requirement for growth (Tatchell 1986). This claim is based on the observation that temperature sensitive mutants which arrest at the nutrient starvation site of start rapidly arrest cell growth (i.e. protein synthesis) at the restrictive temperature, while mutants which arrest at the pheromone site continue growing for a relatively long time at the restrictive temperature (Hartwell et al. 1973; Bedard et al. 1981; Martegani et al. 1984). At present however, it cannot be excluded that the RAS-adenylate cyclase pathway is also directly required for cell cycle progression in addition to its requirement for cell growth. It is worthwhile pointing out that in yeast, cell growth and cell cycle progression are intimately linked (Pringle and Hartwell 1981). The cell growth phase is an essential part of the yeast cell cycle. Yeast cells have to reach a certain critical cell size before they can start with a new cell cycle and this requirement is also checked at the start point (Johnston et al. 1977; Carter & Jagadish 1978). Baroni et al. (1989) have reported evidence that lower or higher activity of the RAS-adenylate cyclase pathway respectively diminishes or enhances the critical cell size.

Is there a connection between nutrient control of the RAS-adenylate cyclase pathway and nutrient control of progression over the start point of the cell cycle?

The discovery that cAMP-deficient yeast cells arrest at exactly the same point in the cell cycle as nutrient-starved yeast cells has led to the idea that nutrient starvation would cause cAMP depletion thereby blocking progression over the start point of the cell cycle (Jacquet & Camonis 1985; Dumont et al. 1989; Engelberg et al. 1989; Gibbs & Marshall 1989; Malone 1990). Only one experimental result in favor of this idea however has been published. Sulfate starvation was shown to decrease the cAMP level while readdition enhanced the cAMP level. Hence, it was proposed that sulfate triggers progression over the start point by virtue of its stimulating effect on cAMP synthesis (Shin et al. 1987b). According to Unger and Hartwell (1976) the signal for sulfate starvation would be generated near the end of the sulfate assimilation pathway, at or beyond the formation of methionyl-tRNA. There is no evidence however linking methionyl-tRNA, or amino-acyl tRNA in general, to cAMP metabolism.

It has been suggested that nitrogen-source availability is also sensed by the RAS-adenylate cyclase pathway (Boy-Marcotte et al. 1987). However, nitrogen sources do not affect the cAMP level in vegetative yeast cells (Thevelein 1984a; Thevelein & Beullens 1985; François et al. 1988) including nitrogen-starved cells which are refed with a nitrogen-source (K. Hirimburegama & P. Durnez, unpublished results). Although glucose and related rapidly-fermented sugars clearly affect the cAMP level, this effect appears to be confined to the transition period between respirative growth (derepressed cells) and fermentative growth (repressed cells) because of the fact that the glucose-activation pathway of RAS-adenylate cyclase is glucose-repressible. The conclusion that glucose-induced activation of the RAS-adenylate cyclase pathway is dispensable for growth on glucose and more specifically for progression over the nutrient starvation site of 'start' is being supported by recent data of Arkinstall et al. (1991) showing that α-mating factor causes inhibition of glucose-induced cAMP signalling. These authors propose that the effect could be involved in α-mating factor induced arrest at 'start.' However, if glucose-induced activation of the RAS-adenylate cyclase pathway would be vital for progression over the start point of the cell cycle then α-mating factor should arrest cells at the nutrient-starvation/cAMP deficiency site and not downstream at the Cyclin/CDC28 site.

Phosphate addition to phosphate-starved cells causes a significant transient increase in the cAMP level, but in this case interference with the acidification effect on cAMP synthesis is easily possible (K. Hirimburegama, unpublished results). Although it cannot be excluded that a specific phosphate-induced activation pathway of RAS-adenylate cyclase exists, the idea that nutrients trigger progression over the start point of the cell cycle by causing increased cAMP levels is not supported by strong experimental evidence. On the contrary, the initial possibility that rapidly-fermented sugars would act in this way is being contradicted by further research.

In spite of the fact that nitrogen sources do not affect the cAMP level, they cause drastic activity changes at the post-translational level of several enzymes known to be regulated by cAMP-dependent protein phosphorylation, e.g. trehalase (Thevelein 1984a; Thevelein & Beullens 1985; K. Hirimburegama & P. Durnez, unpublished results), phosphofructokinase 2, glycogen synthase and phosphorylase (François et al. 1988). For instance, refeeding of a nitrogen source to nitrogen-starved cells causes very rapid activation of trehalase in a protein synthesis independent way (K. Hirimburegama, unpublished results). This nitrogen-source induced activation also requires the presence of glucose or another rapidly-fermented sugar (Thevelein & Beullens 1985; François et al. 1988), which immediately creates the interesting possibility that the pathway involved could serve as a sensor for both the nitrogen and the sugar. The main and obvious problem is whether trehalase and the other enzymes are only regulated by cAMP-PK or whether other protein kinases could also be involved. Recent experiments with mutants displaying different activity levels of cAMP-PK showed corresponding degrees of nitrogen-induced activation of trehalase and therefore point to cAMP-PK as the mediator (P. Durnez & E. Oris, unpublished results). Addition of phosphate to phosphate-starved cells or sulfate to sulfate-starved cells also triggers rapid activation of trehalase (K. Hirimburegama, unpublished results). It appears therefore that cAMP-PK, and not cAMP, would be the universal integrator of nutrient availability in yeast (at least for cells growing on fermentable sugar). The effect of nitrogen would be introduced at the level of the free catalytic subunits of the kinase, as we have suggested before (Thevelein 1991). The GGS1 (FDP1) protein in its function of glucose sensor might provide the link with sugar availability since it also appears to be required for functioning of the nitrogen signalling pathway (P. Durnez, unpublished results). The glucose concentration required for nitrogen-induced activation of trehalase is similar to the concentration required for glucose-induced activation of cAMP synthesis. This is relatively high: the apparent K_m for activation by glucose is about 20 mM (Beullens et al. 1988; K. Hirimburegama & R. Vergauwen, unpublished results). The dependency of the nitrogen signaling pathway on a relatively high glucose concentration might explain why during the diauxic shift of yeast cells growing on glucose, unbudded cells start to accumulate and many physiological changes occur not after but already well before the glucose in the medium is exhausted (e.g. François et al. 1987; Werner-Washburne et al. 1989; Bataille et al. 1991). The importance of the nitrogen signalling pathway has also been extended by recent results showing that trehalose-6-phosphate synthase and trehalose-6-phosphate phosphatase are inactivated by nitrogen-sources in a glucose-dependent manner (François et al. 1991). In addition, transcriptional regulation of the *CTT1* gene which also appeared to be controlled by the RAS-adenylate cyclase pathway is strongly affected by the combined presence of glucose and a nitrogen source in the medium (Bissinger et al. 1989; Belazzi et al. 1991).

Activation of the free catalytic subunits of cAMP-PK by the nitrogen source pathway provides an explanation for the existence of cell cycle mutants which arrest at the restrictive temperature at the nutrient-starvation/cAMP-PK deficiency site of start, but are not affected in terms of cAMP metabolism itself. They include *cdc33*, which encodes protein synthesis initiation factor 4E (Bedard et al. 1981; Verdier et al. 1989; J. Keleman, unpublished results), the *PRT1* gene product, which is also involved in protein synthesis initiation (Hanic-Joyce et al. 1987; Linder & Prat 1990) and *cdc60*, which encodes cytoplasmic leucyl-tRNA synthetase (S. Hohmann, R. Singer & J. Keleman, unpublished results). Other temperature sensitive mutants in amino-acyl tRNA synthetases, i.e. methionyl- (Unger & Hartwell 1976; Chatton et al. 1987), isoleucyl- (Hartwell & McLaughlin 1968), and maybe also valyl- (Chatton et al. 1988) and histidyl- (Natsoulis et al. 1986) tRNA synthetase show the same behaviour. These results appear to point to the initiation of protein synthesis as a crucial component of the nitrogen signalling pathway (Fig. 1). Protein synthesis itself is not involved as inferred from the fact that its inhibition does not prevent nitrogen-source induced activation of cAMP-PK (as followed by activation of trehalase) (Thevelein & Beullens 1985; J. Keleman, unpublished results). The suppression of both the prolonged G_1 phase at the permissive temperature and the specific start arrest at the restrictive temperature of the *cdc33* mutant by deletion of *bcyl* (Brenner et al. 1988) supports involvement of protein synthesis initiation in a pathway leading to activation of protein kinase A.

Possible involvement of Ca^{2+} as second messenger for progression over the start point of the yeast cell cycle is rather unclear. Iida et al. (1990) showed that intracellular Ca^{2+}-depletion caused transient G_1 arrest in wild type cells but not in mutants with an overactive adenylate cyclase pathway or with constitutively high activity of cAMP-PK. These results point to the possible presence of a Ca^{2+}-requiring step upstream of cAMP-PK in the nutrient-induced signaling pathway. Uno et al. (1988) showed that PIP_2 hydrolysis is essential to proceed through the G_1 phase of the cell cycle. Results from the same group showed that PIP_2 formation is under positive control by cAMP-PK (Kato et al. 1989). The relationship with nutrient-induced signaling however is not clear. It is not known for instance whether PIP_2-depleted cells arrest at the nutrient starvation site or the pheromone site of 'start'.

It has been claimed that growth control by nutrients could also be carried out by an unknown cAMP-independent pathway (e.g. Futcher 1990; Broach 1991). The paper by Cameron et al. (1988) which reported cAMP-independent control of sporulation, glycogen metabolism and heat shock resistance in *S. cerevisiae*, has caused much confusion in this respect. The evidence in this paper was clearly limited to the above mentioned phenotypic properties. cAMP-independent nutrient control of yeast metabolism is well known, e.g. the main glucose repression pathway does not involve the RAS-adenylate cyclase pathway (Matsumoto et al. 1982). Hence, one glucose-repressible protein in the pathways controlling sporulation, glycogen metabolism and heat shock resistance (in yeast the latter is mainly determined by the level of trehalose: Wiemken 1990) would be enough to introduce cAMP-independent nutrient control of the phenomenon. In fact, it has already been demonstrated that glycogen synthase is not only regulated by phosphorylation but that it is also under glucose repression control (François et al. 1987). In addition, yeast cells contain many other cAMP-independent nutrient-controlled regulatory pathways, which could be responsible for nutrient control of the described phenotypic characteristics. Unfortunately, the paper by Cameron et al. (1988) has been taken as evidence that yeast cells lacking a regulated cAMP-PK show normal growth control and by extension that it is possible to delete the RAS - adenylate cyclase - cAMP-PK pathway while retaining the over-all growth behavior of wild type cells. This is certainly not the case. There is not a single strain available that lacks the RAS - adenylate cyclase - cAMP-PK pathway and that shows the same growth and starvation behaviour as wild type strains.

Downstream of cAMP-dependent protein kinase: what is the link between the nutrient and pheromone sites in start?

Progression over the pheromone site is controlled by the well-known CDC28 protein kinase (generally known in other eukaryotes as cdc2 kinase). The activity of CDC28 is dependent upon G_1 cyclins. These cyclins are synthesized specifically in G_1 and rapidly degraded afterwards presumably because of the PEST sequences they contain in their C-terminus (Reed 1991). Nutrient depletion prevents accumulation of G_1 cyclins and causes inactivation of the CDC28 protein kinase. According to Reed (1991) the reduced rates of protein synthesis associated with starvation are probably responsible for the failure to accumulate G_1 cyclins, since G_1 cyclins are intrinsically unstable. Because of the failure to accumulate the cyclins, the CDC28 protein kinase would remain inactivated and the cells arrested in G_1. This explanation is unlikely since in that case nutrient-deprived cells should arrest at the CDC28 site of the cell cycle and not at an earlier site, the *TPK*-encoded protein kinase site. It appears more likely that synthesis of G_1 cyclins is dependent on a previous phosphorylation event by the *TPK*-encoded protein kinase earlier in the cell cycle. That dominant *CLN2* alleles (which are truncated versions of CLN2, lacking the sequences causing metabolic instability) confer an inability to arrest in G_1 in response to nutrient limitation (Hadwiger et al. 1989) does not contradict the phosphorylation requirement. The proteins encoded by these alleles are stable and therefore continuously present throughout the cell cycle. Shutdown of phosphorylation-dependent G_1 cyclin synthesis during nutrient starvation does not cause cell cycle arrest at the nutrient site because the requirement to continue with the next step of the cell cycle is already fulfilled. The experiments with the dominant CLN2 alleles show very clearly that no other factor dependent on protein phosphorylation by the *TPK*-encoded protein kinase is required other than the synthesis of G_1 cyclin. Hence, based on this experiment one can indeed conclude that nutrient availability is ultimately integrated into the over-all decision to continue the cell cycle at the level of the G_1 cyclins (Reed 1991).

Yeast cells can only start with a new cell cycle when the cell has reached the proper size. In strains with dominant, overactivating alleles of *CLN2*, the cell size requirement is overruled. The cells start with a new S phase shortly after mitosis and since the main growth phase of the yeast cell is in G_1, the cells become continuously smaller. Cell size information therefore has to be integrated at or before the Cyclins (Reed 1991). When nitrogen-starved yeast cells are fractionated according to cell size by elutriation and the different fractions refed with nitrogen, large cells start budding appreciably earlier than small cells. Apparently the small cells first have to grow to full size before they can continue with the new cell cycle. In all cases however, refeeding of nitrogen causes immediate activation of trehalase indicating that the block due to insufficient cell size is not situated in the nitrogen signalling pathway which is responsible for activation of *TPK*-encoded protein kinase (H. Mergelsberg, unpublished results).

Only two components have been identified which could be involved in the pathway between *TPK*-encoded protein kinase and the G_1 cyclins: the *SCH9* and *YAK1* gene products, which are both protein kinases themselves. The SCH9 protein kinase shows homology to the *TPK*-encoded protein kinases and is able to rescue lethality caused by deletion of the three *TPK* genes. Deletion of SCH9 causes slower growth with a longer G_1 phase (Toda et al. 1988). If the SCH9 protein kinase is a downstream component of the signaling pathway to cyclin synthesis, there should be at least one other protein kinase acting in parallel. Deletion of the *YAK1* gene partially rescues lethality caused by a triple deletion of the *TPK* genes. Deletion of YAK1 itself however does not produce a phenotype while overexpression of YAK1 is only able to cause growth arrest in strains with reduced activity of protein kinase A (Garrett & Broach 1989; Garrett et al. 1991). Apparently the YAK1 protein kinase shuts off a pathway able to take over the function of the *TPK*-encoded protein kinase. Since deletion of *YAK1* is unable to suppress the temperature-sensitive start arrest of the *cdc33* mutant (Garrett et al. 1991), as opposed to a deletion of *BCY1* (Brenner et al. 1988), the TPK-replacing pathway which is set free by deletion of *YAK1* seems to be at least subordinate to the protein kinase A pathway or might even be nonfunctional in strains with normal protein kinase A activity.

Conclusions

The main conclusion presented in this review is that the RAS-adenylate cyclase pathway does not serve as the mediator of nutrient-control of growth and cell cycle progression in yeast. This pathway is only involved in signaling the level of fermentable sugar during the transition from the derepressed state to the repressed state. In this case, a 'general glucose sensing complex', composed of at least a low-affinity glucose carrier, a sugar kinase and the *GGS1* (= *FDP1*) gene product is proposed as the initial activator of the RAS-adenylate cyclase pathway and as the universal activator in all other cases of glucose-induced regulatory phenomena in yeast. During growth of glucose-repressed cells on glucose the main function of the RAS-adenylate cyclase complex is to provide a basal level of cAMP. This basal level causes liberation of part of the catalytic subunits of cAMP-PK. A glucose-dependent nitrogen-induced signaling pathway, which specifically involves the initiation of protein synthesis, is suggested to activate the free, and only the free, catalytic subunits of cAMP-PK. cAMP-PK rather than cAMP therefore would serve as the universal integrator of nutrient availability in yeast.

Acknowledgements

I wish to thank J.C. Argüelles, J. Becher dos Passos, M. Beullens, E. Boy-Marcotte, R. Brandao, B. Bulaya, P. Durnez, K. Hirimburegama, S. Hohmann, M. Jacquet, A.W.H. Jans, K. Mbonyi, J. Ramos, L. Sierkstra, R. Singer, L. Van Aelst, M. Vanhalewyn and F. Zimmermann for the fruitful co-operation. Research carried out in our laboratory was supported by the Belgian National Fund for Scientific Research (FGWO, 'Kom op tegen Kanker'), the Belgian National Lottery, the Research Fund of the K.U.Leuven and the North Atlantic Treaty Organization.

References

Alonso A, Pascual C, Herrera L, Gancedo JM & Gancedo C (1984) Metabolic imbalance in a *Saccharomyces cerevisiae* mutant unable to grow on fermentable hexoses. Eur. J. Biochem. 138: 407–411

Argüelles JC, Mbonyi K, Van Aelst L, Vanhalewyn M, Jans AWH & Thevelein JM (1990) Absence of glucose-induced cAMP signaling in the *Saccharomyces cerevisiae* mutants *cat1* and *cat3* which are deficient in derepression of glucose-repressible proteins. Arch. Microbiol. 154: 199–205

Arkinstall SJ, Papasavvas SG, & Payton MA (1991) Yeast α-mating factor receptor-linked G-protein signal transduction suppresses RAS-dependent activity. FEBS Lett. 284: 123–128

Banuelos M & Fraenkel DG (1982) *Saccharomyces carlsbergensis fdp* mutant and futile cycling of fructose-6-phosphate. Mol. Cell. Biol. 2: 921–929

Barbacid M (1987) *ras* Genes. Ann. Rev. Biochem. 56: 779–827

Baroni MD, Martegani E, Monti P & Alberghina L (1989) Cell size modulation by *CDC25* and *RAS2* genes in *Saccharomyces cerevisiae*. Mol. Cell. Biol. 9: 2715–2723

Bataille N, Regnacq M & Boucherie H (1991) Induction of a heat-shock-type response in *Saccharomyces cerevisiae* following glucose limitation. Yeast 7: 367–378

Becher dos Passos J, Vanhalewyn M, Brandao RL, Castro IM, Nicoli JR & Thevelein JM (1992) Glucose-induced activation of plasma membrane H^+-ATPase in mutants of the yeast *Saccharomyces cerevisiae* affected in cAMP metabolism, cAMP-dependent protein phosphorylation and the initiation of glycolysis. Biochim. Biophys. Acta. (in press)

Bedard DP, Johnston GC & Singer RA (1981) New mutations in the yeast *Saccharomyces cerevisiae* affecting completion of 'start'. Curr. Genet. 4: 205–214

Belazzi T, Wagner A, Wieser R, Schanz M, Adam G, Hartig A & Ruis H (1991) Negative regulation of transcription of the *Saccharomyces cerevisiae* catalase T (*CTT1*) gene by cAMP is mediated by a positive control element. EMBO J. 10: 585–592

Beullens M & Thevelein JM (1990) Investigation of transport-associated phosphorylation of sugar in yeast mutants (*snf3*) lacking high-affinity glucose transport and in a mutant (*fdp1*) showing deficient regulation of initial sugar metabolism. Curr. Microbiol. 21: 39–46

Beullens M, Mbonyi K, Geerts L, Gladines D., Detremerie K, Jans AWH & Thevelein JM (1988) Studies on the mechanism of the glucose-induced cAMP-signal in glycolysis- and glucose repression-mutants of the yeast *Saccharomyces cerevisiae*. Eur. J. Biochem. 172: 227–231

Bissinger PH, Wieser R, Hamilton B & Ruis H (1989) Control of *Saccharomyces cerevisiae* catalase T gene (*CTT1*) expression by nutrient supply via the *ras*-cyclic AMP pathway. Mol. Cell. Biol. 9: 1309–1315

Boutelet F, Petitjean A & Hilger F (1985) Yeast *cdc35* mutants are defective in adenylate cyclase and are allelic with *cyr1* mutants while *CAS1*, a new gene, is involved in the regulation of adenylate cyclase. EMBO J. 4: 2635–2641

Boy-Marcotte E, Garreau H & Jacquet M (1987) Cyclic AMP controls the switch between division cycle and resting state programs in response to ammonium availability in *Saccharomyces cerevisiae*. Yeast 3: 85–93

Breitenbach-Schmitt I, Schmitt HD, Heinisch J & Zimmermann FK (1984) Genetic and physiological evidence for the existence of a second glycolytic pathway in yeast parallel to the phosphofructokinase-aldolase reaction sequence. Mol. Gen. Genet. 195: 536–540

Brenner C, Nakayama N, Goebl M, Tanaka K, Toh-E A & Matsumoto K (1988) *CDC33* encodes mRNA Cap-binding protein eIF-4E of *Saccharomyces cerevisiae*. Mol. Cell. Biol. 8: 3556–3559

Breviario D, Hinnebusch A, Cannon J, Tatchell K & Dhar R (1986) Carbon source regulation of *RAS1* expression in *Saccharomyces cerevisiae* and the phenotypes of ras2 cells. Proc. Natl. Acad. Sci. USA 83: 4152–4156

Broach JR (1991) RAS genes in *Saccharomyces cerevisiae*: signal transduction in search of a pathway. Trends Genet. 7: 28–33

Broek D, Samiy N, Fasano O, Fujiyama A, Tamanoi T, Northup J & Wigler M (1985) Differential activation of yeast adenylate cyclase by wild-type and mutant RAS proteins. Cell 41: 763–769

Broek D, Toda T, Michaeli T, Levin L, Birchmeier C, Zoller M, Powers S & Wigler M (1987) The *S. cerevisiae CDC25* gene product regulates the RAS/adenylate cyclase pathway. Cell 48: 789–799

Cameron S, Levin L, Zoller M & Wigler M (1988) cAMP-independent control of sporulation, glycogen metabolism, and heat shock resistance in *S. cerevisiae*. Cell 53: 555–566

Camonis JH, Kalékine M, Gondré B, Garreau H, Boy-Marcotte E & Jacquet M (1986) Characterization, cloning and sequence analysis of the *CDC25* gene which controls the cyclic AMP level of *Saccharomyces cerevisiae*. EMBO J. 5: 375–380

Cannon JF, Gitan R & Tatchell K (1990) Yeast cAMP-dependent protein kinase regulatory subunit mutations display a variety of phenotypes. J. Biol. Chem. 265: 11897–11904

Carter BLA & Jagadish MN (1978) The relationship between cell size and cell division in the yeast *Saccharomyces cerevisiae*. Exp. Cell Res. 112: 15–24

Casperson GF, Walker N, Brasier AR & Bourne HR (1983) A guanine nucleotide-sensitive adenylate cyclase in the yeast *Saccharomyces cerevisiae*. J. Biol. Chem. 258: 7911–7914.

Charlab R, Oliveira DE & Panek A (1985) Investigation of the relationship between *sst1* and *fdp* mutations in yeast and their effect on trehalose synthesis. Brazilian J. Med. Biol. Res. 18: 447–454

Chatton B, Winsor B, Boulanger Y & Fasiolo F (1987) Cloning and characterization of the yeast methionyl-tRNA synthetase mutation *mes1*. J. Biol. Chem. 262: 15094–15097

Chatton B, Walter P, Ebel J-P, Lacroute F & Fasiolo F (1988) The yeast *VAS1* gene encodes both mitochondrial and cytoplasmic valyl-tRNA synthetases. J. Biol. Chem. 263: 52–57

Cherry JR, Johnson TR, Dollard C, Sushter JR & Denis CL (1989) Cyclic AMP-dependent protein kinase phosphorylates and inactivates the yeast transcriptional activator ADR1. Cell 56: 409–419

Coote PJ, Cole MB & Jones MV (1991) Induction of increased thermotolerance in *Saccharomyces cerevisiae* may be triggered by a mechanism involving intracellular pH. J. Gen. Microbiol. 137: 1701–1708

DeFeo-Jones D, Scolnick EM, Koller R & Dahr R (1983) *ras*-Related gene sequences identified and isolated from *Saccharomyces cerevisiae*. Nature 306: 707–709

DelaFuente G (1970) Specific inactivation of yeast hexokinase induced by xylose in the presence of a phosphoryl donor substrate. Eur. J. Biochem. 16: 240–243

De Vendittis E, Vitelli A, Zahn R & Fasano O (1986) Suppression of defective RAS1 and RAS2 functions in yeast by an adenylate cyclase activated by a single amino acid change. EMBO J. 5: 3657–3663

Dhar R, Nieto A, Koller R, DeFeo-Jones D & Scolnick E (1984) Nucleotide sequence of two *ras*H related-genes isolated from the yeast *Saccharomyces cerevisiae*. Nucleic Acids Res. 12: 3611–3618

Dumont JE, Jauniaux JC & Roger PP (1989) The cyclic AMP-mediated stimulation of cell proliferation. TIBS 14: 67–71

Engelberg D, Perlman R & Levitzki A (1989) Transmembrane signalling in *Saccharomyces cerevisiae*. Cellular Signalling 1: 1–7

Entian KD, Droll L & Mecke D (1983) Studies on rapid reversible and non-reversible inactivation of fructose-1,6-bisphosphatase and malate dehydrogenase in wild-type and glycolytic block mutants of *Saccharomyces cerevisiae*. Arch. Microbiol. 134: 187–192

Fasano O, Crechet JB, De Vendittis E, Zahn R, Feger G, Vitelli A & Parmeggiani A (1988) Yeast mutants temperature-sensitive for growth after random mutagenesis of the chromosomal *RAS2* gene and deletion of the *RAS1* gene. EMBO J. 7: 3375–3383

Fedor-Chaiken M, Deschenes RJ & Broach JR (1990) *SRV2*, a gene required for RAS activation of adenylate cyclase in yeast. Cell 61: 329–340

Fernandez R, Herrero P & Moreno F (1985) Inhibition and inactivation of glucose-phosphorylating enzymes from *Saccharomyces cerevisiae* by D-xylose. J. Gen. Microbiol. 131: 2705–2709

Field J, Vojtek A, Ballester R, Bolger G, Colicelli J, Ferguson K, Gerst J, Kataoka T, Michaeli T, Powers S, Riggs M, Rodgers L, Wieland I, Wheland B & Wigler M (1990) Cloning and characterization of CAP, the *S. cerevisiae* gene encoding the 70 kd adenylyl cyclase-associated protein. Cell 61: 319–327

Fraenkel DG (1982) Carbohydrate metabolism. In: Strathern JN, Jones EW & Broach JR (Eds) The Molecular Biology of the Yeast *Saccharomyces*. Metabolism and Gene Expression (pp 1–37). Cold Spring Harbory Laboratory, Cold Spring Harbor

Fraenkel DG (1985) On *RAS* gene function in yeast. Proc. Natl. Acad. Sci. USA 82: 4740–4744

François J, Eraso P & Gancedo C (1987) Changes in the concentration of cAMP, fructose-2,6-bisphosphate and related metabolites and enzymes in *Saccharomyces cerevisiae* during growth on glucose. Eur. J. Biochem. 164: 369–373

François J, Villanueva ME & Hers HG (1988) The control of glycogen metabolism in yeast. 1. Interconversion *in vivo* of glycogen synthase and glycogen phosphorylase induced by glucose, a nitrogen source or uncouplers. Eur. J. Biochem. 174: 551–559

François J, Neves M-J & Hers H-G (1991) The control of trehalose biosynthesis in *Saccharomyces cerevisiae*: evidence for a catabolite inactivation and repression of trehalose-6-phosphate synthase and trehalose-6-phosphate phosphatase. Yeast 7: 575–587

François J, Van Schaftingen E & Hers HG (1984) The mechanism by which glucose increases fructose-2,6-bisphosphate concentration in *Saccharomyces cerevisiae*. A cyclic-AMP-dependent activation of phosphofructokinase 2. Eur. J. Biochem. 145: 187–193

Franzusoff AJ & Cirillo VP (1982) Uptake and phosphorylation of 2-deoxy-D-glucose by wild-type and single-kinase strains of *Saccharomyces cerevisiae*. Biochim. Biophys. Acta 688: 295–304

Frascotti G, Baroni D & Martegani E (1990) The glucose-induced polyphosphoinositides turnover in *Saccharomyces cerevisiae* is not dependent on the CDC25-RAS mediated signal transduction pathway. FEBS Lett. 274: 19–22

Futcher AB (1990) Yeast cell cycle. Curr. Opinion Cell Biol. 2: 246–251

Gancedo C & Schwerzmann K (1976) Inactivation by glucose of phosphoenolpyruvate carboxykinase from *Saccharomyces cerevisiae*. Arch. Microbiol. 109: 221–225

Gancedo JM & Gancedo C (1979) Inactivation of glyconeogenic enzymes in glycolytic mutants of *Saccharomyces cerevisiae*. Eur. J. Biochem. 101: 455–460

Garrett S & Broach J (1989) Loss of Ras activity in *Saccharomyces cerevisiae* is suppressed by disruptions of a new kinase gene, *YAK1*, whose product may act downstream of the cAMP-dependent protein kinase. Genes & Dev. 3: 1336–1348

Garrett S, Menold MM & Broach JR (1991) The *Saccharomyces cerevisiae YAK1* gene encodes a protein kinase that is induced by arrest early in the cell cycle. Mol. Cell. Biol. 11: 4045–4052

Gerst JE, Ferguson K, Vojtek A, Wigler M & Field J (1991) CAP is a bifunctional component of the *Saccharomyces cerevisiae* adenylyl cyclase complex. Mol. Cell. Biol. 11: 1248–1257

Gibbs JB & Marshall MS (1989) The *ras* oncogene – an important regulatory element in lower eucaryotic organisms. Microbiol. Rev. 53: 171–185

Gilman AG (1987) G Proteins : transducers of receptor-generated signals. Annu. Rev. Biochem. 56: 615–649

Hadwiger JA, Wittenberg C, Richardson HE, Lopes MD & Reed SI (1989) A family of cyclin homologs that control the G_1 phase in yeast. Proc. Natl. Acad. Sci. USA 86: 6255–6259

Hanic-Joyce PJ, Johnston GC & Singer RA (1987) Regulated arrest of cell proliferation mediated by yeast *prt1* mutations. Exp. Cell Res. 172: 134–145

Hartwell LH (1974) *Saccharomyces cerevisiae* cell cycle. Bacteriol. Rev. 38: 164–198

Hartwell LH & McLaughlin CS (1968) Mutants of yeast with temperature-sensitive isoleucyl-tRNA synthetases. Poc. Natl. Acad. Sci. USA 59: 422–428

Hartwell LH, Mortimer RK, Culotti J & Culotti M (1973) Genetic control of the cell division cycle in yeast: genetic analysis of cdc mutants. Genetics 74: 267–286

Hartwell LH, Culotti J, Pringle JR & Reid BJ (1974) Genetic control of the cell division cycle in yeast. Science 183: 46–51

Hohmann S, Huse K, Valentin E, Mbonyi K, Thevelein JM & Zimmermann FK (1992) Glucose-induced regulatory defects in the *Saccharomyces cerevisiae* growth initiation mutant *byp1* and identification of *MIG1* as a partial suppressor. J. Bacteriol. (in press)

Holzer H (1984) Mechanism and function of reversible phosphorylation of fructose-1,6-bisphosphatase in yeast. In: Cohen P (Ed) Molecular Aspects of Cellular Regulation, Vol 3 (pp 143–154). Elsevier, Amsterdam

Hottiger T, Boller T & Wiemken A (1989) Correlation of trehalose content and heat resistance in yeast mutants altered in the RAS/adenylate cyclase pathway: is trehalose a thermoprotectant? FEBS Lett. 255: 431–434

Iida H, Sakaguchi S, Yagawa Y & Anraku Y (1990) Cell cycle control by Ca^{2+} in *Saccharomyces cerevisiae*. J. Biol. Chem. 265: 21216–21222

Jacquet M & Camonis J (1985) Contrôle du cycle de division cellulaire et de la sporulation chez *Saccharomyces cerevisiae* par le système de l'AMP cyclique. Biochimie 67: 35–43

Jaspers HTA & Van Steveninck J (1975) Transport-associated phosphorylation of 2-deoxy-D-glucose in *Saccharomyces fragilis*. Biochim. Biophys. Acta 406: 370–385

Johnston GC, Pringle JR & Hartwell LH (1977) Coordination of growth with cell division in the yeast *Saccharomyces cerevisiae*. Exp. Cell Res. 105: 79–98

Kaibuchi K, Miyajima A, Arai K & Matsumoto K (1986) Possible involvement of RAS-encoded proteins in glucose-induced inositolphospholipid turnover in *Saccharomyces cerevisiae*. Proc. Natl. Acad. Sci. USA 83: 8172–8176

Kataoka T, Powers S, McGill C, Fasano O, Strathern J, Broach J & Wigler M (1984) Genetic analysis of yeast *RAS1* and *RAS2* genes. Cell 37: 437–445

Kataoka T, Broek D & Wigler M (1985a) DNA sequence and characterization of the *S. cerevisiae* gene encoding adenylate cyclase. Cell 43: 493–505

Kataoka T, Powers S, Cameron S, Fasano O, Goldfarb M, Broach J & Wigler M (1985b) Functional homology of mammalian and yeast *RAS* genes. Cell 40: 19–26

Kato H, Uno I, Ishikawa T & Takenawa T (1989) Activation of phosphatidylinositol kinase and phosphatidyl-4-phosphate kinase by cAMP in *Saccharomyces cerevisiae*. J. Biol. Chem. 264: 3316–3321

Kim J-H & Powers S (1991) Overexpression of RPI1, a novel inhibitor of the yeast ras-cyclic AMP pathway, down-regulates normal but not mutationally activated ras functions. Mol. Cell. Biol. 11: 3894–3904

Kruckeberg AL & Bisson LF (1990) The *HXT2* gene of *Saccharomyces cerevisiae* is required for high-affinity glucose transport. Mol. Cell. Biol. 10: 5903–5913

Leao C & Van Uden N (1984) Effects of ethanol and other alkanols on passive proton influx in the yeast *Saccharomyces cerevisiae*. Biochim. Biophys. Acta 774: 43–48

Liao H & Thorner J (1980) Yeast mating pheromone α factor inhibits adenylate cyclase. Proc. Natl. Acad. Sci. USA 77: 1898–1902

Linder P & Prat A (1990) Baker's yeast, the new work horse in protein synthesis studies – analyzing eukaryotic translation initiation. Bio Essays 12: 519–526

Londesborough J & Lukkari TM (1980) The pH and temperature dependence of the activity of the high K_m cyclic nucleotide phosphodiesterase of baker's yeast. J. Biol. Chem. 255: 9262–9267

Malone RE (1990) Dual regulation of meiosis in yeast. Cell 61: 375–378

Marshall MS, Gibbs JB, Scolnick EM & Sigal IS (1987) Regulatory function of the *Saccharomyces cerevisiae* RAS C-terminus. Mol. Cell. Biol. 7: 2309–2315

Martegani E, Vanoni M & Baroni M (1984) Macromolecular synthesis in the cell cycle mutant *cdc25* of budding yeast. Eur. J. Biochem. 144: 205–210

Martegani E, Baroni M & Wanoni M (1986) Interaction of cAMP with the CDC25-mediated step in the cell cycle of budding yeast. Exp. Cell Res. 162: 544–548

Matsumoto K, Uno I, Toh-e A, Ishikawa T & Oshima Y (1982) Cyclic AMP may not be involved in catabolite repression in *Saccharomyces cerevisiae* : evidence from mutants capable of utilizing it as an adenine source. J. Bacteriol. 150: 277–285

Matsumoto K, Uno I & Ishikawa T (1983) Initiation of meiosis in yeast mutants defective in adenylate cyclase and cyclic AMP-dependent protein kinase. Cell 32: 417–423

Matsumoto K, Uno I & Ishikawa T (1985) Genetic analysis of the role of cAMP in yeast. Yeast 1: 15–24

Mazon MJ, Gancedo JM & Gancedo C (1982) Inactivation of yeast fructose-1,6-bisphosphatase. In vivo phosphorylation of the enzyme. J. Biol. Chem. 257: 1128–1130

Mbonyi K, Beullens M, Detremerie K, Geerts L & Thevelein JM (1988) Requirement of one functional *RAS* gene and inability of an oncogenic *ras*-variant to mediate the glucose-induced cAMP signal in the yeast *Saccharomyces cerevisiae*. Mol. Cell. Biol. 8: 3051–3057

Mbonyi K, Van Aelst L, Argüelles JC, Jans AWH & Thevelein JM (1990) Glucose-induced hyperaccumulation of cAMP and absence of glucose repression in yeast strains with reduced activity of cAMP-dependent protein kinase. Mol. Cell. Biol. 10: 4518–4523

Meredith SA & Romano AH (1977) Uptake and phosphorylation of 2-deoxy-D-glucose by wild type and respiration-deficient baker's yeast. Biochim. Biophys. Acta 497: 745–759

Mitts MR, Grant DB & Heideman W (1990) Adenylate cyclase in *Saccharomyces cerevisiae* is a peripheral membrane protein. Mol. Cell. Biol. 10: 3873–3883

Mitts MR, Bradshaw-Rouse J & Heideman W (1991) Interactions between adenylate cyclase and the yeast GTPase-activating protein IRA1. Mol. Cell. Biol. 11: 4591–4598

Müller D & Holzer H (1981) Regulation of fructose-1,6-bisphosphatase in yeast by phosphorylation/dephosphorylation. Biochem. Biophys. Res. Commun. 103: 926–933

Munder T & Küntzel H (1989) Glucose-induced cAMP signaling in *Saccharomyces cerevisiae* is mediated by the CDC25 protein. FEBS Lett. 242: 341–345

Nakajima H, Oshima I, Yashiro M, Yoda K, Yamasaki M & Tamura G (1987) Novel *Saccharomyces cerevisiae* temperature-sensitive *cyr1* mutant which accumulates glycogen particles. Agric. Biol. Chem. 51: 2679–2689

Natsoulis G, Hilger F & Fink GR (1986) The *HTS1* gene encodes both the cytoplasmic and mitochondrial histidine tRNA synthetases of *S. cerevisiae*. Cell 46: 235–243

Nikawa J, Cameron S, Toda T, Ferguson KW & Wigler M (1987a) Rigorous feedback control of cAMP levels in *Saccharomyces cerevisiae*. Genes & Dev. 1: 931–937

Nikawa J, Sass P & Wigler M (1987b) Cloning and characterization of the low-affinity cyclic AMP phosphodiesterase gene of *Saccharomyces cerevisiae*. Mol. Cell. Biol. 7: 3629–3636

Panek AD, Sampaio AL, Braz GC, Baker SJ & Mattoon JR (1979) Genetic and metabolic control of trehalose and glycogen synthesis. New relationships between energy reserves, catabolite repression and maltose utilization. Cell. Mol. Biol. 25: 345–354

Panek AC, François J & Panek AD (1988) New insights into a mutant of *Saccharomyces cerevisiae* having impaired sugar uptake and metabolism. Curr. Gen. 13: 15–20

Pardee AB (1974) A restriction point for control of normal animal cell proliferation. Proc. Natl. Acad. Sci. USA 71: 1286–1290

Perlman R, Eilam Y, Padan E, Simchen G & Levitzki A (1989) Rapid intracellular alkalinization of *Saccharomyces cerevisiae* Mata cells in response to Alpha-Factor requires the Cdc25 gene product. Cellular Signalling 1: 577–586

Petitjean A, Hilger F & Tatchell K (1990) Comparison of thermosensitive alleles of the *Cdc25* gene involved in the cAMP metabolism of *Saccharomyces cerevisiae*. Genetics 124: 797–806

Piper P (1990) Interdependence of several heat shock gene activations, cyclic AMP decline and changes at the plasma membrane of *Saccharomyces cerevisiae*. A. v. Leeuwenhoek 58: 195–201

Postma PW & Lengeler JW (1985) Phosphoenolpyruvate:carbohydrate phosphotransferase system of bacteria. Microbiol. Rev. 49: 232–269

Powers S, Kataoka T, Fasano O, Goldfarb M, Strathern J, Broach J & Wigler M (1984) Genes in *S. cerevisiae* encoding proteins with domains homologous to the mammalian ras proteins. Cell 36: 607–612

Praekelt UM & Meacock PA (1990) HSP12, a new small heat shock gene of *Saccharomyces cerevisiae*: analysis of structure, regulation and function. Mol. Gen. Genet. 223: 97–106

Pringle JR & Hartwell LH (1981) The *Saccharomyces cerevisiae* cell cycle. In : Strathern JN, Jones EW & Broach JR (Eds) The Molecular Biology of the Yeast *Saccharomyces*. Metabolism and Gene Expression (pp 97–142). Cold Spring Harbor Laboratory, Cold Spring Harbor

Reed SI (1980) The selection of *S. cerevisiae* mutants defective in the start event of cell division. Genetics 95: 561–577

Reed SI (1991) G1-specific cyclins: in search of an S-phase-promoting factor. Tr. Genet. 7: 95–99

Resnick RJ & Racker E (1988) Phosphorylation of the *RAS2* gene product by kinase A inhibits the activation of yeast adenylyl cylase. Proc. Natl. Acad. Sci. USA 85: 2474–2478

Robinson LC, Gibbs JB, Marshall MS, Sigal IS & Tatchell K (1987) cdc25 : a component of the Ras-adenylate cyclase pathway in *Saccharomyces cerevisiae*. Science 235: 1218–1221

Rose M, Entian KD, Hofmann L & Vogel RF (1988) Irreversible inactivation of *Saccharomyces cerevisiae* fructose-1,6-bisphosphatase independent of protein phosphorylation at Ser11. FEBS Lett. 241: 55–59

Rose M, Albig W & Entian KD (1991) Glucose repression in

Saccharomyces cerevisiae is directly associated with hexose phosphorylation by hexokinase-PI and hexokinase-PII. Eur. J. Biochem. 199: 511–518

Saier MH Jr, Wu L-F & Reizer J (1990) Regulation of bacterial physiological processes by three types of protein phosphorylating systems. Tr. Biochem. Sci. 15: 391–395

Sass P, Field J, Nikawa J, Toda T & Wigler M (1986) Cloning and characterization of the high-affinity cAMP phosphodiesterase of *S. cerevisiae*. Proc. Natl. Acad. Sci. USA 83: 9303–9307

Schuddemat J, Van den Broek PJA & Van Steveninck J (1986) Effect of xylose incubation on the glucose transport system in *Saccharomyces cerevisiae*. Biochim. Biophys. Acta 861: 489–493

Shin D-Y, Matsumoto K, Iida H, Uno I & Ishikawa T (1987a) Heat shock response of *Saccharomyces cerevisiae* mutants altered in cyclic AMP-dependent protein phosphorylation. Mol. Cell. Biol. 7: 244–250

Shin D-Y, Uno I & Ishikawa T (1987b) Control of the G1 – G0 transition and G0 protein synthesis by cyclic AMP in *Saccharomyces cerevisiae*. Curr. Genet. 12: 577–582

Sols A (1976) The Pasteur effect in the allosteric era. In: Kornberg A & Ochoa S (Eds) Reflections on Biochemistry (pp 199–206). Pergamon Press, Oxford

Suoranta K (1985) Cyclic AMP phosphodiesterase activities in growing cells of baker's yeast (*Saccharomyces cerevisiae*). J. Cyclic Nucleot. Prot. Phosphor. Res. 10: 121–127

Suoranta K & Londesborough J (1984) Purification of intact and nicked forms of a zinc-containing, Mg^{+} dependent, low Km cyclic AMP phosphodiesterase from baker's yeast. J. Biol. Chem. 259: 6964–6971

Tanaka K, Matsumoto K, Toh-e A (1988) Dual regulation of the expression of the polyubiquitin gene by cyclic AMP and heat shock in yeast. EMBO J. 7: 495–502

Tanaka K, Matsumoto K & Toh-e A (1989) Ira1, an inhibitory regulator of the RAS – cyclic AMP pathway in *Saccharomyces cerevisiae*. Mol. Cell. Biol. 9: 757–768

Tanaka K, Nakafuku M, Satoh T, Marshall MS, Gibbs JB, Matsumoto K, Kaziro Y & Toh-e A (1990a) *S. cerevisiae* genes *IRA1* and *IRA2* encode proteins that may be functionally equivalent to mammalian ras GTPase activating protein. Cell 60: 803–807

Tanaka K, Nakafuku M, Tamanoi F, Kaziro Y, Matsumoto K & Toh-e A (1990b) *IRA2*, a second gene of *Saccharomyces cerevisiae* that encodes a protein with a domain homologous to mammalian *ras* GTPase-activating protein. Mol. Cell. Biol. 10: 4303–4313

Tatchell K (1986) *RAS* genes and growth control in *Saccharomyces cerevisiae*. J. Bacteriol. 166: 364–367

Tatchell K, Chaleff DT, DeFeo-Jones D, Scolnick EM (1984) Requirement of either of a pair of *ras*-related genes of *Saccharomyces cerevisiae* for spore viability. Nature 309: 523–527

Tatchell K, Robinson LC & Breitenbach M (1985) *RAS2* of *Saccharomyces cerevisiae* is required for gluconeogenic growth and proper response to nutrient limitation. Proc. Natl. Acad. Sci. USA 82: 3785–3789

Thevelein JM (1984a) Cyclic-AMP content and trehalase activation in vegetative cells and ascospores of yeast. Arch. Microbiol. 138: 64–67

Thevelein JM (1984b) Regulation of trehalose mobilization in fungi. Microbiol. Rev. 48: 42–59

Thevelein JM (1988) Regulation of trehalase activity by phosphorylation – dephosphorylation during developmental transitions in fungi. Exp. Mycol. 12: 1–12

Thevelein JM (1991) Fermentable sugars and intracellular acidification as specific activators of the RAS adenylate cyclase signalling pathway in yeast – the relationship to nutrient-induced cell cycle control. Mol. Microbiol. 5: 1301–1307

Thevelein JM & Beullens M (1985) Cyclic AMP and the stimulation of trehalase activity in the yeast *Saccharomyces cerevisiae* by carbon sources, nitrogen sources and inhibitors of protein synthesis. J. Gen. Microbiol. 131: 3199–3209

Toda T, Uno I, Ishikawa T, Powers S, Kataoka T, Broek D, Cameron S, Broach J, Matsumoto K & Wigler M (1985) In yeast, Ras proteins are controlling elements of adenylate cyclase. Cell 40: 27–36

Toda T, Cameron S, Sass P, Zoller M, Scott JD, McBullen B, Hurwitz M, Krebs EG & Wigler M (1987a) Cloning and characterization of *BCY1*, a locus encoding a regulatory subunit of the cyclic AMP-dependent protein kinase in *Saccharomyces cerevisiae*. Mol. Cell. Biol. 7 : 1371–1377

Toda T, Cameron S, Sass P, Zoller M & Wigler M (1987b) Three different genes in *Saccharomyces cerevisiae* encode the catalytic subunits of the cAMP-dependent protein kinase. Cell 50: 277–287

Toda T, Cameron S, Sass P & Wigler M (1988) *SCH9*, a gene of *Saccharomyces cerevisiae* that encodes a protein distinct from, but functionally and structurally related to cAMP-dependent protein kinase catalytic subunits. Genes & Dev. 2: 517–527

Unger MW & Hartwell LH (1976) Control of cell division in *Saccharomyces cerevisiae* by methionyl-tRNA. Proc. Natl. Acad. Sci. USA 73: 1664–1668

Uno I, Fukami K, Kato H, Takenawa T & Ishikawa T (1988) Essential role for phosphatidylinositol 4,5-biphosphate in yeast cell proliferation. Nature 333, 188–190

Van Aelst L, Boy-Marcotte E, Camonis JH, Thevelein JM & Jacquet M (1990) The C-terminal part of the *CDC25* gene product plays a key role in signal transduction in the glucose-induced modulation of cAMP level in *Saccharomyces cerevisiae*. Eur. J. Biochem. 193: 675–680

Van Aelst L, Hohmann S, Zimmermann K, Jans AWH & Thevelein JM (1991a) A yeast homologue of the bovine lens fibre MIP gene family complements the growth defect of a *Saccharomyces cerevisiae* mutant on fermentable sugars but not its defect in glucose-induced RAS-mediated cAMP signalling. EMBO J. 10: 2095–2104

Van Aelst L, Jans AWH & Thevelein JM (1991b) Involvement of the *CDC25* gene product in the signal transmission pathway of the glucose-induced RAS-mediated cAMP signal in the yeast *Saccharomyces cerevisiae*. J. Gen. Microbiol. 137: 341–349

Van de Poll KW & Schamhart DHJ (1977) Characterization of a regulatory mutant of fructose-1,6-diphosphatase in *Saccharomyces carlsbergensis*. Mol. Gen. Genet. 154: 61–66

Van de Poll KW, Kerkenaar A & Schamhart DHJ (1974) Isolation of a regulatory mutant of fructose-1,6-diphosphatase in Saccharomyces carlsbergensis. J. Bacteriol. 117: 965–970

Van der Plaat JB (1974) Cyclic 3,'5' – adenosine monophosphate stimulates trehalose degradation in bakers' yeast. Biochem. Biophys. Res. Commun. 56: 580–587

Van Steveninck J (1968) Transport and transport – associated phosphorylation of 2-deoxy-D-glucose in yeast. Biochim. Biophys. Acta 163: 386–394

Verdier JM, Camonis JH & Jacquet M (1989) Cloning of *CDC33*: a gene essential for growth and sporulation which does not interfere with cAMP production of *Saccharomyces cerevisiae*. Yeast 5: 79–90

Vojtek A, Haarer B, Field J, Gerst J, Pollard D, Brown S & Wigler M (1991) Evidence for a functional link between profilin and CAP in the yeast *S. cerevisiae*. Cell 66: 497–505

Weitzel G, Pilatus U & Rensing L (1987) The cytoplasmic pH, ATP content and total protein synthesis rate during heat-shock protein inducing treatments in yeast. Exp. Cell Res. 170: 64–79

Werner-Washburne M, Becker J, Kosic-Smithers J & Craig EA (1989) Yeast Hsp70 RNA levels vary in response to the physiological status of the cell. J. Bacteriol. 171: 2680–2688

Wiemken A (1990) Trehalose in yeast, stress protectant rather than reserve carbohydrate. A. v. Leeuwenhoek 58: 209–217

Antonie van Leeuwenhoek **62**: 131–153, 1992.

Genetic approaches to the study of mitochondrial biogenesis in yeast

M. Bolotin-Fukuhara[1] & L.A. Grivell[2]
[1] *Laboratoire de Génétique Moléculaire, Institut de Génétique et Microbiologie, Université Paris-Sud, Centre d'Orsay, 91405 Orsay, France;* [2] *Section for Molecular Biology, Department of Molecular Cell Biology, University of Amsterdam, Kruislaan 318, 1098 SM, Amsterdam, The Netherlands*

Key words: mitochondrial DNA, mutational analysis, nucleo-mitochondrial interactions, gene expression, membrane assembly, respiratory deficiency

Abstract

In contrast to most other organisms, the yeast *Saccharomyces cerevisiae* can survive without functional mitochondria. This ability has been exploited in genetic approaches to the study of mitochondrial biogenesis. In the last two decades, mitochondrial genetics have made major contributions to the identification of genes on the mitochondrial genome, the mapping of these genes and the establishment of structure-function relationships in the products they encode. In parallel, more than 200 complementation groups, corresponding to as many nuclear genes necessary for mitochondrial function or biogenesis have been described. Many of the latter are required for post-transcriptional events in mitochondrial gene expression, including the processing of mitochondrial pre-RNAs, the translation of mitochondrial mRNAs, or the assembly of mitochondrial translation products into the membrane. The aim of this review is to describe the genetic approaches used to unravel the intricacies of mitochondrial biogenesis and to summarize recent insights gained from their application.

Introduction

As in other eukaryotic cells, biosynthesis of functional mitochondria in yeast is a complex affair, with the genes for the many hundreds of proteins required to construct the organelle being divided over nuclear and mitochondrial DNAs. Consistent with its size, nuclear DNA contributes the lion's share to this joint venture, coding both for structural components of the organelle and for parts of its genetic apparatus. The contribution of the mitochondrial genome is modest by comparison, but nevertheless essential, consisting of key subunits of mitochondrial respiratory enzymes, together with the remaining parts of the genetic system.

As might be expected, mutations in either genome can lead to either mitochondrial dysfunction, or to an inability to assemble a functional organelle. In most organisms, this is ultimately a lethal event. However, the fact that yeast cells with defective mitochondria can fall back on the energy produced by the fermentation of sugars means that genetics can be used to identify the components encoded by each of the genomes and to dissect the regulatory mechanisms responsible for balancing their synthesis. Here we describe the genetic approaches that have been used to unravel the intricacies of mitochondrial biogenesis and summarize recent insights gained from their application. For a more detailed account of developments in this field, the reader is referred to full-length reviews by Tzagoloff & Myers (1986), Costanzo & Fox (1990), Grivell (1989), Tzagoloff & Dieckmann (1990).

The genetics of mitochondrial biogenesis: a plain man's guide

1. Mutations in mtDNA

Three types of mitochondrial mutation have been used in the analysis of mitochondrial biogenesis:

1.1. Rho⁻(ϱ⁻) petite mutants
These are deletion mutants of mtDNA, that can arise at high frequency either spontaneously by recombination across short repeated sequences in the genome, or as the result of the action of acridine and phenanthridine dyes. The deletion usually removes more than 50% of the wild type mtDNA sequence, while some mutants may retain as little as 0.1%. The amount of mtDNA in rho⁻ cells is the same as in wild type cells, regardless of the size of the deletion. Amplification of the remaining segment is responsible for this and the resulting molecules thus contain tandem repeats of this segment. Rho⁻ petite mutants lack the capacity for mitochondrial protein synthesis, presumably because one or more genes essential for this process have been hit. Petite mutants are still able to mate with wild type cells and to transfer the information contained in their mtDNA to wild type mtDNA. The results of such crosses, when combined with physical mapping data, were instrumental in the construction of the first genetic maps of yeast mtDNA (see Dujon (1981) and refs. therein).

1.2. Antibiotic- or inhibitor-resistant mutants
Compounds that specifically inhibit mitochondrial functions, such as protein synthesis, electron transport or oxidative phosphorylation, block the growth of yeast cells on non-fermentable media. Resistant mutants can be induced, or arise spontaneously and are easily selected. These mutants result from point mutations or small deletions and, like rho⁻ mutants, were extremely useful in initial mapping of the yeast mitochondrial genome.

1.3. Mit- mutants
These are respiratory-deficient, defective in the synthesis of one or more components of either the mitochondrial respiratory chain, or the mitochondrial protein synthetic machinery. Mutants in the former group can be distinguished from the phenotypically similar rho⁻ mutants by the fact that they retain mitochondrial protein synthesis (Tzagoloff et al. 1975; Slonimski & Tzagoloff 1976). Mutants in the latter class (sometimes referred to as syn⁻) are of necessity conditional, since mitochondrial protein synthesis is essential for the maintenance of wild type mtDNA (Myers & Tzagoloff 1985). Both types of mutant owe their phenotype either to point mutations, or deletions ranging in size from a few to several thousand base pairs. As described by Kotylak et al. (1976), mit⁻ mutant collections can be established in strains carrying the *op1* mutation in one of the genes for the adenine nucleotide translocator in yeast mitochondria (*AAC2*; Lawson & Douglas 1988). This has the advantage of preventing the accumulation of rho⁻ mutants in populations of mit⁻ cells, since combination of rho⁻ and *op1* is lethal, probably due to an inability of the cell to energize the mitochondrial membrane with cytosolically generated ATP (Kovacova et al. 1968). Maintenance of mit⁻ mutant collections in strains carrying the *op1* mutation will obviously select against mutants in components of the translational machinery due to the tendency of such mutants to throw off rho⁻ cells. To enrich for these mutants, a procedure involving selection for conditional mutants and crosses to selected rho⁻ mutants to screen for restoration of wild type function has been used with success (Bolotin-Fukuhara et al. 1977).

2. Mutations in nuclear DNA

Two classes of mutant have been of value in the identification of nuclear genes involved in mitochondrial biogenesis:

2.1. Pet *mutants*
These are respiratory-deficient as a result of a mutation in a nuclear gene encoding a mitochondrial component, or a protein required for the synthesis, transport or modification of such a component. The vast majority of *pet* mutations studied so far are recessive. They are distinguishable from the

more abundant class of cytoplasmic (rho^-) petite mutants by virtue of the fact that after crossing to a tester strain lacking mtDNA (rho^o), the diploid cells are capable of growth on non-fermentable carbon sources. Of some 2000 independent *pet* strains so far characterized, 1700 have been unambiguously assigned to 215 complementation groups (see Tzagoloff & Dieckmann (1990) for review). The distribution of these with respect to their involvement in the synthesis of major mitochondrial components is shown in Table 1. Although impressively large, there are several reasons to think that this collection may still be incomplete:

1. Mutations in some genes lead to instability of mtDNA and hence to rapid accumulation of cytoplasmic (rho^-) petites. Among such mutations are those affecting synthesis and function of components of the mitochondrial translation machinery and replication of mtDNA. Mutations so far identified in these functions are few in number and tend to be highly leaky.
2. Some mutations may not result in a discernable respiratory defect. Among these are mutations in (a) genes involved in the fine tuning of gene expression, (b) genes for certain Krebs cycle enzymes, (c) duplicated genes and (d) genes whose function only becomes apparent in combination with a particular mitochondrial genetic background, e.g. the presence of intron(s) whose processing is dependent on the action of the nuclear-encoded gene product.

Table 1. Genes for mitochondrial proteins identified by analysis of nuclear pet mutations (adapted from Tzagoloff & Dieckmann, 1990).

Protein/enzyme/complex	Compl. groups
Cytochrome c oxidase	40
QH2-cytochrome c reductase	23
ATP synthase	9
Translational machinery	17
Other	21
Unidentified	105
Total	215

2.2. Second-site suppressor mutants

As in the study of other genetic systems, the isolation of second-site mutants displaying phenotypic suppression of a first (usually mitochondrial) mutation has yielded valuable information on the interaction of nuclear and mitochondrial gene products during mitochondrial biogenesis. Although originally limited to a small group of so-called *NAM* mutants (Nuclear Accommodation of Mitochondria, Dujardin et al. 1980), affected in RNA splicing, this class of mutants now defines many genes, including those involved in mitochondrial transcription, other forms of RNA processing and translation.

3. Recombinant DNA techniques

3.1. The pleasures and pitfalls of transformation-complementation

Subsequent to characterization of a *pet* or *NAM*-type mutant, the mutated gene is in principle identifiable by application of recombinant DNA technology. Transformation of the mutant with a DNA library derived from wild type or suppressor strains should allow complementation of the deficiency and thus isolation of the corresponding gene. There are, however, many cases known in which the cloned sequences have been found not to correspond to the gene defined by the original mutation. For example, in the complementation assays usually employed to screen for suppression of splicing deficiency, restoration of even low levels of splicing will give rise to respiratory-sufficiency. Positive colonies can thus arise not only by direct restoration of mitochondrial splicing activity, but indirectly, for example by changes in translation, energy metabolism, or compensatory changes in the synthesis or import of other proteins involved in the splicing reaction. In both these and other situations, this type of suppression is more liable to occur when the complementing sequences are present in multiple copies, presumably reflecting a need for gross overproduction of the protein involved. Such cases are recognizable by virtue of the fact that a single copy of the gene complements incompletely, while inactivation of the chromosomal copy of the cloned sequences fails to produce the original mutant phenotype.

Although multi-copy suppression is usually regarded as a pitfall to be avoided, it has unexpectedly uncovered some of the more ingenious strategies employed by the yeast cell to bypass various mutational blocks. As such, therefore, it can be regarded as a valid tool for genetic analysis of the interrelationships that exist between events such as transcription (both nuclear and mitochondrial) and translation, translation and RNA processing, RNA processing and energy metabolism (see section on RNA processing). New relationships between organellar and cytosolic protein synthesis may also be revealed by this approach, as illustrated by the case of the *TIF1* and *TIF2* genes, characterised by Linder & Slonimski (1989) as redundant genes for the yeast cytoplasmic initiation factor eIF4A. When present in high copy number, they allow the suppression of missense mutations in the mitochondrial gene for subunit III of cytochrome c oxidase. The mechanism of this effect is as yet unknown.

3.2. Allotopic gene expression

Valuable though the study of various mutations in mtDNA has been, the lack of success of standard nucleic acid transformation procedures to introduce DNA into the organelle has formed a frustrating restriction on our ability to test ideas on many features of mitochondrial gene expression. This restriction has been partly overcome by the use of what has been termed allotopic gene expression (Nagley & Devenish 1989). Gene synthesis and/or repeated cycles of mutagenesis are used to create copies of mitochondrial genes that make use of standard codon assignments. After fusion to sequences coding for the appropriate targeting signals, these are introduced into the nucleus and synthesis, import and mitochondrial function of the encoded proteins are followed. The technical difficulties of such an approach should not be underestimated, in particular when components of the respiratory chain are involved. Efficient import and assembly of such proteins is critically dependent on the choice of targeting sequence and the availability of partially assembled complexes capable of accommodating the newly imported protein (Law et al. 1990).

3.3. Biolistic transformation

This recently developed technique makes use of high-velocity microprojectiles to deliver DNA into the mitochondrion (Fox et al. 1988). Transformation frequencies are at present low and unpredictable (Butow & Fox 1990), but the use of specially adapted strains, refined bombardment protocols (Williams et al. 1991) and/or better vectors should improve prospects for the application of *in vitro* mutagenesis to the *in vivo* study of a wide range of mitochondrial genetic functions, including the role of cis-acting sequence elements in RNA processing and translation.

4. Diverse genetic tricks: complementation and cytoduction

4.1. Complementation

The high rate of recombination of mtDNA and the rapid mitotic segregation of mitochondrial genomes in genetically mixed zygotes mean that it is impossible to maintain mitochondrial genomes in a heteroplasmic state long enough to carry out complementation analysis at the level of colony growth. However, short term complementation studies are possible:

1. *Zygotic gene rescue*: expression of the repeated sequences retained by rho$^-$ petite mutant mtDNAs cannot be studied directly, since these mutants lack the capacity for mitochondrial protein synthesis. They can, however, be studied in zygotes formed in crosses between the petite and a suitable rho$^+$, or mit$^-$ tester strain, since their high copy number allows them to compete successfully with those of the tester mtDNA for transcription and translation. The method of zygotic gene rescue was originally developed for the genetic localisation of sequences encoding *VAR1*, using petite and tester strains encoding electrophoretically distinguishable forms of the protein (Butow et al. 1978). Although it is likely that this technique will ultimately be replaced by

approaches based on biolistic transformation, it is of value in cases in which it is desirable to study expression of mtDNA segments in a completely homologous genetic system. It can be applied as long as a mit$^-$ tester strain, either deleted for the sequences under study, or at least unable to make the product specified by them, is available.

2. *Zygotic complementation*: the success of zygotic gene rescue shows that even short-lived heteroplasmy is sufficient for functional complementation. Use has been made of this fact to identify the tRNA synthesis locus in yeast mtDNA (Martin & Underbrink-Lyon 1981) and to assign different mit$^-$ mutations to complementation groups in the mosaic genes for cytochrome *b* and cytochrome *c* oxidase subunit I (summarized in Dujon 1981). In the case of complementing mutations, respiratory activity (measured as oxygen uptake) is recovered within 5–8 hours after zygote formation and is not related to the extent of recombination. Non-complementing mutants acquire respiration only after some 15–20 hours, the time at which wild type recombinants begin to accumulate.
3. *rho$^+$/rho$^-$ heterozygotes*: although inherently unstable, some heteroplasmic combinations can be maintained by selection. In these cases, the cell contains rho$^-$ and rho$^+$ genomes, both of which are required for respiration, but which are unable to form stable recombination products by homologous recombination. Such heterozygotes were first picked up as mitochondrial suppressors of *pet* mutants disturbed in mitochondrial translation (see Mueller et al. 1984 and section on mitochondrial translation below). Their use in other situations is limited by the requirement for selective maintenance of respiring cells.

4.2. Cytoduction

An extremely valuable technique that allows the construction of haploid strains carrying different types of (mutant) mtDNA in combination with a particular (mutant) nuclear background by exploiting the property of *kar1–1* mutants to delay nuclear fusion after cytoplasmic fusion during mating (Fox et al. 1991). Transfer of mitochondrial genotype is achieved by using a *kar1* rho° strain with the appropriate nuclear genotype as partner in the cross with the mitochondrial mutant under study. A fraction of the resulting zygotic buds are haploids with the nucleus of one parent and the mtDNA of the other. Because non-respiring diploids do not sporulate, cytoduction is the method of choice for the introduction of mit$^-$ mutations into different nuclear genetic backgrounds. The technique has also revolutionized the study of nuclear genes involved in mitochondrial RNA splicing by making possible the construction of families of mutant strains lacking some or all mitochondrial introns (Séraphin et al. 1987a).

Replication, recombination and repair of mtDNA

Despite the fact that a specific mitochondrial DNA polymerase was identified many years ago, study of the replication of yeast mtDNA has so far been held back by difficulties in the isolation or visualization of replicating molecules. Likely reasons for these difficulties have been the large size of the DNA and a high frequency of recombination. The first of these problems may be circumventable by application of pulsed-field, or related type of electrophoresis techniques for the separation of molecules of large size, or complex topologies (cf. Maleszka et al. 1991)

First indications for the existence of multiple replication origins in yeast mtDNA came from the study of rho$^-$ mutants. As mentioned above, these mutants, when crossed to a cell containing wild-type mtDNA, are able to transmit their genotype to the zygotic progeny. This phenomenon (termed suppressivity) occurs at a frequency that is characteristic for individual rho$^-$ mutants (Ephrussi et al. 1955). It has been suggested that suppressivity is due to the recombinational spread of defective genomes into the wild type mtDNA population (Perlman & Birky 1974). Present indications are, however, that it more probably results from the replication advantage held by a petite mtDNA by virtue of its possession of an amplified set of replication origins. This idea was formulated as early as 1969 (Carnevali et al. 1969) and is lent support by the finding that hypersuppressive rho⁻ mutants (producing more than 99 % of rho⁻ cells in the progeny) share elements that resemble replication origins in other organisms in terms of both sequence motifs and predicted secondary structure (de Zamaroczy et al. 1979; Blanc & Dujon 1980). Depending on strain, 7–8 of such ori sequences are present in yeast mtDNA (de Zamaroczy et al. 1984; Faugeron-Fonty 1974). Of these, four (ori 1,2,3 and 5) are probably functional *in vivo*, since they are able to initiate DNA synthesis through RNA priming (Baldacci & Bernardi 1982). For ori3, functionality is also implied by the finding of a point mutation conferring temperature sensitivity on replication of mtDNA (R. Zelikson & M. Bolotin-Fukuhara, unpubl. obs.). Ori 1,2 and 7 may be dispensable, since they can be deleted without obvious deleterious effects (Piskur 1988a,b).

Despite these data, however, it is still not clear how many ori sequences really are functional in wild type mtDNA *in vivo*, nor how mitochondrial replication proceeds. Experimental approaches aimed at answering these questions have involved looking for mutations affecting ori function and for nuclear genes required for replication. Complementation of one nuclear mutant, deficient in replication of mtDNA (Genga et al. 1986), has led to the identification of the gene for the mitochondrial DNA polymerase (*MIP1*; Foury 1989). This 143.5 kD protein exhibits sequence similarities with both eukaryotic/viral DNA polymerases and reverse transcriptase (Blanco et al. 1991) and possesses a 3′-5′ exonuclease activity (F. Foury & S. Vanderstraeten, pers. commun.). In another study, two nuclear suppressors of ts-ori3 mutation have been identified, thus opening the way to the identification of additional elements of the mitochondrial replication apparatus (R. Zelikson & M. Bolotin-Fukuhara, unpubl. obs.).

Little is so far known about repair of yeast mtDNA, although the suspicion exists that pathways operative in other genetic systems may not be operative (Waters & Moustacchi 1974). Foury and her co-workers have described the properties of *pif1* mutants, which emerged from a screen for deficiencies in mitochondrial recombination behaviour (Foury & Kolodynski 1983). These mutants are also defective in the repair of mtDNA after UV irradiation or treatment with Mn^{2+} or ethidium bromide and lose mtDNA at high temperature. The *PIF1* gene has been cloned (Foury & Lahaye 1987). Its product contains a number of sequence motifs common to a group of DNA helicases and displays sequence homology to subunits B, C and D of the Rec BCD complex of *E.coli*. By analogy with the role of this complex in *E. coli*, the *PIF1* gene product may act in a recombination/repair pathway that is sensitive to recombinogenic signals formed by DNA topology, rather than specific sequences (Foury & van Dyck 1985). As a DNA helicase, it may also participate in DNA replication and/or stabilization (Lahaye et al. 1991).

The isolation and cloning of both *MIP1* and *PIF1* genes was difficult. In the case of *MIP1*, loss of function immediately converts the cell into a rho° state, while for *PIF1*, progress was hampered by the difficulties involved in characterizing a recessive mutation that can only be detected in the diploid state. For both mutants, genetic tricks including cytoduction (Rickwood et al. 1988; Fox et al. 1991) and use of a partial suppressor (Genga et al. 1986) were necessary for their characterization and this will probably also hold for mutations in proteins mediating other steps in recombination, replication or repair. Indeed, the current scarcity of mutants in these processes may well be explicable in terms of the failure to appreciate this fact.

Transcription of mitochondrial genes

A single, unusually simple RNA polymerase is responsible for both the transcription of coding sequences and priming of DNA synthesis in yeast mitochondria (Schinkel & Tabak 1989). The catalytic subunit of the enzyme is the product of the *RPO41* gene, a 145 kDa protein with a surprisingly high degree of sequence similarity to the DNA-directed RNA polymerases of bacteriophages T3 and T7 (Masters et al. 1987). During transcription, this subunit is associated with a second protein, a specificity factor, necessary for accurate initiation of transcription. The identity of this specificity factor has been to some extent controversial, with reports of a 43 kDa protein associated with the catalytic subunit itself and a 70 kDa protein capable of direct binding to the promoter (Schinkel et al. 1988; Ticho et al. 1988; Wilcoxen et al. 1988). The identification of the 43 kDa subunit as the product of the *MTF1* gene (Jang & Jaehning 1991) has, however, brought this controversy a step nearer resolution. Identified initially by its ability to complement a temperature-sensitive mutant defective in mitochondrial transcription, *MTF1* can act as a multi-copy suppressor of a ts-mutation in the catalytic subunit, while its inactivation leads to the rapid loss of mtDNA (Lisowsky & Michaelis 1988, 1989). Close examination of the *MTF1* sequence reveals similarities to bacterial sigma factors in domains identified with −10 promoter recognition, promoter melting and holoenzyme stability, strongly suggestive of similarities in the mechanism of promoter recognition (Jang & Jaehning 1991).

The search for ts-mutations in mitochondrial transcription has turned up a second gene (*RF1320*; *MTF2*), which encodes a protein of 51 kDa (Lisowsky 1990). This gene is, rather surprisingly, identical with *NAM1*, a sequence cloned on the basis of its ability to act as a high copy number suppressor of splicing defects caused by certain mitochondrial intron mutations (Ben Asher et al. 1989) (see below). The properties of cells carrying a disrupted *NAM1* sequence suggest that the gene encodes a factor specifically required for translation of the *COX1* mRNA. Its involvement in transcription may thus reflect the existence of independent functional domains within the protein. Just as for the 70 kDa promoter-binding protein identified by Ticho et al. (1988), direct evidence for interaction of the *MTF2* product with the catalytic subunit of RNA polymerase still has to be obtained. The further characterization of both proteins is eagerly awaited.

RNA processing

RNA processing is an important step in the expression of all genes in yeast mtDNA. Modifications undergone by precursor RNAs include splicing, cleavage of individual transcripts from multi-cistronic precursors, trimming and in the case of tRNAs, -CCA addition and base modification. The major features of most of these reactions have been covered in recent reviews (Grivell 1989; Costanzo & Fox 1990). We will therefore focus here on recent developments and on the specific contributions made by genetic approaches.

1. mRNA processing

Most yeast mitochondrial mRNAs are generated by cleavage from multi-cistronic precursors. Stable 3′-termini appear to be generated by cleavage at the dodecamer motif AAUAAUAUUCUU (Osinga et al. 1984). Sequences marking the sites of stable 5′-termini have not been defined. Controlled cleavage at the dodecamer has in two cases been suggested as a means of modulating gene expression (Zhu et al. 1987; Bordonné et al. 1988). The further characterization of the protein responsible for this activity is thus of some interest, as is also the analysis of the product of the *SUV3* gene, identified by its ability to suppress effects of deletion of the dodecamer sequence downstream of the var1 gene (Conrad-Webb et al. 1990). Although unlikely to be the processing endonuclease itself, the *SUV3* gene product appears to be involved in several post-transcriptional processes, including RNA stability, splicing and translation.

2. mRNA stability

Mitochondrial transcripts differ in their relative stabilities, but still too little is known of sequence determinants and enzymes responsible for (in)stability. Genetic studies have led to the identification of two nuclear genes affecting stability of specific mRNAs: *CBP1* for cytochrome *b* and *AEP2* for ATP synthase subunit 9 (Dieckmann & Mittelmeier 1987; Payne et al. 1991). In both cases, mRNA for these proteins are strongly reduced in level or absent, despite the fact that the respective genes are transcribed and normal levels of precursor transcripts observed. The products of both genes could be either nucleases that specifically generate the 5′-terminus of the mature mRNA, or they could be RNA-binding proteins capable of stabilizing the pre-mRNAs with 5′-extensions until the final cut has taken place. In both instances biochemical studies of the interaction between protein and (pre-)mRNA are required to establish the mode of action.

3. tRNA processing and modification

As in bacteria, 5′-maturation of yeast mitochondrial tRNAs is carried out by an RNAse P enzyme containing an RNA moiety (Hollingsworth & Martin 1986). This RNA is encoded by the mitochondrial genome, as demonstrated by a series of elegant genetic studies, involving the techniques of rho^- mapping and zygotic induction (see above). Less is known about processing events at the 3′-termini, apart from the demonstration of an endonuclease activity in crude mitochondrial extracts (Chen & Martin 1988). However, the recent characterization of nuclear mutations capable of compensating for tRNA-encoded deficiency in 3′-end maturation (Valens et al. 1991) should lead to the isolation of the gene for this enzyme in the near future.

Genetic approaches have also led to the identification of *MOD5*, *TRM1* and *TRM2*, nuclear genes responsible for modification of mitochondrial tRNAs (Najarian et al. 1987; Ellis et al. 1989; Hopper et al. 1982). Interestingly, all these gene products perform the same modifications on nuclear-encoded tRNAs and have a dual localization in the cell (see below).

4. RNA splicing

Three genes in yeast mtDNA are interrupted by introns, many of which have been demonstrated to possess RNA cleavage activity *in vitro*. The introns can be divided into two main groups (I and II) on the basis of a number of criteria, the most important of which are the differences in their predicted secondary structure and the mechanism by which catalysis of splicing occurs (Michel et al. 1989; Cech 1990). Several introns contain long open reading frames (ORFs), translatable by readthrough of ribosomes from an upstream exon. The ORFs present in group I introns form a relatively homogeneous group of basic proteins, sharing conserved sequence motifs both with each other and with reading frames downstream of the *COX2*, *COX3* and *ATP6* genes, but not with other known proteins (Hensgens et al. 1983; Séraphin et al. 1987b). The ORFs of group II introns form a distinct group and display significant similarity to reverse transcriptases (Michel & Lang 1985).

4.1. Intron-encoded proteins: role in intron excision and mobility

As long ago as 1977, Slonimski and his co-workers predicted from an impressive genetic analysis that the gene for apocytochrome *b* would be split and that some of the introns would encode separate functions essential for the synthesis of this protein (Slonimski et al. 1978; Dujon 1979). Since then, the concept of intron-encoded RNA maturases (Lazowska et al. 1980) has become a familiar one and further studies have implicated the products of other intron-encoded proteins derived from both group I and group II introns in RNA splicing (reviewed by Dujon 1981). The biochemical basis of maturase action in promoting splicing is, however, still unknown. In the same period of time, genetic evidence has also accumulated for a role of certain intron-encoded proteins in the apparently unrelated phenomenon of intron transposition. Biochemically, this process is much better understood, since it is directly attributable to the ability of certain of the ORF products to act as site-specific DNA endonucleases (Sargueil 1991; Nakagawa et al. 1991). The intron-encoded enzymes can cleave specific exon-exon junctions, thus triggering a gene-conversion event in which intron sequences are inserted into a previously unsplit gene (reviewed by Dujon 1989). Given the high degree of sequence similarity amongst group I ORFs, it is not clear what endows a specific ORF product with RNA splicing or DNA endonuclease activities. Two observations suggest that both activities are closely related . First, when genetically-engineered versions of proteins encoded by the aI4 and bI4 introns have been produced in E. coli, both bind to DNA with high affinity and introduce double-strand breaks therein (Goguel et al. 1989; Perea et al. 1990). Second, only a single point mutation (*MIM2*) is required to confer RNA maturase activity on the aI4 endonuclease (Dujardin et al. 1982). On the other hand, the Endo.SceI characterized by Nakagawa et al. (1991) lacks maturase activity and when the aI4 and bI4 intron products are synthesized in yeast and imported into mitochondria, only the bI4 intron product is capable of promoting splicing (Perea et al. 1990). Furthermore, in domain swapping experiments between the two proteins, none of about 40 chimaeric forms possessing RNA maturase activity was active as a DNA endonuclease, strongly suggesting that the two activities are mutually exclusive.

4.2. Nuclear-encoded proteins

Alongside intron encoded-proteins, systematic genetic analysis has unearthed several additional proteins involved in mitochondrial splicing *in vivo* (see Lambowitz & Perlman (1990) for a review). Encoded by nuclear genes, some appear to be required for only a single intron, while others have been implicated in the splicing of one or both groups of introns. With a certain risk of oversimplification, three main groups of factors can be distinguished (see also Table 2):

1. *Proteins likely to be directly involved in splicing.* This group includes the *NAM2* , *MRS1/PET157*, *MRS2*, *PET54* and *CBP2* proteins. *NAM2*, *MRS1* and *MRS2* were picked up by complementation cloning of nuclear suppressor mutations of intronic splicing defects. *PET54* and *CBP2* emerged from the direct complementation cloning of pet mutants disturbed in the synthesis of *COX3* and *COB* respectively. With the exception of *CBP2*, none of the proteins have been biochemically characterized. *CBP2* enhances splicing of the group I intron bI5 *in vitro* by reducing the dependence of catalytic activity of the RNA on GTP and Mg^{2+}-ions whilst leaving the basic transesterification mechanism of self-splicing unaltered (Gampel et al. 1989). *NAM2* and *PET54* also function in mitochondrial protein synthesis, but their roles in splicing appear to be distinct from those in translation (Herbert et al. 1990; Valencik & McEwen 1991). *NAM2* encodes the mitochondrial leucyl aminoacyl-tRNA synthetase and may resemble the *cyt-18* gene product (tyrosyl aminoacyl-tRNA synthetase) in *N. crassa* mitochondria in the way it promotes splicing. However, *cyt-18* acts on a collection of group I introns and is by itself sufficient for splicing (Kittle et al. 1991). *NAM2*, in contrast, acts on only two closely related introns (bI4 and aI4) and is dependent for its action on either of the proteins encoded by these introns (Dujardin et al. 1983). Alongside its role in the initiation of translation on the *COX3* mRNA, *PET54* is specifically required for excision of the aI5b intron in the COX1 gene. Genetic studies suggest that different domains of the proteins are required for each of these functions (Valencik & McEwen 1991). *MRS2* appears relatively specific for group II introns, but must have an additional function, since intron-less $MRS2^0$ mutants fail to respire (Wiesenberger et al. 1992).
2. *Proteins required for RNA maturase synthesis.* *MSS18*, *MSS116*, *MSS51* and *MRF1* belong to this group. All were picked up by biochemical analysis and complementation cloning of pet mutants defective in the synthesis of one or more mitochondrial translation products. The mutant phenotypes are ascribable to blocks on RNA splicing. As with the above group of genes, available evidence suggests that the gene isolated corresponds to that defined by the original mutation. Gene disruptants display a respiratory-deficient phenotype in a strain lacking introns, indicating an involvement of each of the proteins in a process other than splicing. In each case, this process appears to be mitochondrial translation and it has thus been suggested that these proteins might resemble the *NAM2* or

PET54 products in having a dual function, or that there are mechanistic links between the processes of splicing and translation (Grivell & Schweyen 1989; Ben Asher et al. 1989; Ahsen et al. 1991). At this stage, however, we prefer a more mundane explanation: that the defect in splicing results from an inability to synthesise one or more intron-encoded RNA maturases (cf. Pel et al. 1991). In the case of *MSS51*, which may encode a gene-specific initiation factor (Decoster et al. 1990), the defects in splicing are predictably gene-, but not intron-specific. In contrast, defects in splicing resulting from mutations in *MRF1*, coding for the mitochondrial peptide chain release factor, are both gene and intron-specific. The basis of this effect is not fully understood, but may be due to context-dependent recognition of the termination codon in the reading frames involved (H.J. Pel, M.J. Maat & L.A. Grivell, unpubl. obs.).

3. *Pseudo-suppressors of splicing defects*. Typified by *MRS3* and *MRS4* , *NAM1*, *NAM7* and *NAM8*, which were picked up as suppressors of splicing deficiencies, this group differs from the others in comprising genes that only display suppression when present on high copy number vectors. These sequences are genetically distinct from the mutations they suppress and their disruption does not lead to splicing deficiency. The initial failure to recognize the fact of non-identity has given rise to a certain amount of confusion in nomenclature (e.g. the *NAM1* gene cloned by Ben Asher et al. (1989) is not identical to the locus originally defined by Groudinsky et al. (1981) by genetic techniques). However, in later studies, multi-copy complementation has been consciously applied as a technique for the identification of links between RNA splicing and other mitochondrial processes (Koll et al. 1987). As Table 2 shows, the harvest so far is an intriguing one, including as it does mitochondrial carrier proteins (*MRS3*, *MRS4*)and a factor apparently involved in both transcription and translation (*NAM1/MTF2*).

Table 2. Proteins affecting splicing of mitochondrial introns in yeast.

Gene	Other function(s)	Intron(s) involved	Reference
Mitochondrial DNA			
Intronic reading frames (RNA maturases)	DNA endonuclease (intron transposition)	aI1, aI2, aI4, bI2, bI4	a, b, c
Nuclear DNA			
CBP2		bI5	d
MRF1	Polypeptide release factor	aI1 (aI2)	e
MSS18		aI5β	f
MSS51	Translational initiation	aI1, aI2, aI4, aI5	g
MSS116	Translational initiation	aI1, aI5α/β, bI1, bI2/3	h
MRS1 (*PET157*)		bI3, aI5β	i, j
MRS2		group II	k
MRS3,4	Cation carrier		l
SUP-101		bI1	m
NAM1	Transcription/RNA stability	groups I and II	n, o
NAM2	mt Leu-tRNA synthetase	bI4	p
NAM7			n, q
NAM8			n
PET54	Translational initiation	aI5β	r

a: Dujon 1981; b: Dujon 1989; c: Perea et al. 1990; d: Gampel et al. 1989; e: Pel H.J., M.J. Maat & L.A. Grivell, unpubl. obs.; f: Séraphin et al. 1988; g: Decoster et al. 1990; h: Séraphin et al. 1989; i: Kreike et al. 1987; j: Bousquet et al. 1990; k: Wiesenberger et al. 1992; l: Wiesenberger et al. 1991; m: Schmelzer & Schweyen 1986; n: Ben-Asher et al. 1989; o: Lisowsky & Michaelis 1989; p: Herbert et al. 1988; q: Altamura et al. 1992; r: Valencik et al. 1989.

Translation

The mitochondrial translational machinery in most organisms is unusual (Benne & Sloof 1987) and that of yeast is no exception. A different genetic code, a low number of tRNAs, unconventional rRNAs, a lack of a typical 5S rRNA, mRNAs with long leader and trailer sequences, a large number of what appear to be mRNA-specific translation factors are just some of the features that distinguish this machinery from typical prokaryote or eukaryotic cell sap systems. Study of these features should give valuable insight into the problems of rRNA and tRNA design, ribosome structure, decoding and mRNA selection. Current points of interest in translation concern the structure, function and evolutionary conservation of individual components of the translational machinery; details of mRNA selection and subsequent steps in translational initiation and the mechanism(s) of translational regulation. Each of these points is discussed below.

1. Structure-function relationships in rRNAs and tRNAs

The mutational analysis of the genes for rRNAs and tRNAs has been facilitated by the fact that, in contrast to other genetic systems, genes for each occur in only one copy per mitochondrial genome. Mutations conferring antibiotic resistance, or heat-/cold sensitivity to mitochondrial protein synthesis have turned out to lie within the rRNA genes and identify rRNA domains involved in the peptidyl transferase centre and in translational fidelity (Sor & Fukuhara 1982 1984; Kruszewska & Slonimski 1984; Weiss-Brummer et al. 1987; Daignan-Fornier et al. 1988; Sakai et al. 1991). Second-site nuclear or mitochondrial suppressors of some of these mutations have also been isolated (Julou et al. 1984; Daignan-Fornier & Bolotin-Fukuhara 1988) and these are likely to give much needed information on RNA-protein interactions within the ribosome and possibly also on long-range folding interactions within the rRNAs themselves.

Similar mutational approaches have been applied to tRNA genes (ElElj-Frighi et al. 1991 and references therein)and these have yielded a number of mutants deficient in amino-acylation and a single mutant defective in 3′-end maturation. In this last case, selection of nuclear suppressors should be of value in the identification of enzymes involved in processing.

2. Shared proteins

It is in the translational apparatus that the cost to the cell of maintaining a separate organellar genetic system is most obvious. Only the rRNAs and tRNAs and a single protein of the small ribosomal subunit (*VAR1*) are encoded by the mitochondrial genome. The remainder, including specific mitochondrial versions of aminoacyl-tRNA synthetases, initiation, elongation and release factor(s) and 70 or so ribosomal proteins are encoded by nuclear genes. Surprisingly, not all components are uniquely mitochondrial, since a select number of aminoacyl-tRNA synthetases and tRNA modification enzymes are shared between mitochondrial and cell sap tRNAs (Natsoulis et al. 1986; Chatton et al. 1988; Hopper et al. 1982; Ellis et al. 1989; Gillman et al. 1991). The advantages of this arrangement are far from obvious, while on the debet side problems abound: (1) sharing of enzymes places limitations on their evolutionary flexibility: interactions must be maintained with two substrates that differ widely in sequence and in their rates of evolutionary change; (2) the shared enzymes are required to function in different cellular compartments in which the prevailing ionic conditions are likely to be different; (3) synthesis of the two forms of the protein, carrying different topogenic signals has to be finely balanced to achieve correct dosage of each protein in the two cellular compartments. So why should sharing be maintained? Economy, never a safe bet where Nature is concerned, is unattractive in view of the small number of proteins shared. The arrangement may thus simply reflect an evolutionary accident whose origins lie in the strongly conserved tertiary structure of the tRNAs involved, or more intriguingly, it may be part of a regulatory circuit that links the biosynthetic capacities of mitochondrial and cytosolic systems, thus relating mitochondrial to cellular metabolism.

3. Initiation of translation

Of all the steps in mitochondrial gene expression, initiation of translation is probably the most poorly understood. Unlike their cell sap counterparts, for which cap recognition and leader scanning are important steps in initiation (Kozak 1989), mRNAs in yeast mitochondria are uncapped and most possess extremely long untranslated leader and trailer sequences of high average A+U content (>95%). These leader sequences may be difficult for a ribosome to scan: they often contain both short open reading frames and short G+C rich clusters capable of forming highly stable secondary structures. On the other hand, if mitochondrial ribosomes are like bacterial ribosomes in being able to bind directly to the initiator codon region (Gualerzi & Pon 1990), it is far from clear how the initiator AUG is distinguished from other potential start sites. Sequences complementary to a region of the small subunit rRNA are present in most mRNAs (Li et al. 1982), but these are found at different sites relative to the AUG start codon in the various leaders and an mRNA lacking this putative ribosome binding site is still active in protein synthesis (Costanzo & Fox 1988).

Classical genetic approaches have so far yielded little information on translational initiation: only one mit$^-$ mutant apparently impaired in this process has been described and the reason for this impairment is not known (Ooi et al. 1987). The construction of mutant strains by means of biolistic transformation is likely to be more informative. Studies so far, which show that a mutant version of the *COX3* gene, in which the initiator AUG has been replaced by AUA, is still translatable, suggest a role for sequences and/or structures around the initiator codon (Folley & Fox 1991). Formal proof of this will, however, require the demonstration that protein synthesis indeed initiates at the correct site.

4. PET factors, mRNA selection and regulation of translation

As first shown by Fox and co-workers, individual mRNAs in yeast mitochondria each require nuclear-coded activator proteins for their translation (see Fox 1986; Costanzo & Fox 1990 and refs therein). Certain nuclear pet$^-$ mutants fail to accumulate individual mitochondrial translation products, even though the corresponding mRNA is present in relatively normal amounts, suggesting a block in translation. The mutations are recessive, thus defining products which serve to activate translation (see Table 3). A common feature of these activators is that they depend for their action on the 5′ untranslated regions of mitochondrial mRNAs, as shown by the study of mitochondrial suppressors of these pet$^-$ mutations. These suppressor mutants are heteroplasmic, carrying a rearranged mitochondrial genome alongside the wild type genome. The rearrangement places the 5′-leader of another mitochondrial mRNA upstream of the coding sequence of the affected product, thus placing its synthesis under the control of a new set of nuclear factors. The simplest explanation for the ability of such rearranged molecules to restore translation is that the PET factors interact directly with the 5′-leader sequence. Such an interaction still has to be demonstrated directly by biochemical means, however.

Nuclear suppressors of *pet* mutations affecting translation also exist. In the case of *PET122* (required for *COX3* translation), the mutations in two such suppressors have turned out to be located in the genes for two proteins of the small subunit of the mitochondrial ribosome (Haffter et al. 1990; McMullin et al. 1990; Haffter et al. 1991), suggesting that the *PET122* product in some way mediates an interaction between the *COX3* mRNA and the mitochondrial ribosome.

How do the PET factors stimulate translational initiation? Some speculations are possible, based on sequence comparisons and localisation studies. The sequence comparisons show that the factors *PET111*, *PET54*, *CBS1* and *CBP6* display significant sequence similarity both to each other and to the eukaryotic initiation factors eIF2α or eIF2β

(Grivell 1989). These factors function in charging the small ribosomal subunit with GTP and initiator Met-tRNA prior to the association with mRNA and in yeast direct evidence for their involvement in AUG start codon recognition has been obtained (Donahue et al. 1988; Cigan et al.,1989). Thus at least these PET factors may function in start site selection. How they do this is not clear. The PET factors that activate translation of *COX3* mRNA depend for their action on a site that is located at least 172 nucleotides upstream of the initiator AUG (Costanzo & Fox 1988), so that long-range interactions have to be invoked to juxtapose the two regions.

Localisation studies have been carried out on the PET factors 494, 54 and 122, required for translation of *COX3* mRNA. All three proteins, which apparently function as a complex, are associated with the mitochondrial inner membrane (Costanzo & Fox 1990). This has led to the idea that these proteins act by bringing the *COX3* mRNA together with the small ribosomal subunit to the membrane, where synthesis of *COX3* can occur near to the site of assembly into cytochrome c oxidase. Localisation studies on the *CBS1* and *CBS2* factors have led to a similar proposal for cytochrome *b* synthesis (Michaelis et al. 1991).

Table 3. Nuclear gene products required for translation of specific mitochondrial mRNAs in yeast.

Nuclear gene product	mRNA affected	Reference
MSS51	*COX1*	a
PET111	*COX2*	b
PET54, PET494, PET122	*COX3*	c, d
CBS1, CBS2, CBP6	*COB*	e, f
AEP1	*ATP9*	g

a: Decoster et al. 1990; b: Poutre et al. 1987; c: Costanzo et al. 1986; d: Kloeckener-Gruissem et al. 1988; e: Rödel 1986; f: Dieckmann & Tzagoloff 1985; g: Payne et al. 1991.

Transcription of nuclear genes for mitochondrial proteins

1. Factors and motifs

Unlike most other eukaryotes, *S. cerevisiae* can tailor the level of mitochondrial biosynthesis to its specific needs in response to its environment. Oxygen, haem and carbon source are the prime regulatory factors, exerting control mainly at the level of transcription (Grivell 1989; Forsburg & Guarente 1989a). Genetic approaches to unravelling the signal transduction pathways mediating these responses have made use of schemes involving the selection of mutants displaying aberrant expression of nuclear genes for mitochondrial proteins (see Zitomer and Lowry 1992 and references therein). This work has led to the demarcation of sequence elements responsible for the activation or repression of transcription and to the identification of protein factors capable of recognizing such elements and mediating the effects on transcription. As detailed below, many of the mutations affecting these latter factors have turned out to possess highly pleiotropic phenotypes. This suggests that the factors participate in more than one regulatory pathway and that the regulation of mitochondrial biosynthesis is coupled to a variety of cellular functions, including carbohydrate and nitrogen metabolism, mating type response and cell morphology. A good example of the level of complexity is provided by the *CYC1* gene, encoding iso-1-cytochrome c. Figure 1 shows that this gene is equipped with two upstream regions capable of activating transcription, namely UAS1 and UAS2. Factors interacting directly or indirectly with these elements are also shown.

UAS1 is mainly responsible for mediating response to oxygen, most likely by sensing changes in the level of intracellular haem (Pfeifer et al. 1989). This sequence, which consists of two sub-domains, is the target for the cooperative binding (Kim et al. 1990) of two molecules of HAP1, a protein with several zinc-finger motifs and several other sequence features that have led to the suggestion that it may act as a sensor of the redox state of the cell (Creusot et al. 1988). The factor activates in a

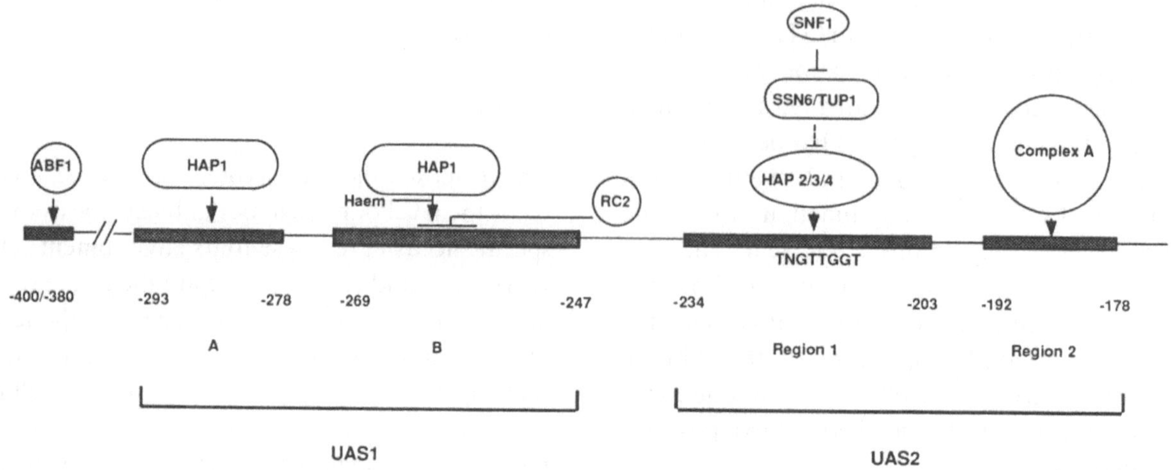

Fig. 1. Regulatory elements upstream of *CYC1*, the gene for iso-1 cytochrome c.
Schematic representation of the upstream region of the *CYC1* gene, showing regulatory elements and the factors that interact with them. Based on data presented by Olesen et al. (1987), Pfeifer et al. (1987), Forsburg & Guarente (1988, 1989b), Dorsman et al. (1990) and Wright & Poyton (1990). Distances, which are not drawn exactly to scale, are given in base pairs relative to the start site of RNA synthesis. See text for further explanation.

haem-dependent fashion the transcription of *CYC1* and of many other genes, including those for non-haemoproteins and at least one non-mitochondrial protein.

As mentioned above, genes encoding mitochondrial proteins are subject to catabolite repression and *CYC1* is no exception. UAS2 is mainly responsible for mediating this control. This sequence also consists of two sub-domains, binding sites for a complex containing the transcriptional activators HAP2,3 and 4 and the poorly characterized complex A respectively (Forsburg & Guarente 1988; Forsburg & Guarente 1989b). Like HAP1, the HAP2,3,4 complex binds to the promoter regions of a large number of nuclear genes coding for mitochondrial proteins. The consensus sequence for binding is TNATTGGT, an element that contains the CCAAT motif common to many mammalian promoters and HAP2 and 3 possess sufficient sequence similarity with their mammalian CCAAT-box binding counterparts to allow the formation of functional hybrid complexes (Chodosh et al. 1988). Domain swap experiments suggest that HAP2 and HAP3 contain domains that together form a binding site for DNA, while HAP4 is responsible for activation of transcription (Olesen & Guarente 1990). Synthesis of both HAP2 and HAP4 is subject to catabolite repression (Pinkham & Guarente 1985; Forsburg & Guarente,1989b), so that induction of gene expression in response to glucose depletion can thus at least in part be attributed to changes in the level of these factors. If *hap2* or *hap3* mutations lead to impaired nitrogen metabolism, while *hap4* mutations do not, as suggested by Forsburg & Guarente (1989a), then HAP2 and 3 may constitute general transcriptional regulators in several metabolic pathways, with specificity perhaps being conferred by different HAP4-like subunits.

2. C-source control of mitochondrial biogenesis: relationship to other catabolite-repressible pathways

Although the interactions sketched above account in broad lines for the transcriptional responses of the *CYC1* gene, there are several indications that the picture is not yet complete: mutant studies show that *CYC1*, like many other catabolite-repressible genes, falls under control of the *SNF1* gene. Mutations in this gene, which encodes a protein kinase, prevent induction of *CYC1* on a shift to non-fermentable media (Wright & Poyton 1990).

Interestingly, initial indications for this control came more than 10 years ago from the pioneer studies of Rothstein and Sherman that led to the identification of the *CYC8* and *CYC9* loci (Rothstein & Sherman 1980), now known to be allelic to *SSN6* and *TUP1* respectively (Trumbly 1988; Williams & Trumbly 1990). Both *ssn6* and *tup1* mutants are highly pleiotropic, displaying a tendency to flocculate, altered permeability and morphology and defects in mating type response alongside the constitutive expression of genes normally repressible by glucose and/or haem. The mutations therefore probably define general regulatory factors with roles in several signal transduction pathways, including those of catabolite repression and haem activation/repression. Recent biochemical work shows that the products of both genes are located exclusively in the nucleus and are physically associated with each other in a complex (Williams et al. 1991). *ssn6* mutations restore high levels of gene expression in cells carrying mutational defects in the *SNF1* protein kinase and this has been interpreted as support for the idea that the *SSN6* gene product acts as a negative regulator. Its site of action in the signal transduction pathway mediating catabolite repression is likely to be downstream of that for the *SNF1* product, possibly at the level of assocation with and modification of the activity of DNA-binding proteins (Schultz & Carlson 1987). Regulation by the *SNF1/SSN6/TUP1* pathway of *CYC1* and other genes for mitochondrial proteins may thus be either direct via effects on the activity of the HAP2,3,4 complex or indirect via the synthesis of HAP2 and HAP4, since the genes for both factors are themselves catabolite-repressible (Pinkham & Guarente 1985; Forsburg & Guarente 1989a).

3. Mitochondrial biogenesis and cell growth

Band-shift studies show that alongside the HAP factors, other proteins bind to the *CYC1* promoter, or to sequences upstream of it. Two of these simply modulate response to either HAP1 (factor RC2; Arcangioli and Lescure 1985), or HAP2,3,4 (complex A; Forsburg & Guarente 1989a). A third factor may, however, influence transcription via a distinct mechanism. It has been identified as ABF1, a zinc-finger DNA binding protein with a large number of potential targets in the yeast genome and the ability to either activate or repress transcription, dependent on binding site context and interaction with other factors (Dorsman et al. 1990; Diffley, this issue). The role of ABF1 in CYC1 transcription has not been established, but in the case of the QCR8 gene encoding a subunit of the mitochondrial ubiquinol-cytochrome c reductase, the factor appears to be important for both basal-level and induced transcription and for a rapid escape from catabolite repression (J.H. de Winde & L.A. Grivell, unpubl. obs.). We have previously suggested that ABF1, together with other global factors such as CPF1 and RAP1, may be part of a regulatory circuit that links the rate of a variety of biosynthetic processes to cell growth (Grivell et al. 1990). In this context, it may be significant that binding sites for ABF1 are present in the promoter regions of genes for several regulatory factors, including SSN6 and TUP1.

4. Nucleo-mitochondrial cross-talk

Given that the rate of mitochondrial biogenesis is coupled to cell growth, the cell must in some way be able to monitor mitochondrial mass during division and growth. How this is done is not known. The idea that the mitochondrion can signal changes in gene expression within the nucleus is a relatively old one, but there is as yet no firm support for it. In yeast, it has been argued that such a pathway cannot be quantitatively important, since synthesis of most nuclear-coded components of mitochondrial respiratory complexes is not dependent on the presence of an intact mitochondrial genome (Myers & Tzagoloff 1985). Despite this, several groups have reported that the levels of expression of a number of nuclear DNA sequences- some of them unrelated to mitochondrial biogenesis- change in response to defects in, or loss of the mitochondrial genome (Van Loon et al. 1982; Parikh et al. 1987; Butow et al. 1988; Partaledis & Mason 1988; Farrell et al. 1990; Liao et al. 1991).

For respiratory-deficient versus wild type cells, such changes (which include activation as well as repression) may simply be attributable to changes in energy balance within the cell and consequently, in levels of ATP or GTP. In contrast, the transcriptional differences observed between rho° and rho^- or mit^- cells could apparently only be correlated with the loss of mtDNA from the former. In these cases, it has been speculated that nucleus and mitochondrion may share common transcriptional specificity factors and that nuclear transcription responds to an altered distribution of such factors over the two compartments (Marczynski et al. 1989). Experimental support for this idea has yet to be obtained.

Mitochondrial protein import and assembly

Recent reviews on the topic of mitochondrial protein import have dealt with the identification of signals responsible for directing proteins to the mitochondrion and with the main features of the import process itself (Pfanner & Neupert 1990; Schatz 1991). So far, about 10 components of the import machinery have been identified, mainly by means of biochemical studies. Genes for several of these have been cloned and sequenced and have turned out to be essential for cell viability, a feature entirely consistent with the importance of the mitochondrion for the proper functioning of metabolic pathways required for cell survival. This fact no doubt accounts for the limited success of early genetic approaches to the identification of import-deficient mutants: these depended primarily on the selection of cells displaying precursor accumulation, or respiratory-deficiency (Yaffe & Schatz 1984; Smith & Yaffe 1991; Atencio & Yaffe 1992) and resulted in the identification of subunits of the matrix-localised processing protease and heat shock proteins thought to play a role in inducing an import-competent conformation in precursor proteins. More recent approaches, that make use of mislocalization of the product of a reporter gene to isolate conditional or leaky mutants in import functions, may help fill the gaps in our knowledge of the components that catalyze translocation of proteins through the membrane (Pollock et al. 1988; Maarse et al. 1990).

Mutant studies have been instrumental in identifying genes for subunits of some of the respiratory complexes and have helped define pathways for their assembly. In addition, *pet* mutations have identified gene products, which although not components of the complexes themselves are essential for the assembly of ATP synthase and the respiratory complexes (Wu & Tzagoloff 1989; Bousquet et al. 1991; Nobrega et al. 1990; Ackerman & Tzagoloff 1990a,b; Bowman et al. 1991). These proteins have been proposed to act as complex-specific chaperones but their mechanism of action awaits clarification.

Conclusions and prospects

The mitochondrial genetics of yeast have evolved considerably in two decades. Initially directed at a definition of the rules for recombination and genomic segregation, formal genetics have made major contributions to the identification of genes on the mitochondrial genome, the mapping of these genes and the establishment of structure-function relationships in the products they encode. In parallel, more than 200 complementation groups, corresponding to as many nuclear genes necessary for mitochondrial function or biogenesis have been described. Many of the latter are required for post-transcriptional events in mitochondrial gene expression, including the processing of mitochondrial pre-RNAs, the translation of mitochondrial mRNAs, or the assembly of mitochondrial translation products into the membrane.

What still remains to be done and how can genetic approaches help?

1. Nuclear genes involved in mitochondrial biogenesis

Although it has been argued that with the 200 or so genes identified so far few other functions remain to be defined, there are, as reviewed in the introductory section, several reasons not to accept this statement just yet. Efforts to identify new gene products should therefore be continued. In addition to the approaches so far used, new techniques may be required for the isolation of mutant classes under-represented in collections so far. Of especial value may be (a) the deliberate selection of multicopy suppressors of both mitochondrial and nuclear mutations in order to identify interacting components in regulatory networks within the mitochondrion; (b) transposon mutagenesis, which may allow the detection of genes whose disruption causes the cell to rapidly convert to rho^-, thus permitting enrichment for mutants in functions affecting replication, recombination or stability of mtDNA (cf. Daignan-Fornier & Bolotin-Fukuhara 1989) and finally, (c) the use of fusion gene libraries to identify families of co-regulated genes (Daignan-Fornier et al., unpubl. results). Characterization of these proteins may shed light on similar processes in human mitochondria and thus lead to a better understanding of the molecular basis of defects in mitochondrial function and the role of such defects in human degenerative diseases.

2. Site-directed mutagenesis of mtDNA

With a few exceptions, sequences signalling RNA (in)stability or processing are not known, while only the vaguest ideas exist on the identity of sequences responsible for directing translational initiation. These should be identifiable by systematic site-directed mutagenesis, followed by biolistic transformation of mitochondria. Selection of second-site suppressors of the resulting mutants should allow identification of (nuclear-coded) proteins that interact with these sites.

Biolistic transformation should also permit the study of structure-function relationships in mitochondrially encoded proteins. Of interest are both components of the respiratory complexes and the biochemically more intractable products, such as the intron-encoded maturases and the few unassigned reading frames still remaining.

3. Signal transduction in the regulation of mitochondrial biogenesis

Although most nuclear genes encoding mitochondrial proteins are regulated at the level of transcription, too little is known of the signals or factors that modulate transcriptional activity in response to environment or growth rate. For a handful of genes, cis-acting elements and DNA-binding proteins mediating control by haem, or carbon source have been identified. These factors interact with more general regulatory pathways in which the SNF gene products and the global transcriptional factors such as ABF1 and CPF1 are the main figures. The nature of these (protein-protein) interactions deserve study and genetics can make an important contribution to this, as they have in other signal transduction pathways in yeast (see reviews by Konopka & Fields and Thevelein, this volume).

4. Chaperones and membrane assembly

The idea of molecular chaperones- proteins that act as molecular work-benches to promote unfolding or re-folding of other proteins before or after their translocation across a membrane-, is now a familiar one, even though little is known about their mode of action. Less familiar is the idea of chaperones acting at the level of the assembly of the respiratory complexes themselves. *ABC1*, *CBP3*, *COX10*, *ATP10–12* are products that were identified genetically as being required for assembly of functional cytochrome bc1 complex, cytochrome oxidase, or ATPase, without being themselves a constituent of these complexes. How they achieve this is worthy of further study and no doubt the isolation of second-site suppressors of mutations in these genes will be a useful first step in this process.

Acknowledgements

We are grateful to Amy Bednarz, Peter Dekker, Michiel Meijer, Wietse Mulder, Herman Pel, Martijn Rep and Han de Winde for useful discussions and critical comments on the manuscript. Collaboration between our two laboratories was greatly stimulated by an award from MRT (Réseaux Européens de Laboratoires 41821011). Research in our respective groups was also supported by the CNRS (grant to URA.D.1354), the Netherlands Foundation for Chemical Research and the Netherlands Organization for the Advancement of Research (NWO).

References

Ackerman SH & Tzagoloff A (1990a) *ATP10*, a yeast nuclear gene required for the assembly of the mitochondrial F1-F0 complex. J. Biol. Chem. 265: 9952–9959

Ackerman SH & Tzagoloff A (1990b) Identification of 2 nuclear genes (*ATP11, ATP12*) required for assembly of the yeast F1-ATPase. Proc. Natl. Acad. Sci. USA 87: 4986–4990

Ahsen U von, Davies J & Schroeder R (1991) Antibiotic inhibition of group I ribozyme function. Nature 353: 368–370

Altamura N, Groudinsky O, Dujardin G & Slonimski PP (1992) *NAM7* nuclear gene encodes a novel member of a family of helicases with zinc ligand motif and is involved in mitochondrial functions in *S. cerevisiae*. J. Mol. Biol. (in press)

Arcangioli B & Lescure B (1985) Identification of proteins involved in the regulation of yeast iso-1-cytochrome c expression by oxygen. EMBO J. 4: 2627–2633

Atencio DP & Yaffe MP (1992) *MAS5*, a yeast homolog of *DnaJ* involved in mitochondrial protein import. Mol. Cell. Biol. 12: 283–291

Baldacci G & Bernardi G (1982) Replication origins are associated with transcription initiation sequences in the mitochondrial genome of yeast. EMBO J. 1: 987–994

Ben Asher E, Groudinsky O, Dujardin G, Altamura N, Kermorgant M & Slonimski, P.P. (1989) Novel class of nuclear genes involved in both mRNA splicing and protein synthesis in *Saccharomyces cerevisiae* mitochondria. Mol. Gen. Genet. 215: 517–528

Benne R & Sloof P (1987) Evolution of the mitochondrial protein synthetic machinery. Biosystems 21: 51–68

Blanc H & Dujon B (1980) Replicator origins of the yeast mitochondrial DNA responsible for suppressiveness. Proc. Natl. Acad. Sci. USA 77: 3942–3946

Blanco L, Bernad A & Salas M (1991) MIP1 DNA polymerase of *S. cerevisiae*: structural similarity with the *E. coli* DNA polymerase I-type enzymes. Nucleic Acids Res. 19: 955

Bolotin-Fukuhara M, Faye G & Fukuhara H (1977) Temperature-sensitive respiratory-deficient mitochondrial mutations: isolation and genetic mapping. Mol. Gen. Genet. 152: 295–305

Bordonné R, Dirheimer G & Martin RP (1988) Expression of the oxi1 and maturase-related RF1 genes in yeast mitochondria. Curr. Genet. 13: 227–233

Bousquet I, Dujardin G, Poyton RO & Slonimski PP (1990) Two group I mitochondrial introns in the cob-box and coxI genes require the same *MRS1/PET157* nuclear gene product for splicing. Curr. Genet. 18: 117–124

Bousquet I, Dujardin G & Slonimski PP (1991) *ABC1*, a novel yeast nuclear gene has a dual function in mitochondria: it suppresses a cytochrome b mRNA translation defect and is essential for the electron transfer in the bc1 complex. EMBO J. 10: 2023–2031

Bowman S, Ackerman SH, Griffiths DE & Tzagoloff A (1991) Characterization of *ATP12*, a yeast nuclear gene required for the assembly of the mitochondrial F1-ATPase. J. Biol. Chem. 266: 7517–7523

Butow RA & Fox TD (1990) Organelle transformation: shoot first, ask questions later. Trends in Biochem. Sci., 15: 465–468

Butow RA, Strausberg RL, Vincent RD, Paulson LD & Perlman PS (1978) Analysis of structural genes on mitochondrial DNA. In: Bacila M, Horecker BL & Stoppani AOM (Eds) Biochemistry and Genetics of Yeast (pp 403–411). Academic Press, New York

Butow RA, Docherty R & Parikh VS (1988) A path from mitochondria to the yeast nucleus. Phil. Trans. R. Soc. Lond. B 319: 127–133

Carnevali F, Morpurgo G & Tecce G (1969) Cytoplasmic DNA from petite colonies of *Saccharomyces cerevisiae*: a hypothesis on the nature of the mutation. Science 163: 1331–1333

Chatton B, Walter P, Ebel J-P, Lacroute F & Fasiolo F (1988) The yeast *VAS1* gene encodes both mitochondrial and cytoplasmic valyl-tRNA synthetases. J. Biol. Chem. 263: 52–57

Cech TR (1990) Self-splicing of group I introns. Ann. Rev. Biochem. 59: 543–568

Chen J-Y & Martin NC (1988) Biosynthesis of tRNA in yeast mitochondria. An endonuclease is responsible for the 3′-processing of tRNA precursors. J. Biol. Chem. 263: 13677–13682

Chodosh LA, Olesen J, Hahn S, Baldwin AS, Guarente L & Sharp PA (1988) A human and yeast CCAAT-binding protein have heterologous subunits that are functionally interchangeable. Cell 53: 25–35

Cigan AM, Pabich EK, Feng L & Donahue TF (1989) Yeast translation initiation suppressor *sui2* encodes the alpha subunit of eukaryotic initiation factor 2 and shares sequence identity with the human alpha subunit. Proc. Natl. Acad. Sci. USA 86: 2784–2788

Conrad-Webb H, Perlman PS, Zhu H & Butow RA (1990) The nuclear *Suv3-1* mutation affects a variety of post-transcriptional processes in yeast mitochondria. Nucleic Acids Res. 18: 1369–1376

Costanzo M, Seaver EC & Fox TD (1986) At least two nuclear gene products are specifically required for translation of a single yeast mitochondrial mRNA. EMBO J. 5: 3637–3641

Costanzo MC & Fox TD (1988) Specific translational activation by nuclear gene products occurs in the 5′ untranslated leader of a yeast mitochondrial mRNA. Proc. Natl. Acad. Sci. USA 85: 2677–2681

Costanzo MC & Fox TD (1990) Control of mitochondrial gene expression in *Saccharomyces cerevisiae*. Ann. Rev. Genet. 24: 91–113

Creusot F, Verdière J, Gaisne M & Slonimski PP (1988) The *CYP1* (*HAP1*) regulator of oxygen dependent gene expression in yeast. 1. Overall organisation of the protein displays several novel structural motifs. J. Mol. Biol. 204: 263–276

Daignan-Fornier B & Bolotin-Fukuhara B (1988) Mutational study of the rRNA in yeast mitochondria: the functional importance of T1696 in the large rRNA gene. Nucleic Acids Res. 16: 9299–9306

Daignan-Fornier B, Su I, Contamine V & Bolotin-Fukuhara B (1988) A genetic approach to molecular interactions in theyeast mitochondrial ribosome. In: Tuite M, Picard M & Bolotin-Fukuhara M (Eds) NATO-ASI series, Vol. H14 Genetics of Translation (pp 159–169). Springer Verlag, Berlin, Heidelberg

Daignan-Fornier B & Bolotin-Fukuhara B (1989) Functional exploration of the yeast (*S. cerevisiae*) genome: use of a mini-Mu transposon to analyze randomly cloned sequences. Yeast 5: 259–269

Decoster E, Simon M, Hatat D & Faye G (1990) The *MSS51* gene product is required for translation of the *COX1* mRNA in yeast mitochondria. Mol. Gen. Genet. 224: 111–118

Dieckmann CL & Tzagoloff A (1985) Assembly of the mitochondrial membrane system. *CBP6*, a yeast nuclear gene necessary for the synthesis of cytochrome b. J. Biol. Chem. 260: 1513–1520

Dieckmann CL & Mittelmeier TM (1987) Nuclearly-encoded CBP1 interacts with the 5′-end of mitochondrial cytochrome b pre-mRNA. Curr. Genet. 12: 391–397

Diffley JFX (1992) Global regulators of chromosome function in yeast. Antonie van Leeuwenhoek 62: 25-33 (this issue)

Donahue TD, Cigan AM, Pabich EK & Castilho Valavicius B (1988) Mutations at a Zn(II) finger motif in the yeast eIF-2beta gene alter ribosomal start-site selection during the scanning process. Cell 54: 621–632

Dorsman JC, Van Heeswijk WC & Grivell LA (1990) Yeast general transcription factor GFI-sequence requirements for binding to DNA and evolutionary conservation. Nucleic Acids Res. 18: 2769–2776

Dujardin G, Pajot P, Groudinsky O, Slonimski PP (1980) Long range control circuits within mitochondria and between nucleus and mitochondria. I. Methodology and Phenomenology of suppressors. Mol. Gen. Genet. 179: 469–482

Dujardin G, Jacq C & Slonimski PP (1982) Single base substitution in an intron oxidase gene compensates splicing defects of the cytochrome b gene. Nature 298: 628–632

Dujardin G, Labouesse M, Netter P & Slonimski PP (1983) Genetic and biochemical studies of the nuclear suppressor *NAM2*: extraneous activation of a latent pleiotropic maturase. In: Schweyen RJ, Wolf K & Kaudewitz F (Eds) Mitochondria 1983. Nucleo-Mitochondrial Interactions (pp 233–250). de Gruyter

Dujon B (1979) Mutants in a mosaic gene reveals functions for introns. Nature 282: 777–778

Dujon B (1981) Mitochondrial genetics and functions. In: Strathern JN, Jones EW & Broach JR (Eds) The Molecular Biology of the Yeast *Saccharomyces*. Life Cycle and Inheritance (pp 505–635). Cold Spring Harbor Laboratory

Dujon B (1989) Group I introns as mobile genetic elements: facts and mechanistic speculations – a review. Gene 82: 91–114

Elelj-Fridhi N, Pallier C, Zelikson R, Guetari M & Bolotin-Fukuhara M (1991) Mutational studies of the major tRNA region of the *S. cerevisiae* mitochondrial genome. Curr. Genet. 19: 301–308

Ellis SR, Hopper AK & Martin NC (1989) Amino-terminal extension generated from an upstream AUG codon increases the efficiency of mitochondrial import of yeast N2, N2-dimethylguanosine-specific tRNA methyltransferases. Mol. Cell. Biol. 9: 1611–1620

Ephrussi B, Margerie-Hottinguer H de & Roman H (1955) Suppressiveness: a new factor in the genetic determinism of the synthesis of respiratory enzymes in yeast. Proc. Natl. Acad. Sci. USA 41: 1065–1071

Farrell LE, Trawick JD & Poyton RO (1990) Mitochondrial-nuclear interactions: transcription of nuclear cox genes in yeast is reduced in cells that lack a mitochondrial genome. In: Quagliariello E, Papa S, Palmieri F & Saccone C (Eds) Structure, Function and Biogenesis of Energy Transfer Systems (pp 131–134). Elsevier, Amsterdam

Faugeron-Fonty G, Le Van Kim C, Zamaroczy M de, Goursot R & Bernardi G (1984) A comparative stuy of the ori sequences from the mitochondrial genomes of twenty wild-type yeast strains. Gene 32: 459–473

Folley LS & Fox TD (1991) Site-directed mutagenesis of a *Saccharomyces cerevisiae* mitochondrial translation initiation codon. Genet. 129: 659–668

Forsburg SL & Guarente L (1988) Mutational analysis of upstream activation sequence 2 of the *CYC1* gene of *Saccharomyces cerevisiae*: a HAP2-HAP3-responsive site. Mol. Cell. Biol. 8: 647–654

Forsburg SL & Guarente L (1989a) Communication between mitochondria and the nucleus in regulation of cytochrome genes in the yeast *Saccharomyces cerevisiae*. Ann. Rev. Cell Biol. 5: 153–180

Forsburg SL & Guarente L (1989b) Identification and characterization of HAP4: a third component of the CCAAT-bound HAP2/HAP3 heteromer. Genes & Dev. 3: 1166–1178

Foury F (1989) Cloning and sequencing of the nuclear gene *MIP1* encoding the catalytic subunit of the yeast mitochondrial DNA polymerase. J. Biol. Chem. 264: 20552–20560

Foury F & Kolodynski J (1983) *Pif* mutation blocks recombination between mitochondrial r^+ and r^- genomes having tandemly arrayed repeat units in *Saccharomyces cerevisiae*. Proc. Natl. Acad. Sci. USA 80: 5345–5349

Foury F & Van Dyck E (1985) A *PIF*-dependent recombinogenic signal in the mitochondrial DNA of yeast. EMBO J. 4: 3525–3530

Foury F & Lahaye A (1987) Cloning and sequencing of the *PIF* gene involved in repair and recombination of yeast mitochondrial DNA. EMBO J. 6: 1441–1449

Fox TD (1986) Nuclear gene products required for translation of specific mitochondrially coded mRNAs in yeast. Trends in Genet. 2: 97–100

Fox TD, Sanford JC & McMullin TW (1988) Plasmids can stably transform yeast mitochondria lacking endogenous mtDNA. Proc. Nat. Acad. Sci. USA 85: 7288–7292

Fox TD, Folley LS, Mulero JJ, McMullin TW, Thorsness PE, Hedin LO & Costanzo MC (1991) Analysis and manipulation of yeast mitochondrial genes. In: Guthrie C & Fink GR (Eds) Guide to Yeast Genetics and Molecular Biology. Meth. in Enzymol. 194 (pp 149–165). Academic Press, New York

Gampel A, Nishikimi M & Tzagoloff A (1989) CBP2 protein promotes in vitro excision of a yeast mitochondrial group I Intron. Mol. Cell. Biol. 9: 5424–5433

Genga A, Bianchi L & Foury F (1986) A nuclear mutant of *Saccharomyces cerevisiae* deficient in mitochondrial DNA replication and polymerase activity. J. Biol. Chem. 261: 9328–9332

Gillman EC, Slusher LB, Martin NC & Hopper AK (1991) *MOD5* initiation sites determine N6-isopentenyladenosine modification of mitochondrial and cytoplasmic tRNA. Mol. Cell. Biol. 11: 2382–2390

Goguel V, Bailone A, Devoret R & Jacq C (1989) The b14 RNA mitochondrial maturase of *Saccharomyces cerevisiae* can stimulate intra-chromosomal recombination in *Escherichia coli*. Mol. Gen. Genet. 216: 70–74

Grivell LA (1989) Nucleo-mitochondrial interactions in mitochondrial biogenesis. Eur. J. Biochem. 182: 477–493

Grivell LA & Schweyen RJ (1989) RNA splicing in yeast mitochondria: taking out the twists. Trends in Genet. 5: 39–41

Grivell LA, Dorsman JC, Gozdzicka-Jozefiak A, van Heeswijk WC & de Winde JH (1990) Generalized growth control of mitochondrial biogenesis in yeast. In: Quagliariello E, Papa S, Palmieri F & Saccone C (Eds) Structure, Function and Biogenesis of Energy Transfer Systems(pp 193–200). Elsevier, Amsterdam

Groudinsky O, Dujardin G & Slonimski PP (1981) Long range control circuits within mitochondria and between nucleus and mitochondria. II. Genetic and biochemical analyses of suppressors which selectively alleviate the mitochondrial intron mutations. Mol. Gen. Genet. 184: 493–503

Gualerzi CO & Pon CL (1990) Initiation of messenger RNA translation in prokaryotes. Biochem. 29: 5881–5889

Haffter P, Mcmullin TW & Fox TD (1990) A genetic link between a messenger RNA-specific translational activator and the translation system in yeast mitochondria. Genet. 125: 495–503

Haffter P, McMullin TW & Fox TD (1991) Functional interactions among two yeast mitochondrial ribosomal proteins and an mRNA-specific translational activator. Genetics 127: 319–326

Hensgens LAM, Bonen L, De Haan M, Van der Horst G & Grivell LA (1983) Two intron sequences in yeast mitochondrial *COX1* gene: homology among URF-containing introns and strain-dependent variation in flanking exons. Cell 32: 379–389

Herbert CJ, Labouesse M, Dujardin G & Slonimski PP (1988) The NAM2 proteins from *S. cerevisiae* and *S. douglasii* are mitochondrial leucyl-tRNA synthetases, and are involved in mRNA splicing. EMBO J. 7: 473–483

Herbert CJ, Ben Asher E, Bousquet I, Dujardin G, Groudinsky O, Kermorgant M, Labouesse M & Slonimski PP (1990) Yeast mitochondrial pre-mRNA splicing: genes involved in both protein synthesis and splicing. In: Quagliariello E, Papa S, Palmieri F & Saccone C (Eds). Structure, Function and Biogenesis of Energy Transfer Systems (pp 201–204). Elsevier, Amsterdam

Hollingsworth MJ & Martin NC (1986) RNase P activity in the mitochondria of *Saccharomyces cerevisiae* depends on both mitochondrion and nucleus-encoded components. Mol. Cell. Biol. 6: 1058–1064

Hopper AK, Furukawa AH, Pham HD & Martin NC (1982) Defects in modification of cytoplasmic and mitochondrial transfer RNAs are caused by single nuclear mutations. Cell 28: 543–550

Jang SH & Jaehning JA (1991) The yeast mitochondrial RNA polymerase specificity factor, MTF1, is similar to bacterial s factors. J. Biol. Chem. 266: 22671–22677

Julou C, Contamine V, Sor F & Bolotin-Fukuhara M (1984) Mitochondrial ribosomal RNA genes of yeast: their mutations and a common nuclear suppressor. Mol. Gen. Genet. 193: 275–279

Kim KS, Pfeifer K, Powell L & Guarente L (1990) Internal deletions in the yeast transcriptional activator HAP1 have opposite effects at two sequence elements. Proc. Natl. Acad. Sci. USA 87: 4524–4528

Kittle JD jr, Mohr G, Gianelos JA, Wang H & Lambowitz AM (1991) The *Neurospora* mitochondrial tyrosyl-tRNA synthetase is sufficient for group I intron splicing in vitro and uses the carboxy-terminal tRNA-binding domain along with other regions. Genes & Dev. 5: 1009–1021

Kloeckener-Gruissem B, McEwen JE & Poyton RO (1988) Identification of a third nuclear protein-coding gene required specifically for post-transcriptional expression of the *COX3* gene in *Saccharomyces cerevisiae*. J. Bacteriol. 170: 1399–1402

Koll H, Schmidt C, Wiesenberger G & Schmelzer C (1987) Three nuclear genes suppress a yeast mitochondrial splice defect when present in high copy number. Curr. Genet. 12: 503–509

Konopka J & Fields S (1992) The pheromone signal pathway in *Saccharomyces cerevisiae*. A. v. Leeuwenhoek 62: 95-108 (this issue)

Kotylak Z & Slonimski PP (1976) Joint control of cytochromes a and b by a unique mitochondrial DNA region comprising four genetic loci. In: Saccone C & Kroon AM (Eds) The Genetic Function of Mitochondrial DNA(pp 143–154). Elsevier, Amsterdam

Kovacova V, Irmlerova J & Kovac L (1968) Oxidative phosphorylation in yeast IV. Combination of a nuclear mutation affecting oxidative phosphorylation with cytoplasmic mutation to respiratory deficiency. Biochim. Biophys. Acta 162: 157–163

Kozak M (1989) The scanning model for translation: an update. J. Cell Biol. 108: 229–241

Kreike J, Schulze M, Ahne F & Lang BF (1987) A yeast nuclear gene, *MRS1*, involved in mitochondrial RNA splicing: nucleotide sequence and mutational analysis of two overlapping open reading frames on opposite strands. EMBO J. 6: 2123–2129

Kruszewska A & Slonimski PP (1984) Mitochondrial and nuclear mitoribosomal suppressors that enable misreading of ochre codons in yeast mitochondria. I. Isolation, localization and allelism of suppressors. Curr. Genet. 9: 1–10

Lahaye A, Stahl H, Thines-Sempoux D & Foury F (1991) PIF1: a DNA helicase in yeast mitochondria. EMBO J. 10: 997–1007

Lambowitz AM & Perlman PS (1990) Involvement of aminoacyl-tRNA synthetases and other proteins in group I and group II intron splicing. Trends in Biochem. Sci. 15: 440–444

Law RHP, Devenish RJ & Nagley P (1990) Assembly of imported subunit 8 into the ATP synthase complex of isolated yeast mitochondria. Eur. J. Biochem. 188: 421–429

Lawson JE & Douglas MG (1988) Separate genes encode functionally equivalent ADP/ATP carrier proteins in *Saccharomyces cerevisiae*. Isolation and analysis of *AAC2*. J. Biol. Chem. 263: 14812–14818

Lazowska J, Jacq C & Slonimski PP (1980) Sequence of introns and flanking exons in wild-type and *box3* mutants of cytochrome *b* reveals an interlaced splicing protein coded by an intron. Cell 22: 333–348

Li M, Tzagoloff A, Underbrink-Lyon K & Martin NC (1982) Identification of the paromomycin-resistance mutation in the 15S rRNA gene of yeast mitochondria. J. Biol. Chem. 257: 5921–5928

Liao X, Small WC, Srere PA & Butow RA (1991) Intramitochondrial functions regulate nonmitochondrial citrate synthase (CIT2) expression in *Saccharomyces cerevisiae*. Mol. Cell. Biol. 11: 38–46

Linder P & Slonimski PP (1989) An essential yeast protein, encoded by duplicated genes *TIF1* and *TIF2* and homologous to the mammalian translation initiation factor eIF-4A, can suppress a mitochondrial missense mutation. Proc. Natl. Acad. Sci. USA 86: 2286–2290

Lisowsky T (1990) Molecular analysis of the mitochondrial transcription factor *MTF2* of *Saccharomyces-cerevisiae*. Mol. Gen. Genet. 220: 186–190

Lisowsky T & Michaelis G (1988) A nuclear gene essential for mitochondrial replication suppresses a defect of mitochondrial transcription in *Saccharomyces cerevisiae*. Mol. Gen. Genet. 214: 218–223

Lisowsky T and Michaelis G (1989) Mutations in the genes for mitochondrial RNA polymerase and a second mitochondrial transcription factor of *Saccharomyces cerevisiae*. Mol. Gen. Genet. 219: 125–128

Maarse AC, van Tuinen EJ, Meijer M & Grivell LA (1990) Identification of yeast mutants with defects in mitochondrial protein import. Yeast 6: S456

Maleszka R, Skelly PJ & Clark-Walker GD (1991) Rolling circle replication of DNA in yeast mitochondria. EMBO J. 10: 3923–3929

Marczynski GT, Schultz PW & Jaehning JA (1989) Use of yeast nuclear DNA sequences to define the mitochondrial RNA polymerase promoter in vitro. Mol. Cell. Biol. 9: 3193–3202

Martin NC & Underbrink-Lyon K (1981) A mitochondrial locus is necessary for the synthesis of mitochondrial tRNA in the yeast *Saccharomyces cerevisiae*. Proc. Natl. Acad. Sci. USA 78: 4743–4747

Masters BS, Stohl LL & Clayton DA (1987) Yeast mitochondrial RNA polymerase is homologous to those encoded by bacteriophages T3 and T7. Cell 51: 89–99

McMullin TW, Haffter P & Fox TD (1990) A novel small-subunit ribosomal protein of yeast mitochondria that interacts functionally with an mRNA-specific translational activator. Mol. Cell. Biol. 10: 4590–4595

Michaelis U, Körte A. & Rödel G (1991) Association of cytochrome b translational activator proteins with the mitochondrial membrane: implications for cytochrome b expression in yeast. Mol. Gen. Genet. 230: 177–185

Michel F & Lang BF (1985) Mitochondrial class II introns encode proteins related to the reverse transcriptases of retroviruses. Nature 316: 641–643

Michel F, Umesono K & Ozeki H (1989) Comparative and functional anatomy of group II catalytic introns – a review. Gene 82: 5–30

Mueller PP, Reif MK, Zonghou S, Sengstag C, Mason TL & Fox TD (1984) A nuclear mutation that post-transcriptionally blocks accumulation of a yeast mitochondrial gene product can be suppressed by a mitochondrial gene rearrangement. J. Mol. Biol. 175: 431–452

Myers AM, Pape LK & Tzagoloff A (1985) Mitochondrial protein synthesis is required for maintenance of intact mitochondrial genomes in Saccharomyces cerevisiae. EMBO J. 4: 2087–2092

Nakagawa K, Morishima N & Shibata T (1991) A maturase-like subunit of the sequence-specific endonuclease Endo.SceI from yeast mitochondria. J. Biol. Chem. 266: 1977–1984

Nagley P & Devenish RJ (1989) Leading organellar proteins along new pathways: the relocation of mitochondrial and chloroplast genes to the nucleus. Trends in Biochem. Sci. 14: 31–35

Natsoulis G, Hilger F & Fink GR (1986) The *HTS1* gene encodes both the cytoplasmic and mitochondrial histidine tRNA synthetases of S. cerevisiae. Cell 46: 235–243

Nobrega MP, Nobrega FG & Tzagoloff A (1990) *COX10* codes for a protein homologous to the orf1 product of *Paracoccus denitrificans* and Is required for the synthesis of yeast cytochrome oxidase. J. Biol. Chem. 265: 14220–14226

Olesen JT & Guarente L (1990) The HAP2 subunit of yeast CCAAT transcriptional activator contains adjacent domains for subunit association and DNA recognition – model for the HAP2/3/4 complex. Genes & Dev. 4: 1714–1729

Olesen J, Hahn S & Guarente L (1987) Yeast HAP2 and HAP3 activators both bind to the CYC1 upstream activation site, UAS2 in an interdependent manner. Cell 51: 953–961

Ooi BG, Lukins HB, Linnane AW & Nagley P (1987) Biogenesis of mitochondria: a mutation in the 5′-untranslated region of yeast mitochondrial *oli1* mRNA leading to impairment in translation of subunit 9 of the mitochondrial ATPase complex. Nucleic Acids Res. 15: 1965–1977

Osinga KA, De Vries E, Van der Horst G & Tabak HF (1984) Processing of yeast mitochondrial messenger RNAs at a conserved dodecamer sequence. EMBO J. 3: 829–834

Parikh VS, Morgan MM, Scott R, Clements LS & Butow RA (1987) The mitochondrial genotype can influence nuclear gene expression in yeast. Science 235: 576–580

Partaledis JA & Mason TL (1988) Structure and regulation of a nuclear gene in *Saccharomyces cerevisiae* that specifies MRP13, a protein of the small subunit of the mitochondrial ribosome. Mol. Cell. Biol. 8: 3647–3660

Payne MJ, Schweizer E & Lukins HB (1991) Properties of two nuclear mutants affecting expression of the mitochondrial oli1 gene of *Saccharomyces cerevisiae*. Curr. Genet. 19: 343–351

Pel HJ, Tzagoloff A & Grivell LA (1991) The identification of 18 nuclear genes required for the expression of the yeast mitochondrial gene encoding cytochrome c oxidase subunit 1. Curr. Genet. (in press)

Perea J, Delahodde A, Goguel V, Hatat D, Sargueil B & Jacq C (1990) Functions of intron-encoded proteins from yeast mitochondrial genome. RNA maturase and DNA endonuclease. In: Quagliariello E, Papa S, Palmieri F & Saccone C (Eds) Structure, Function and Biogenesis of Energy Transfer Systems (pp 205–208). Elsevier, Amsterdam

Perlman PS & Birky CW jr (1974) Mitochondrial genetics in bakers' yeast: a molecular mechanism for recombinational polarity and suppressiveness. Proc. Natl. Acad. Sci. USA 71: 4612–4616

Pfanner N & Neupert W (1990) The mitochondrial protein import apparatus. Ann. Rev. Biochem. 59: 331–353

Pfeifer K, Arcangioli B & Guarente L(1987) Yeast HAP1 activator competes with the factor RC2 for binding to the upstream activator site UAS1 of the CYC1 gene. Cell 49: 9–18

Pfeifer K, Kim K-S, Kogan S & Guarente L (1989) Functional dissection and sequence of yeast HAP1 activator. Cell 56: 291–301

Pinkham JL & Guarente L (1985) Cloning and molecular analysis of the *HAP2* locus: a global regulator of respiratory genes in *Saccharomyces cerevisiae*. Mol. Cell. Biol. 5: 3410–3416

Piskur J (1988a) Transmission of yeast mitochondrial loci to progeny is reduced when nearby intergenic regions containing ori sequences are deleted. Mol. Gen. Genet. 214: 425–432

Piskur J (1988b) A 5 kb intergenic region containing ori1 in the mitochondrial DNA of *Saccharomyces cerevisiae* is dispensable for expression of the respiratory phenotype. FEBS Lett. 229: 145–149

Pollock RA, Hartl F-U, Cheng MY, Ostermann J, Horwich A & Neupert W (1988) The processing peptidase of yeast mitochondria: the two co-operating components MPP and PEP are structurally related. EMBO J. 7: 3493–3500

Poutre C & Fox TD (1987) *PET111*, a *Saccharomyces cerevisiae* nuclear gene required for translation of the mitochondrial mRNA encoding cytochrome c oxidase subunit II. Genetics 115: 637–647

Rödel G (1986) Two yeast nuclear genes, *CBS1* and *CBS2*, are required for translation of mitochondrial transcripts bearing the 5′-untranslated COB leader. Curr. Genet. 11: 41–45

Rothstein R & Sherman F (1980) Genes affecting the expression of cytochrome c in yeast: genetic mapping and genetic interactions. Genetics 94: 871–889

Sakai H, Stiess R & Weiss-Brummer B (1991) Mitochondrial mutations restricting spontaneous translational frameshift suppression in the yeast Saccharomyces cerevisiae. Mol. Gen. Genet. 227: 306–317

Sargueil B, Delahodde A, Hatat D, Lazowska J & Jacq C (1991) A new specific DNA endonuclease activity in yeast mitochondria. Mol. Gen. Genet. 255: 340–341

Schatz G (1991) Transport of proteins into mitochondria. The Harvey Lectures, Series 85 (pp 109–126). Wiley-Liss Inc

Schinkel AH & Tabak HF (1989) Mitochondrial RNA polymerase: dual role in transcription and replication. Trends in Genet. 5: 149–154

Schinkel AH, Groot-Koerkamp MJA & Tabak HF (1988) Mitochondrial RNA polymerase of *Saccharomyces cerevisiae*: subunit composition and mechanism of promoter recognition. EMBO J. 7: 3255–3262

Schmelzer C & Schweyen RJ (1986) Self-splicing of group II introns in vitro: mapping of the branch point and mutational inhibition of lariat formation. Cell 46: 557–565

Schultz J & Carlson M (1987) Molecular analysis of *SSN6*, a gene functionally related to the *SNF1* protein kinase of *Saccharomyces cerevisiae*. Mol. Cell. Biol. 7: 3637–3645

Séraphin B, Boulet A, Simon M & Faye G (1987a) Construction of a yeast strain devoid of mitochondrial introns and its use to screen nuclear genes involved in mitochondrial splicing. Proc. Natl. Acad. Sci. USA 84: 6810–6814

Séraphin B, Simon M & Faye G (1987b) The mitochondrial reading frame RF3 is a functional gene in *Saccharomyces uvarum*. J. Biol. Chem. 262: 10146–10153

Séraphin S, Simon M & Faye G (1988) *MSS18*, a yeast nuclear gene involved in the splicing of intron a15b of the mitochondrial *cox1* transcript. EMBO J. 7: 1455–1464

Séraphin B, Simon M, Boulet A & Faye G (1989) Mitochondrial splicing requires a protein from a novel helicase family. Nature 337: 84–87

Slonimski PP & Tzagoloff A (1976) Localization in mitochondrial DNA of mutations expressed in a deficiency of cytochrome oxidase and/or coenzyme QH2-cytochrome c reductase. Eur. J. Biochem. 61: 27–41

Slonimski PP, Claisse ML, Foucher M, Jacq C, Kochko A, Lamouroux A, Pajot P, Perrodin G, Spyridakis A & Wambier-Kluppel ML (1978) Mosaic organization and expression of the mitochondrial DNA region controlling cytochrome c reductase and oxidase. In: Bacila M, Horecker BL & Stoppanni AOM (Eds) Biochemistry and Genetics of Yeasts (pp 391–401). Academic Press, New York

Smith BJ & Yaffe MP (1991) A mutation in the yeast heat-shock factor gene causes temperature-sensitive defects in both mitochondrial protein import and the cell cycle. Mol. Cell. Biol. 11: 2647–2655

Sor F & Fukuhara H (1982) Identification of two erythromycin resistance mutations in the mitochondrial gene coding for the large ribosomal RNA in yeast. Nucleic Acids Res. 21: 6571–6577

Sor F & Fukuhara H (1984) Erythromycin and spiramycin resistance mutations of yeast mitochondria: nature of the rib2 locus in the large ribosomal RNA gene. Nucleic Acids Res. 12: 8313–8318

Thevelein JM (1992) The RAS-adenylate cyclase pathway and cell cycle control in *Saccharomyces cerevisiae*. A. v. Leeuwenhoek 62: 109-130 (this issue)

Ticho BS & Getz GS (1988) The characterization of yeast mitochondrial RNA polymerase. A monomer of 150,000 daltons with a transcription factor of 70,000 daltons. J. Biol. Chem. 263: 10096–10103

Trumbly RJ (1988) Cloning and characterization of the *CYC8* gene mediating glucose repression in yeast. Gene 73: 97–111

Tzagoloff A & Dieckmann CL (1990) *Pet* genes of *Saccharomyces cerevisiae*. Microbiol. Rev. 54: 211–225

Tzagoloff A & Myers AM (1986) Genetics of mitochondrial biogenesis. Ann. Rev. Biochem. 55: 249–285

Tzagoloff A, Akai A, Needleman RB & Zulch G (1975) Assembly of the mitochondrial membrane system. Cytoplasmic mutants of *Saccharomyces cerevisiae* with lesions in enzymes of the respiratory chain and the mitochondrial ATPase. J. Biol. Chem. 250: 8236–8242

Valencik ML & McEwen JE (1991) Genetic evidence that different functional domains of the *PET54* gene product facilitate expression of the mitochondrial genes *COX1* and *COX3* in *Saccharomyces cerevisiae*. Mol. Cell. Biol. 11: 2399–2409

Valencik ML, Kloeckener-Gruissem B, Poyton RO & McEwen JE (1989) Disruption of the yeast nuclear *PET54* gene blocks excision of mitochondrial intron al5β from pre-mRNA for cytochrome c oxidase subunit I. EMBO J. 8: 3899–3904

Valens M, Rinaldi T, Daignan-Fornier B & Bolotin-Fukuhara M (1992) Identification of nuclear genes which participate to mitochondrial translation in *S. cerevisiae*. Biochimie (in press)

Van Loon APGM, de Groot RJ, Van Eyk E, Van der Horst GTJ & Grivell LA (1982) Isolation and characterization of genes coding for subunits of the yeast ubiquinol-cytochrome c reductase complex. Gene 20: 323–327

Waters R & Moustacchi E (1974) The fate of ultraviolet-induced pyrimidie dimers in the mitochondrial DNA of *Saccharomyces cerevisiae* following various post-irradiation cell treatments. Biochim. Biophys. Acta 366: 241–250

Weiss-Brummer B, Sakai H & Kaudewitz F (1987) A mitochondrial frameshift suppressor (+) of the yeast S. cerevisiae maps in the mitochondrial 15S rRNA locus. Curr. Genet. 11: 295–301

Wiesenberger G, Link TA, Ahsen U von, Waldherr M & Schweyen RJ (1991) *MRS3* and *MRS4*, two suppressors of mtRNA splicing defects in yeast, are new members of the mitochondrial carrier family. J. Mol. Biol. 217: 23–37

Wiesenberger G, Waldherr M & Schweyen RJ (1992) The nuclear gene *MRS2* is essential for the excision of group II introns from yeast mitochondrial transcripts in vivo. J. Biol. Chem. (in press)

Wilcoxen SE, Peterson CR, Winkley CS, Keller MJ & Jaehning JA (1988) Two forms of *RP041*-dependent RNA polymerase. Regulation of the RNA polymerase by glucose-repression may control yeast mitochondrial gene expression. J. Biol. Chem. 263: 12346–12351

Williams FE & Trumbly RJ (1990) Characterization of *TUP1*, a mediator of glucose repression in *Saccharomyces cerevisiae*. Mol. Cell. Biol. 10: 6500–6511

Williams FE, Varanasi U & Trumbly RJ (1991) The *CYC8* and *TUP1* proteins involved in glucose repression in *Saccharomyces cerevisiae* are associated in a protein complex. Mol. Cell. Biol.11: 3307–3316

Williams RS, Johnston SA, Riedy M, DeVit MJ, McElligot SG & Sanford JC (1991) Introduction of foreign genes into tissues of living mice by DNA-coated microprojectiles. Proc. Natl. Acad. Sci. USA 88: 2726–2730

Wright RM & Poyton RO (1990) Release of two *Saccharomyces cerevisiae* cytochrome genes, *COX6* and *CYC1*, from glucose repression requires the *SNF1* and *SSN6* gene products. Mol. Cell. Biol. 10: 1297–1300

Wu M & Tzagoloff A (1989) Identification and characterization of a new gene (*CBP3*) required for the expression of yeast coenzyme QH2-cytochrome c reductase. J. Biol. Chem. 264: 11122–11130

Yaffe MP & Schatz G (1984) Two nuclear mutations that block mitochondrial protein import in yeast. Proc. Natl. Acad. Sci. USA 81: 4819–4823

de Zamaroczy M, Baldacci G & Bernardi G (1979) Putative origins in the mitochondrial genome of yeast. FEBS Lett. 108: 429–432

de Zamaroczy M, Faugeron-Fonty G, Baldacci G, Coursot R & Bernardi G (1984) The ori sequences of the mitochondrial genome of a wild-type yeast strain: number, location, orientation and structure. Gene 32: 439–457

Zhu H, Macreadie IG & Butow RA (1987) RNA processing and expression of an intron-encoded protein in yeast mitochondria: role of a conserved dodecamer sequence. Mol. Cell. Biol. 7: 2530–2537

Zitomer RS & Lowry CV (1992) Regulation of gene expression by oxygen in *Saccharomyces cerevisiae*. Microbiol. Rev. (in press)

MIX
Papier aus verantwortungsvollen Quellen
Paper from responsible sources
FSC® C105338

If you have any concerns about our products,
you can contact us on
ProductSafety@springernature.com

In case Publisher is established outside the EU,
the EU authorized representative is:
Springer Nature Customer Service Center GmbH
Europaplatz 3, 69115 Heidelberg, Germany

Printed by Libri Plureos GmbH
in Hamburg, Germany